LECTURE NOTES
IN PURE AND APPLIED MATHEMATICS

1. *N. Jacobson,* Exceptional Lie Algebras
2. *L.-Å. Lindahl and F. Poulsen,* Thin Sets in Harmonic Analysis
3. *I. Satake,* Classification Theory of Semi-Simple Algebraic Groups
4. *F. Hirzebruch, W. D. Newmann, and S. S. Koh,* Differentiable Manifolds and Quadratic Forms (out of print)
5. *I. Chavel,* Riemannian Symmetric Spaces of Rank One (out of print)
6. *R. B. Burckel,* Characterization of C(X) Among Its Subalgebras
7. *B. R. McDonald, A. R. Magid, and K. C. Smith,* Ring Theory: Proceedings of the Oklahoma Conference
8. *Y.-T. Siu,* Techniques of Extension of Analytic Objects
9. *S. R. Caradus, W. E. Pfaffenberger, and B. Yood,* Calkin Algebras and Algebras of Operators on Banach Spaces
10. *E. O. Roxin, P.-T. Liu, and R. L. Sternberg,* Differential Games and Control Theory
11. *M. Orzech and C. Small,* The Brauer Group of Commutative Rings
12. *S. Thomeier,* Topology and Its Applications
13. *J. M. Lopez and K. A. Ross,* Sidon Sets
14. *W. W. Comfort and S. Negrepontis,* Continuous Pseudometrics
15. *K. McKennon and J. M. Robertson,* Locally Convex Spaces
16. *M. Carmeli and S. Malin,* Representations of the Rotation and Lorentz Groups: An Introduction
17. *G. B. Seligman,* Rational Methods in Lie Algebras
18. *D. G. de Figueiredo,* Functional Analysis: Proceedings of the Brazilian Mathematical Society Symposium
19. *L. Cesari, R. Kannan, and J. D. Schuur,* Nonlinear Functional Analysis and Differential Equations: Proceedings of the Michigan State University Conference
20. *J. J. Schäffer,* Geometry of Spheres in Normed Spaces
21. *K. Yano and M. Kon,* Anti-Invariant Submanifolds
22. *W. V. Vasconcelos,* The Rings of Dimension Two
23. *R. E. Chandler,* Hausdorff Compactifications
24. *S. P. Franklin and B. V. S. Thomas,* Topology: Proceedings of the Memphis State University Conference
25. *S. K. Jain,* Ring Theory: Proceedings of the Ohio University Conference
26. *B. R. McDonald and R. A. Morris,* Ring Theory II: Proceedings of the Second Oklahoma Conference
27. *R. B. Mura and A. Rhemtulla,* Orderable Groups
28. *J. R. Graef,* Stability of Dynamical Systems: Theory and Applications
29. *H.-C. Wang,* Homogeneous Banach Algebras
30. *E. O. Roxin, P.-T. Liu, and R. L. Sternberg,* Differential Games and Control Theory II
31. *R. D. Porter,* Introduction to Fibre Bundles
32. *M. Altman,* Contractors and Contractor Directions Theory and Applications
33. *J. S. Golan,* Decomposition and Dimension in Module Categories
34. *G. Fairweather,* Finite Element Galerkin Methods for Differential Equations
35. *J. D. Sally,* Numbers of Generators of Ideals in Local Rings
36. *S. S. Miller,* Complex Analysis: Proceedings of the S.U.N.Y. Brockport Conference
37. *R. Gordon,* Representation Theory of Algebras: Proceedings of the Philadelphia Conference
38. *M. Goto and F. D. Grosshans,* Semisimple Lie Algebras
39. *A. I. Arruda, N. C. A. da Costa, and R. Chuaqui,* Mathematical Logic: Proceedings of the First Brazilian Conference

40. *F. Van Oystaeyen,* Ring Theory: Proceedings of the 1977 Antwerp Conference
41. *F. Van Oystaeyen and A. Verschoren,* Reflectors and Localization: Application to Sheaf Theory
42. *M. Satyanarayana,* Positively Ordered Semigroups
43. *D. L. Russell,* Mathematics of Finite-Dimensional Control Systems
44. *P.-T. Liu and E. Roxin,* Differential Games and Control Theory III: Proceedings of the Third Kingston Conference, Part A
45. *A. Geramita and J. Seberry,* Orthogonal Designs: Quadratic Forms and Hadamard Matrices
46. *J. Cigler, V. Losert, and P. Michor,* Banach Modules and Functors on Categories of Banach Spaces
47. *P.-T. Liu and J. G. Sutinen,* Control Theory in Mathematical Economics: Proceedings of the Third Kingston Conference, Part B
48. *C. Byrnes,* Partial Differential Equations and Geometry
49. *G. Klambauer,* Problems and Propositions in Analysis
50. *J. Knopfmacher,* Analytic Arithmetic of Algebraic Function Fields
51. *F. Van Oystaeyen,* Ring Theory: Proceedings of the 1978 Antwerp Conference
52. *B. Kedem,* Binary Time Series
53. *J. Barros-Neto and R. A. Artino,* Hypoelliptic Boundary-Value Problems
54. *R. L. Sternberg, A. J. Kalinowski, and J. S. Papadakis,* Nonlinear Partial Differential Equations in Engineering and Applied Science
55. *B. R. McDonald,* Ring Theory and Algebra III: Proceedings of the Third Oklahoma Conference
56. *J. S. Golan,* Structure Sheaves over a Noncommutative Ring
57. *T. V. Narayana, J. G. Williams, and R. M. Mathsen,* Combinatorics, Representation Theory and Statistical Methods in Groups: YOUNG DAY Proceedings
58. *T. A. Burton,* Modeling and Differential Equations in Biology
59. *K. H. Kim and F. W. Roush,* Introduction to Mathematical Consensus Theory
60. *J. Banas and K. Goebel,* Measures of Noncompactness in Banach Spaces
61. *O. A. Nielsen,* Direct Integral Theory
62. *J. E. Smith, G. O. Kenny, and R. N. Ball,* Ordered Groups: Proceedings of the Boise State Conference
63. *J. Cronin,* Mathematics of Cell Electrophysiology
64. *J. W. Brewer,* Power Series Over Commutative Rings
65. *P. K. Kamthan and M. Gupta,* Sequence Spaces and Series
66. *T. G. McLaughlin,* Regressive Sets and the Theory of Isols
67. *T. L. Herdman, S. M. Rankin, III, and H. W. Stech,* Integral and Functional Differential Equations
68. *R. Draper,* Commutative Algebra: Analytic Methods
69. *W. G. McKay and J. Patera,* Tables of Dimensions, Indices, and Branching Rules for Representations of Simple Lie Algebras
70. *R. L. Devaney and Z. H. Nitecki,* Classical Mechanics and Dynamical Systems
71. *J. Van Geel,* Places and Valuations in Noncommutative Ring Theory
72. *C. Faith,* Injective Modules and Injective Quotient Rings
73. *A. Fiacco,* Mathematical Programming with Data Perturbations I
74. *P. Schultz, C. Praeger, and R. Sullivan,* Algebraic Structures and Applications: Proceedings of the First Western Australian Conference on Algebra
75. *L. Bican, T. Kepka, and P. Nemec,* Rings, Modules, and Preradicals
76. *D. C. Kay and M. Breen,* Convexity and Related Combinatorial Geometry: Proceedings of the Second University of Oklahoma Conference
77. *P. Fletcher and W. F. Lindgren,* Quasi-Uniform Spaces
78. *C.-C. Yang,* Factorization Theory of Meromorphic Functions
79. *O. Taussky,* Ternary Quadratic Forms and Norms
80. *S. P. Singh and J. H. Burry,* Nonlinear Analysis and Applications
81. *K. B. Hannsgen, T. L. Herdman, H. W. Stech, and R. L. Wheeler,* Volterra and Functional Differential Equations

82. *N. L. Johnson, M. J. Kallaher, and C. T. Long,* Finite Geometries: Proceedings of a Conference in Honor of T. G. Ostrum
83. *G. I. Zapata,* Functional Analysis, Holomorphy, and Approximation Theory
84. *S. Greco and G. Valla,* Commutative Algebra: Proceedings of the Trento Conference

Other Volumes in Preparation

COMMUTATIVE ALGEBRA

Proceedings of the Trento Conference

edited by

Silvio Greco
Dipartimento di Matematica
Politecnico di Torino
Torino, Italy

Giuseppe Valla
Istituto Matematico
Università di Genova
Genova, Italy

MARCEL DEKKER, INC. New York and Basel

Library of Congress Cataloging in Publication Data

Trento Conference (1981)
Commutative algebra.

(Lecture notes in pure and applied mathematics ; v. 84)
Conference organized by Centro interuniversitario di ricerca matematica, University of Trento, and held 6/8-13/81 at Villa Monastero, Trento, Italy.
1. Commutative algebra--Congresses. 2. Geometry, Algebraic--Congresses. I. Greco, Silvio. II. Valla Giuseppe, III. Title, IV. Series.
QA251.3.T73 1981 512'.24 82-22028
ISBN 0-8347-1899-2

MARCEL DEKKER, INC.
270 Madison Avenue, New York, New York 10010

Current printing (last digit):
10 9 8 7 6 5 4 3 2 1

PRINTED IN THE UNITED STATES OF AMERICA

PREFACE

This volume contains the proceedings of the Trento meeting on Commutative Algebra held at Villa Monastero, Trento, Italy, from June 8 through June 13, 1981. The meeting was organized by C.I.R.M. (Centro Interuniversitario di Ricerca Matematica), a research institution of the University of Trento.

This book represents a collection of original research papers on topics in commutative algebra and algebraic geometry. Although most of the papers were selected from those submitted by the participants at the conference, some of the papers were prepared after the conference and are based on results that were announced at the meeting.

The main focus of the book is to provide an overview of some of the recent developments in commutative algebra with special reference to algebraic geometry. The original contributions by experts in these fields are directed to university teachers, graduate and post-graduate students, and researchers of pure mathematics. Since division according to subject matter did not seem feasible, the papers are arranged in alphabetical order.

We wish to thank the contributors and all the participants for their active and fruitful cooperation, and the many referees for their invaluable help in editing this volume. Our thanks go also to the Directors of C.I.R.M., Professors M. Miranda and G. Zacher, for the opportunity given us to organize the meeting, and to the Secretary of C.I.R.M., Mr. Micheletti, who brilliantly solved all the organizational problems before and during the meeting. We would also like to thank C.N.R. for their support of some of the participants The publication of the proceedings has been handled promptly by Marcel Dekker. We are very grateful to them and to their editors.

S. Greco
G. Valla

CONTENTS

CONTRIBUTORS

E. Arbarello, Istituto Matematica, Universita di Roma, Roma, Italy

M. Brodmann, Mathematisches Institut, der Universität, Basel, Switzerland

Maria Pia Cavaliere, Istituto di Matematica, Università di Genova, Genova, Italy

C. Ciliberto, Istituto Matematica, Universita di Napoli, Napoli, Italy

Caterina Cumino, Politecnico di Torino, Torino, Italy

David E. Dobbs, University of Tennessee, Knoxville, Tennessee

Marco Fontana, Istituto Matematico, Università di Roma, Roma, Italy

J. Herzog, Universitat Essen Gesamthochschule, Essen, West Germany

Monica Ida, University of Bologna, Bologna, Italy

Mirella Manaresi, University of Bologna, Bologna, Italy

Maria Grazia Marinari, Instituto Matematico, University of Genova, Genova, Italy

Gianfranco Niesi, Istituto di Matematica, Universita de Genova, Genova, Italy

Anna Oneto, Instituto di Matematica, Universita de Genova, Genova, Italy

U. Orbanz, Matematisches Institut der Universitat zu Koln, West Germany

Ferruccio Orecchia, Instituto de Matematica, dell' Università di Napoli, Napoli, Italy

Mario Raimondo, Instituto Matematico, University of Genova, Genova, Italy

Lorenzo Robbiano, Istituto Matematico, Dell' Università di Genova, Genova, Italy

R.Y. Sharp, Department of Pure Mathematics, University of Sheffield, Sheffield, United Kingdom

A. Simis, Departmento de Matematica, Universidade, Federal de Pernambuco, Recife, Brazil

Ezio Stagnaro, Istituto di Matematica Applicata, Facoltà di Ingegneria Università, Padova, Italy

Rosario Strano, Department of Mathematics, University of Catania, Catania, Italy

Grazzia Tamone, Istituto di Matematica, University of Genoa, Genoa, Italy

Giuseppe Valla, Istituto Matematico, Dell' Università di Genova, Genova, Italy

W.V. Vasconcelos, Department of Mathematics, Rutgers University, New Brunswick, New Jersey

Kei-ichi Watanabe, Department of Mathematics, Nagoya Institute of Technology, Nagoya, Japan

Elsa Zatini, Instituto di Matematica, Universita de Genova, Genova, Italy

Adjoint Hypersurfaces to Curves in $\mathbb{P}^r$ Following Petri

E. Arbarello

Istituto di Matematica
-Univ. ROMA-

C. Ciliberto(*)

Istituto di Matematica
-Univ. NAPOLI-

1. Introduction

Let $\Gamma \subset \mathbb{P}^r$ be an irreducible, non degenerate, complete curve of degree n defined over an algebraically closed field k. Let

$$\pi : C \longrightarrow \Gamma$$

be the normalization morphism, and let g be the genus of C. The map π determines an (r+1) - dimensional vector subspace

$$V \subseteq H^0(C, \mathcal{O}_C(D))$$

where D is the pull-back via π of a hyperplane in $\mathbb{P}^r$, and

(*) Both authors are members of the G.N.S.A.G.A. of the Italian C.N.R.

yields the identification $\mathbb{P}^r = \mathbb{P}V^*$.

Let $X_o, \ldots\ldots\ldots\ldots, X_r$ be a basis of V; we set

$$\mathbf{S} = \bigoplus_{m \geq 0} \mathbf{S}_m = k[X_0, \ldots, X_r]$$

The maps

$$\varrho_m : \mathbf{S}_m \longrightarrow H^0(C, \mathcal{O}_C(m\mathbf{D}))$$

have been the object of intensive study. First of all, when Γ is non-singular, it is well known that, for m large enough, ϱ_m is surjective. The question is: <u>what is a lower bound for such</u> m'<u>s</u> ? In [C] Castelnuovo proved that, if $m \geq n - 2$, ϱ_m is surjective. This result was obtained by generically projecting Γ into a plane and then using the completeness of the linear series cut out on a plane curve by adjoint curves of any degree. We learned that recently Lazarsfeld improved this result by showing that ϱ_m is surjective if $m \geq n - r + 1$.

When Γ is a singular curve, the maps ϱ_m are no longer surjective and their analysis presents two aspects. On one hand one should measure exactly to which extent this lack of surjectivity is due to the singularities of Γ, and, on the other, find a suitable way of generalizing, and answering, the question raised in the non-singular case.

To be more precise, one should find a <u>graded</u>, <u>saturated ideal</u>

$$\mathfrak{a}(\Gamma) \subseteq \mathbf{S}$$

depending only on Γ, whose support is the singular locus of Γ and also having the following property: if Δ is the divisor on C defined by $\mathfrak{a}(\Gamma)$, then the maps

$$\varrho_m : \mathfrak{a}(\Gamma)_m \longrightarrow H^0(C, \mathcal{O}_C(m\mathbf{D} - \Delta)) \qquad (1.1)$$

should be surjective for large enough m. The question would then be, as before, to find a lower bound for such m's.

For example, when Γ is a plane curve, $\mathfrak{a}(\Gamma)$ can be taken to be the adjoint ideal (see [A S_1]), and it is a classical result that ϱ_m is surjective for all m.

In the general case Castelnuovo proposed the following construction. First of all let $p_1, \ldots, p_\nu$ be the preimages under π of the singular points of Γ. Take then a generic projection of Γ into a plane, and let Γ' be the image curve. The adjoint ideal $\mathfrak{a}(\Gamma')$ defines a divisor Δ' on C. Set

$$\Delta = \sum_{i}^{\nu} m_{p_i}(\Delta') \cdot p_i$$

where $m_p(\Delta')$ is the multiplicity of Δ' at $p \in C$; define then

$$\mathfrak{a}(\Gamma) = \{ F \in \mathbf{S} : \pi^* F \geq \Delta \}$$

Whit this definition Castelnuovo proves that the maps ϱ_m are surjective for all $m \geq n-2$, thus obtaining, as a particular case, the above mentioned result for non-singular curves.

In the present note, closely following an idea of Petri (see [P]), we propose a different, and perhaps more intrinsic, definition of adjoint ideal $\mathfrak{a}(\Gamma)$, one that, for instance, reflects quite closely the characters of the linear series $|mD|$. This ideal coincides with Castelnuovo's one in some cases, for instance when Γ is smooth, but seems in general to be influenced by the singularities of Γ in a quite different way. We then look at the maps ϱ_m's relative to this ideal and find a number χ such that ϱ_m is surjective for all $m > \chi$. This is what we do in § 2. In § 3 we discuss the behaviour of $\mathfrak{a}(\Gamma)$ for some simple types of singularities. The result of § 2 substantially differs from the ones of Castelnuovo and Lazarsfeld, as we point out discussing some examples in § 4. For instance we shall show that for certain classes of curves with general moduli Petri's bound becomes much sharper than Castelnuovo's and Lazarsfeld's, while for other classes of curves just the contrary occurs.

The disappointing feature is that all the above mentioned bounds are still far from giving the solution of the following very nice conjecture which, we believe, is due to M. Noether and which is perhaps one of the outstanding problems in the theory of curves: let X be an irreducible component of the Hilbert scheme of smooth curves of degree n and genus g in $\mathbb{P}^r$; assume X surjects onto the moduli space M_g and let Γ be a curve corresponding to a general point of X, then the maps ϱ_m are for maximal rank for all m.

2. Derivation of the Ideal

With notation of the introduction, given an invertible sheaf $\mathcal{F}$ on C, we set

$$\gamma(\mathcal{F}) = \bigoplus_{m \in \mathbb{Z}} H^0(C, \mathcal{F}(mD))$$

This is a finitely generated, Cohen-Macaulay, graded S-module, characterized by the property that

$$\gamma(\mathcal{F})^{\sim} = \pi_*(\mathcal{F})$$

(see [S]), and we denote by $F_k(\gamma(\mathcal{F}))$ the k-th Fitting ideal of $\gamma(\mathcal{F})$. The definition of adjoint ideal to Γ implicitely proposed by Petri is the following: the adjoint ideal $\mathfrak{a}(\Gamma)$ is the saturation of the second Fitting ideal of $\gamma(K_C)$, where K_C is the canonical sheaf on C. In other words we first take a minimal free resolution

$$L_1 \xrightarrow{f} L_0 \longrightarrow \gamma(K_C) \longrightarrow 0 \qquad (2.1)$$

of $\gamma(K_C)$ with rk $(L_0) = t$; the second Fitting ideal $F_2(\gamma(K_C))$ of $\gamma(K_C)$ is then generated by the (t-1) - order minors of a matrix representative of f, and

$$\mathfrak{a}(\Gamma) = \{ G \in S : \text{for any } i = 0, \ldots, r \text{ there exists } m \text{ such that } x_i^m G \in F_2(\gamma(K_C)) \}$$

This also means that $\mathfrak{a}(\Gamma)$ is the largest homogeneous ideal of S whose localizations coincide with those of $F_2(\gamma(K_C))$.

First of all we want to show that <u>the support of $\mathfrak{a}(\Gamma)$ is the singular locus Γ_{sg} of Γ</u>. This is a consequence of the following more general

<u>Lemma</u>. - <u>If $\mathcal{F}$ is an invertible sheaf on C, then the support of the second Fitting ideal of $\gamma(\mathcal{F})$ is Γ_{sg}</u>. (2.2)

By a well known theorem of Buchsbaum and Eisenbud (see [BE]), the support of the k-th Fitting ideal of a graded, torsion S-module M coincides with the one of $\Lambda^k M$. In order to prove the Lemma it therefore suffices to show that the support of $\Lambda^2 \gamma(\mathcal{F})$ is Γ_{sg}. Being this a local problem, we may as well assume that $\mathcal{F} = \mathcal{O}_C$. Given a point p on Γ we set $A = \mathcal{O}_{\Gamma, p}$ and let A' denote the integral closure of A. We must then show that p is a singular point for Γ if and only if $\Lambda^2 A' \neq 0$. We only prove the non-trivial implication. Assume that p is a point of multiplicity $\nu > 1$. Let $\mathfrak{m}$ be the maximal ideal of A and let $\mathfrak{m}_1, \ldots, \mathfrak{m}_h$ be the maximal ideals of A', so that

$$\mathfrak{m} A' = \mathfrak{m}_1^{\nu_1} \cdot \ldots \cdot \mathfrak{m}_h^{\nu_h}$$

By [Z S] , vol. II, pg 300, and by the Chinese remainder theorem, we have:

$$\nu = \sum_{1}^{h}{}_{i}\ \nu_h = \dim_k (A'/\mathfrak{m}A')$$

Since $\nu \geq 2$, $\Lambda^2(A'/\mathfrak{m}A')$ is non-zero, and hence $\Lambda^2 A'$ is also non-zero.

The adjoint ideal $\mathfrak{a}(\Gamma)$ defines, by pull-back, a divisor Δ on C. We then have maps ϱ_m as in (1.1); we next want to show that for m large enough these maps are surjective, and, in doing so, we shall actually provide an explicit bound χ such that ϱ_m is surjective whenever $m > \chi$.

To do this, we consider the resolution (2.1), denote by M an (s x t) - matrix representative of f and by λ the maximum degree of minors of order t-1 of M. We also set

$$\mu_m : V \otimes H^0(C, K_C((m-1)\mathbf{D})) \longrightarrow H^0(C, K_C(m\mathbf{D}))$$

These maps are non-trivial for $m \geq -d$, where d is the so called level of D, i.e. the maximum integer such that $H^0(C, K_C(-d\mathbf{D})) \neq 0$. Hence if

$$t_m = \dim(\operatorname{coker} \mu_m)$$

we have

$$t = rk(\mathbf{L}_0) = t_{-d} + \ldots + t_j$$

for some j. It is in fact a fundamental result due to

Castelnuovo (see [C]) that $j \leq 1$ and $j \leq 0$ in case $V = H^{\circ}(C, \mathcal{O}_C(D))$. We now set

$$\chi = \max \{ \lambda , 2\lambda - (2g - 2 + \deg \Delta)/n - 2j \}$$

We shall prove the following

Theorem - When $m > \chi$ the maps ϱ_m are surjective. (2.3)

For any $m \geq -d$, let $y_1^m, \ldots, y_{t_m}^m$ be elements in $H^{\circ}(C, K_C(mD))$ which are linearly independent modulo $\operatorname{Im} \mu_m$; set then

$$(y_1, \ldots, y_t) = (y_1^{-d}, \ldots, y_{t_{-d}}^{-d}, \ldots, y_1^j, \ldots, y_{t_j}^j)$$

and choose a $(t - 1) \times t$ - submatrix A of M with rank equal to $t - 1$. If A_i is the minor obtained by delating the i-th row of A, by Cramer's rule we get

$$A_t \, y_i = A_i \, y_t \tag{2.4}$$

for any $i = 1, \ldots, t$. Let μ be the highest degree of the polynomials A_i and let $W \subseteq S$ be the μ-th graded part of the ideal generated by the A_i's. Relations (2.4) imply that there exists an effective divisor E on C, depending only on A, such that

$$\mathcal{O}_C(\mu D - E) = K_C(jD) \tag{2.5}$$

and that the map

$$\mathbf{W} \longrightarrow H^{\circ}(C, \mathcal{O}_C(\mu \mathbf{D} - \mathbf{E})) \tag{2.6}$$

is surjective. By the very definition of j, we also have that for any $\nu \geq 0$ the maps

$$\mathbf{W} \otimes \mathbf{S}_{\nu} \longrightarrow H^{\circ}(C, \mathcal{O}_C((\mu+\nu)\mathbf{D} - \mathbf{E})) \tag{2.7}$$

are surjective. The various possible choices of the submatrix A lead to subspaces

$$\mathbf{W}^{i} \subseteq \mathbf{S}_{\mu_i} \qquad i = 1, \ldots, N$$

having the property that for each i = 1,...., N, there exists an effective divisor E_i such that (2.5) holds with μ_i, E_i instead of μ,E, and the maps (2.6), (2.7), with μ_i, E_i, W_i instead of μ, E, W, are surjective.

Let now $m \geq \chi$ and let $\nu_1, \ldots, \nu_N$ be such that

$$\mu_i + \nu_i = m > \chi \qquad i = 1, \ldots, N$$

For any k = 1,...., N set

$$\mathbf{F}_k = \text{g.c.d.}(\mathbf{E}_1, \ldots, \mathbf{E}_k)$$

and consider tha maps

$$\alpha_{k,m} : \bigoplus_{i=1}^{k} (\mathbf{W}^{i} \otimes \mathbf{S}_{\nu_i}) \longrightarrow H^{\circ}(C, \mathcal{O}_C(m\mathbf{D} - \mathbf{F}_k))$$

defined by

$$\alpha_{1,m} = \varrho_m |_{W^1 \otimes S_{\nu_1}}$$

$$\alpha_{k,m}\left(\sum_1^k {}_i\, f_i \otimes g_i\right) = \varrho_m(f_1 \otimes g_1) - \sum_2^k {}_i\, \varrho_m(f_i \otimes g_i)$$

We know that $\alpha_{1,m}$ is surjective. Let us prove by induction that, when $m > \bar{\chi}$, $\alpha_{k,m}$ is surjective for any k = 2,..., N. Assume $\alpha_{k-1,m}$ is surjective. We get a commutative diagram

$$\begin{array}{ccc} \bigoplus_{i=1}^{k} (W^i \otimes S_{\nu_i}) & \xrightarrow{\alpha_{k,m}} & H^0(C, \mathcal{O}_C(mD - F_k)) \\ \beta \searrow & & \swarrow \gamma \\ & H^0(C, \mathcal{O}_C(mD - F_{k-1})) \oplus H^0(C, \mathcal{O}_C(mD - E_k)) & \end{array}$$

where $\gamma(f,g) = f - g$. By induction β is surjective. It therefore suffices to prove that γ is surjective. Let

$$B_k = \text{g.c.d.}\ (mD - F_{k-1}, mD - E_{k-1})$$

In order to show that γ is surjective, it is enough to show that

$$H^1(C, \mathcal{O}_C(B_k)) = 0 \tag{2.8}$$

Now

$$B_k = mD - F_{k-1} - E_k + F_k \geqq$$

$$\geqq mD - E_{k-1} - E_k + \Delta$$

since

$$\Delta = \text{g.c.d.}(E_1, \ldots, E_N) \leqq F_k$$

Assuming, as we may, that $\lambda = \mu_N$, since $m > \chi$ we have

$$\deg B_k \geqq mn - 2 \deg E_N + \deg \Delta -$$

$$= mn - 2(\lambda n - 2g + 2 - jn) + \deg \Delta >$$

$$> 2g - 2$$

proving (2.8). Since $\Delta = F_N$, when $k = N$ we get a commutative diagram

$$\begin{array}{ccc} a(\Gamma)_m & \xrightarrow{\varrho_m} & H^0(C, \mathcal{O}_C(mD - \Delta)) \\ \delta \nwarrow & & \nearrow \alpha_{N,m} \\ & \bigoplus_{i=1}^{N} (W^i \otimes S_{\nu_i}) & \end{array}$$

where δ is the obvious map. Being $\alpha_{N,m}$ surjective when $m > \chi$, ϱ_m is also surjective when $m > \chi$.

3. Behavior of the Ideal

In this section we shall analyse the behaviour of the

ideal $\mathfrak{a}(\Gamma)$ in some particular cases.

We keep the notation of Lemma (2.2), and for any ideal $\mathcal{J} \subseteq S$ we denote by $\mathcal{J}^*$ the image of $\mathcal{J}$ in $S/\mathcal{J}(\Gamma)$, where $\mathcal{J}(\Gamma)$ is the ideal of Γ. Let us begin proving that: if char $k \neq 2$ and $p \in \Gamma$ is a double point, then

$$\mathfrak{a}(\Gamma)^*_{\mathfrak{m}} = \mathfrak{c}$$

where $\mathfrak{c}$ is the conductor of A' in A.

By the result of Buchsbaum and Eisenbud previously quoted, we have

$$\begin{aligned}(\operatorname{ann}(\Lambda^2 A'))^{\nu-1} &\subseteq F_2(\gamma(K_C))^*_{\mathfrak{m}} = \\ &= \mathfrak{a}(\Gamma)^*_{\mathfrak{m}} \subseteq \operatorname{ann}(\Lambda^2 A')\end{aligned} \tag{3.1}$$

for any ν-ple point $p \in \Gamma$. Thus in our case we only have to prove that

$$\operatorname{ann}(\Lambda^2 A') = \mathfrak{c}$$

For any $p \in \Gamma$ and for any $h \in \mathbb{N}$, we have

$$\mathfrak{c} \subseteq \operatorname{ann}(\Lambda^h A') \tag{3.2}$$

In fact, if $x_1, \ldots, x_\nu$ generate A' over A and $s \in \mathfrak{c}$, we have

$$s\,(x_{i_1}\wedge \dots \wedge x_{i_h}) = (s x_{i_1})\wedge \dots \wedge x_{i_h} =$$

$$= (s x_{i_1} \dots x_{i_h}) \wedge 1 \wedge \dots \wedge 1 = 0$$

Hence we must show that

$$\mathrm{ann}\,(\Lambda^2 A') \subseteq \mathcal{C}$$

Let 1, x be a minimal system of generators of A'. We have the following exact sequence

$$0 \longrightarrow \mathcal{C} \xrightarrow{\psi} A^2 \xrightarrow{\varphi} A' \longrightarrow 0$$

where

$$\varphi(a_1, a_2) = a_1 + a_2 x$$

$$\psi(b) = (-bx, b)$$

Tensorizing by A' we get the exact sequence

$$\mathcal{C} \otimes A' \xrightarrow{\psi\otimes 1} A^2 \otimes A' \simeq A' \oplus A' \xrightarrow{\varphi\otimes 1} A' \otimes A' \longrightarrow 0$$

Moreover we have the canonical surjective map

$$\pi : A' \otimes A' \longrightarrow \Lambda^2 A'$$

Now $s \in \mathrm{ann}\,(\Lambda^2 A')$ implies $s\,(1 \wedge x) = s \wedge x = 0$.
Being $s \wedge x = \pi(s \otimes x) = \pi(-x \otimes s)$ it is
then $\pi(s \otimes x - x \otimes s) = 0$. By the very definition of wedge product, there exist elements

$$y_i = \alpha_i + \beta_i x \in A',\ \alpha_i, \beta_i \in A,\ i = 1, \dots, n$$

such that

$$
\begin{aligned}
s \otimes x - x \otimes s &= \sum_{1}^{n}{}_{i} (y_i \otimes y_i) = \\
&= \Big(\sum_{1}^{n}{}_{i} \alpha_i^2\Big)(1 \otimes 1) + \Big(\sum_{1}^{n}{}_{i} \alpha_i \beta_i\Big)(1 \otimes x) + \\
&+ \Big(\sum_{1}^{n}{}_{i} \alpha_i \beta_i\Big)(x \otimes 1) + \Big(\sum_{1}^{n}{}_{i} \beta_i^2\Big)(x \otimes x)
\end{aligned}
$$

Then

$$
\begin{aligned}
0 &= s \otimes x - x \otimes s - \Big(\sum_{1}^{n}{}_{i} \alpha_i^2\Big)(1 \otimes 1) - \\
&-\Big(\sum_{1}^{n}{}_{i} \alpha_i \beta_i\Big)(1 \otimes x + x \otimes 1) - \Big(\sum_{1}^{n}{}_{i} \beta_i^2\Big)(x \otimes x) = \\
&= (\varphi \otimes 1)\Big\{(sx,0) - (0,s) - \Big(\sum_{1}^{n}{}_{i} \alpha_i^2\Big)(1,0) - \\
&- \Big(\sum_{1}^{n}{}_{i} \alpha_i \beta_i\Big)\big((x,0) + (0,1)\big) - \Big(\sum_{1}^{n}{}_{i} \beta_i^2\Big)(0,x)\Big\} = \\
&= (\varphi \otimes 1)\Big(sx - \sum_{1}^{n}{}_{i} \alpha_i^2 - \Big(\sum_{1}^{n}{}_{i} \alpha_i \beta_i\Big)x \,, \\
&- s - \sum_{1}^{n}{}_{i} \alpha_i \beta_i - \Big(\sum_{1}^{n}{}_{i} \beta_i^2\Big) x \Big)
\end{aligned}
$$

whence

$$
\begin{aligned}
sx - \sum_{1}^{n}{}_{i} \alpha_i^2 - \Big(\sum_{1}^{n}{}_{i} \alpha_i \beta_i\Big) x &\in \mathbb{C} \\
-s - \sum_{1}^{n}{}_{i} \alpha_i \beta_i - \Big(\sum_{1}^{n}{}_{i} \beta_i^2\Big) x &\in \mathbb{C}
\end{aligned}
$$

This implies

$$s - \sum_{1}^{n}{}_{i} \ \alpha_i \ \beta_i \in \mathfrak{c}$$

$$- s - \sum_{1}^{n}{}_{i} \ \alpha_i \ \beta_i \in \mathfrak{c}$$

and hence $s \in \mathfrak{c}$, since char $k \neq 2$.

The next case we shall analyse is when $p \in \Gamma \subset \mathbb{P}^r$ is a <u>seminormal</u> r-ple point, namely a r-ple point with r independent tangent lines. We shall show that: <u>if</u> $p \in \Gamma$ <u>is a seminormal r-ple point then</u>

$$\mathfrak{a}(\Gamma)^*_{\mathfrak{m}} = \mathfrak{m}^{r-1} = \mathfrak{c}^{r-1}$$

(in this case we have $\mathfrak{c} = \mathfrak{m}$, by [O]).

By (3.2) and Lemma (2.2) we get $\mathrm{ann}(\Lambda^2 A') = \mathfrak{m}$ and then, by (3.1)

$$\mathfrak{a}(\Gamma)^*_{\mathfrak{m}} \supseteq \mathfrak{m}^{r-1}$$

Suppose now that there exists $f \in \mathfrak{a}(\Gamma)^*_{\mathfrak{m}} - \mathfrak{m}^{r-1}$; then there would by some (r-2)-derivative g of f (viewed as a polynomial) such that $g \notin \mathfrak{m}$. This is not possible: in fact $g \in F_r(\gamma(K_C))^*_{\mathfrak{m}}$ and by the result of Buchsbaum-Eisenbud we have

$$F_r(\gamma(K_C))^*_{\mathfrak{m}} \subseteq \mathrm{ann}(\Lambda^r A') = \mathfrak{m}$$

where the last equality follows from (3.2) and from the fact that $\Lambda^r A' \neq 0$, which, in turn, can be shown exactely as we did in Lemma (2.2) for the case r=2.

Let us make some remarks about the above examples. Let $\mathfrak{a}'(\Gamma)$ be Castelnuovo's adjoint ideal to Γ. Since double points are planar singularities we have $\mathfrak{a}'(\Gamma)^*_{\mathfrak{m}} = \mathfrak{c} = \mathfrak{a}(\Gamma)^*_{\mathfrak{m}}$. It seems likely that this is still true for any plane singularity. It is also very easy to check that $\mathfrak{a}'(\Gamma)^*_{\mathfrak{m}} = \mathfrak{a}(\Gamma)^*_{\mathfrak{m}}$ in the case of seminormal singular points. On the other hand, is is not true that $\mathfrak{a}(\Gamma)$ coincides with $\mathfrak{a}'(\Gamma)$, not even "locally", as we are going to see with the following example:

Let $p \in \Gamma \subset \mathbb{P}^3$, with p triple point with three coplanar tangent lines; then

$$\mathfrak{a}(\Gamma)_{\mathfrak{m}_p} = \mathfrak{m}_p^2$$

where $\mathfrak{m}_p$ is the prime ideal corresponding to p in S.

Since the question is local, we may as well assume that Γ is an irreducible quartic lying in the plane $x_0 = 0$ and having an ordinary triple point at p = (0,0,0,1). Looking at the situation in the $\mathbb{P}^2$ with coordinates (X_1, X_2, X_3), we get for the map f in the resolution (2.1) a matrix of the kind:

$$M = \begin{pmatrix} \alpha_1 & \alpha_2 & \alpha_3 \\ \beta_1 & \beta_2 & \beta_3 \\ \gamma_1 & \gamma_2 & \gamma_3 \end{pmatrix}$$

where the α_i's and the β_i's are linear forms, and the γ_i's are quadratic forms in X_1, X_2, X_3 vanishing simply at p (see [A S_2], [S]). The matrix in $\mathbb{P}^3$ is then obtained adding the row (X_o, X_o, X_o) to M, and this proves the assertion.

On the other hand it is very easy to check that in the above case we have

$$\mathfrak{a}'(\Gamma)_{\mathfrak{m}_p} \supset \mathfrak{m}_p^2$$

The question naturally arising from the above remarks is then: what are the relationships between Castelnuovo's and Petri's adjoint ideals, and what is the relationship between them and the conductor ? It would also be interesting to investigate whether the conductor defines a good notion of adjunction, in the sense of the introduction.

Let us finish this section observing that the saturations of the ideal $F_k(\gamma(K_C))$, $k \geq 2$, define subloci of the singular locus of Γ, and precisely a filtration of Γ_{sg} by "increasing singularities". It would perhaps be of some interest to study this filtration.

4. Applications and Conclusions

Keeping fixed the notation of the preceeding sections, we shall now test Petri's bound in a couple of cases.

Let us assume first that $\Gamma \subset \mathbb{P}^r$, $r \geq 3$, satisfies the following conditions:

1) D is of level d = 1 and $V = H^0(C, \mathcal{O}_C(D))$;

2) $\ker \mu_0 = 0$;

3) Γ is smooth.

When $k = \mathbb{C}$, condition 2) implies that Γ belongs to an irreducible component of the Hilbert scheme which surjects on-to the moduli space M_g. In this sense we may say that Γ has general moduli. For a general curve verifying 1), 2) and for $k = \mathbb{C}$, condition 3) is automatically satisfied (unpublished result by J. Harris).

We have

$$t_{-1} = h^0(C, K_C(-D)) = i$$

$$t_0 = g - (r+1)h^0(C, K_C(-D)) = \varrho$$

$$t_h = 0 \quad , \quad h > 0$$

By argueing like in [S], n° 5, and using theorem (3.1) of [A S_2], one easily checks that the matrix M representing the map f in (2.1) is of the kind

$$\begin{pmatrix} \underset{\varrho}{[1]} & \underset{i}{[2]} \end{pmatrix}$$

where $[h]$ is a block of forms in $X_0, \ldots, X_r$, of degree h.

Hence

$$\lambda \leqq 2i + \varrho$$

so that

$$\chi \leqq \max\{2i + \varrho,\ 4i + 2\varrho - 2(g-1)/n\}$$

On the other hand Castelnuovo-Lazarsfeld's bound is here

$$\bar{\chi} = n - r + 1 = ri + \varrho + 1$$

If $\varrho = 0$, namely if there is only a finite number of g^r_n on C, it is $\chi < \bar{\chi}$ as soon as $r \geq 4$, and $\bar{\chi} - \chi$ grows with r.

We now look at the case Γ is non-singular of genus $g \geq 2$, D is non-special and $V = H^{\circ}(C, \mathcal{O}_C(D))$. It is easy to check that here the matrix M is of the type

$$\begin{pmatrix} [1] \\ [2] \end{pmatrix}_g$$

so that

$$g - 1 \leq \lambda \leq 2g - 2$$

and then

$$2g - 4 \leq \chi \leq 4g - 4$$

Here Castelnuovo-Lazarsfeld's bound is

$$\bar{\chi} = n - r + 1 = g + 1$$

so that, as soon as $g \geq 5$, it is certainly $\chi > \bar{\chi}$. The last example suggests that Petri's bound can be convenient only when D is a special divisor.

Let us conclude with a general remark about Petri's bound. Replacing, throughout section n° 2, K_C with any invertible sheaf $\mathcal{F}$ on C, one could get an alternative definition of adjoint ideal and a different bound $\chi_{\mathcal{F}}$. If $\mathcal{F} = \mathcal{O}_C$ we get for instance

$$\chi_{\mathcal{O}_C} = \max\{\lambda_{\mathcal{O}_C}, 2\lambda_{\mathcal{O}_C} + (2(g-1) - \deg \Delta)/n\}$$

where $\lambda_{\mathcal{O}_C}$ is the highest degree of natural generators of $F_2(\gamma(\mathcal{O}_C))$. The question is to evaluate, for each of these bounds, the corresponding $\lambda_{\mathcal{F}}$, which is in general an hard problem. The case $\mathcal{F} = K_C$ turns out to be the more convenient since, by the results of [A S_2], it is possible to estimate the sizes of the matrix M. But it is possible that in particular cases one could get better bounds by taking an ad hoc invertible sheaf $\mathcal{F}$.

REFERENCES

[AS_1] E.ARBARELLO - E.SERNESI, The Equation of a Plane Curve, Duke Math J., 46 (1979), 469-485.

[AS_2] E.ARBARELLO - E.SERNESI, Petri's Approach to the Study of the Ideal Associated to a Special Divisor, Inv. Math., 49 (1978), 99-119.

[BE] P.A.BUCHSBAUM - D.EISENBUD, What Annihilates a Module ? , J.Alg., 47 (1977), 231-243.

[C] G.CASTELNUOVO, Sui multipli di una serie lineare di gruppi di punti appartenenti ad una curva algebrica, Rend. Circ. Mat. Palermo, 7 (1893), 89-110.

[CV] S.CRECO - P.VALABREGA, On the Theory of Adjoints, Springer L.N. in Math n. 732 (1979).

[O] F.ORECCHIA, Points in Generic Position and Conductor of Curves with Ordinary Singularities (to appear).

[P] K.PETRI, Ueber Spazial-kurven I, Math. Ann., 93(1924), 182-209.

[S] E.SERNESI, L'unirazionalità della varietà dei moduli delle curve di genere dodici (to appear)

[ZS] O.ZARISKI - P.SAMUEL, Commutative Algebra, Van Nostrand co. , 1960.

SOME REMARKS ON BLOW-UP AND CONORMAL CONES

M. Brodmann

Mathematisches Institut
der Universität
B A S E L

1. INTRODUCTION

Let W be an irreducible algebraic variety over the complex numbers $\mathbb{C}$ and let $V \subseteq W$ be a closed irreducible subvariety. Let $U \subseteq W$ be an open affine subset which we consider as an algebraic set in some $\mathbb{C}^h$. Let $f \in \mathcal{O}_{\mathbb{C}^h} = \mathbb{C}[X_1,\ldots,X_h]$ and consider the following symmetric differential form on $\mathbb{C}^h$

$$(1.1) \quad d^n f = \sum_{\substack{u_1+\ldots+u_h=n \\ u_1,\ldots,u_h \in \mathbb{Z}^+}} \frac{1}{u_1!\ldots u_h!} \frac{\partial^n f}{\partial^{u_1} X_1 \ldots \partial^{u_h} X_h} (dX_1)^{u_1} \ldots (dX_h)^{u_h} .$$

By $d^n_U f$ we mean the differential form $U \to \mathbb{C}^{\binom{n+h-1}{h-1}}$ defined by $p \to d^n f(p)$ $(p \in U)$. In the same way we define $d^n_{U \cap V} f$. We put $M^{*n}_U = \{d^n_U f \mid f(U) = 0\}$. We say that f vanishes of order ν at $U \cap V$ if $d^{\mu}_{U \cap V} f = 0$ for all $\mu < \nu$ or - equivalently - if all partial derivatives of order $< \nu$ vanish at $U \cap V$. Let $I(U \cap V)$ be the ideal of polynomials vanishing on $U \cap V$. Then - according to Zariski's main Lemma on holomorphic functions - the set of all polynomials vanishing of order ν on $U \cap V$ is given by the symbolic power $I(U \cap V)^{(\nu)}$. Now we define the module of symmetric n-forms on $U \cap V$ with respect to U by

(1.2) $D^{*n}_{U,V} = N^{*n}_{U,V}/M^{*n}_{U} \cap N^{*n}_{U,V}$, where $N^{*n}_{U,V} = \{d^n_{U\cap V} f \mid f \in I(U\cap V)^{(n)}\}$.

$D^{*n}_{U,V}$ is a module over $\mathcal{O}_{U\cap V} = \mathbb{C}[X_1,\dots,X_h]/I(U \cap V)$ which is in fact (up to "natural" isomorphism) independent on the embedding $U \to \mathbb{C}^h$. So by $U \to D^{*n}_{U,V}$ we may define a sheaf $\mathcal{D}^{*n}_{W,V}$ on V , the sheaf of symmetric n-forms on V with respect to W .

Using the multiplication rule

$$(1.3) \qquad d^r(fg) = \sum_{\substack{i+l=r\\ i,l\geq 0}} d^i(f)\, d^l(g) ,$$

we may define a product on $\mathcal{D}^*_{W,V} = \bigoplus_{n\geq 0} \mathcal{D}^{*n}_{W,V}$ in setting $\overline{d^n f}\, \overline{d^m g} = \overline{d^{n+m} fg}$, where $\overline{}^{\bullet}$ stands for the canonical projection $N^{*\bullet}_{U,V} \to D^{*\bullet}_{U,V}$. In this way $\mathcal{D}^*_{W,V}$ becomes a graded algebra over $\mathcal{O}_V$, the algebra of symmetric forms or the symmetric differential algebra of V with respect to W . Clearly $\mathcal{D}^*_{W,V}$ is the right object to study the differential behaviour of V in W . On the other side $\mathcal{D}^*_{W,V}$ is difficult to understand as a geometric object for its own (Recall that $\mathcal{D}^*_{W,V}$ needs not to be noetherian, as is shown in [11]). So one tends to approximate $\mathcal{D}^*_{W,V}$ by a more "geometric" object, namely the associated graded algebra $Gr(\mathcal{J}) = \bigoplus_{n\geq 0} \mathcal{J}^n/\mathcal{J}^{n+1}$, where $\mathcal{J} \subseteq \mathcal{O}_W$ is the ideal of sections vanishing at V . To recall the relation between these two algebras choose U as above. Then there is a $\mathbb{C}$h-homomorphism $I(U \cap V)^{(n)} \xrightarrow{\pi_n} D^{*n}_{U,V}$ given by $f \to d^n_{U\cap V} f$. π_n clearly is onto and its kernel is given by

$$I(U \cap V)^{(n+1)} + I(U) \cap I(U \cap V)^{(n)} .$$

So there is a canonical isomorphism

$$I(U \cap V)^{(n)}/(I(U \cap V)^{(n)} \cap I(U) + I(U \cap V)^{(n+1)}) \xrightarrow{\cong} D^{*n}_{U,V} .$$

So we get a homomorphism of O_V-modules $\Gamma(U, \mathcal{J}^n/\mathcal{J}^{n+1}) =$
$= I(U \cap V)^n/(I(U \cap V)^n \cap I(U) + I(U \cap V)^{n+1}) \xrightarrow{\phi^n_U} D^{*n}_{U,V}$ whose kernel and cokernel are given respectively by
$(I(U \cap V)^{(n)} \cap I(U) + I(U \cap V)^{(n+1)})/(I(U \cap V)^n \cap I(U) + I(U \cap V)^{n+1})$ and
$I(U \cap V)^{(n)}/(I(U \cap V)^n + I(U \cap V)^{(n+1)} + I(U) \cap I(U \cap V)^{(n)})$. So we get

canonical homomorphisms $\phi^n : \mathcal{J}^n/\mathcal{J}^{n+1} \to \mathcal{D}^{*n}_{W,V}$ such that supp ker ϕ^n and supp coker ϕ^n are contained in the (closed) set of those $p \in V$ in which $\mathcal{J}$ is not normally torsion-free. Now we get a homomorphism

$$(1.4) \qquad Gr(\mathcal{J}) \to \mathcal{D}^{*}_{W,V} ,$$

which is an isomorphism if $\mathcal{J}$ is normally torsion-free. This is in particular true if W and V are smooth or - more generally - if W is Cohen Macaulay (CM) and V a locally complete intersection with respect to W .

These remarks show the great importance of the normal torsion-freeness. But in fact there is no simple criterion for this property. So it is natural to try to get information on the sets $Ass(\mathcal{J}^n/\mathcal{J}^{n+1})$ which measure how far away $\mathcal{J}$ is from this property. We relate these sets to the geometric object induced by $Gr(\mathcal{J})$ namely the cone $Spec(Gr(\mathcal{J}))$. This cone is the projecting cone over the exceptional fiber of the blow-up $Bl_W(V)$ of W at V . $Spec(Gr(\mathcal{J}))$ is also called the conormal cone of V with respect to V . This name is motivated by the replacement of $\mathcal{D}^{*}_{W,V}$ by $Gr(\mathcal{J})$ and the fact that $\mathcal{D}^{*1}_{W,V}$ is what we expect the sheaf of conormal fields to be: Namely the sheaf dual to the sheaf of polynomially defined vector fields which are normal to V and tangent to W.

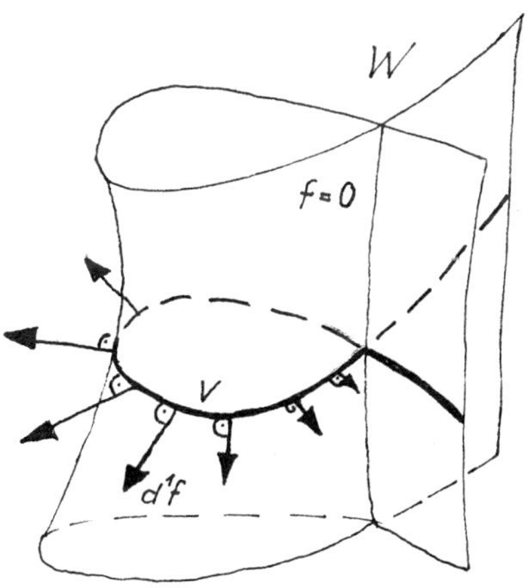

These vector fields may be imagined as sketched. Namely, if f is a section over W which vanishes at V , the corresponding conormal field is given by d^1f . This field really is normal (for the real scalar product) to the hypersurface $f = 0$, thus to V . Similar $\mathcal{J}^n/\mathcal{J}^{n+1}$ is refered to as the n-th conormal sheaf of V with respect to W .

In section 2 we will give a geometric description of the sets $\bigcup_n Ass(\mathcal{J}^n/\mathcal{J}^{n+1})$ and $\lim_{n\to\infty} Ass(\mathcal{J}^n/\mathcal{J}^{n+1})$. For a fixed point $p \in V$ a similar description of $\min_n depth(\mathcal{J}^n_p/\mathcal{J}^{n+1}_p)$ and $\lim_{n\to\infty} depth(\mathcal{J}^n_p/\mathcal{J}^{n+1}_p)$ is given. This gives another improvement of Burch's inequality for the analytic spread.

Once having accepted to replace $\mathcal{D}^*_{W,V}$ by $Gr(\mathcal{J})$ it turns out that even this latter object is too complicated in many cases. This leads to approximate the associated graded ring $Gr(\mathcal{J})$ by the symmetric algebra $S(\mathcal{J}/\mathcal{J}^2)$ of the conormal sheaf. This object in fact only depends on the first order differential behaviour of V in W. Clearly there is a canonical epimorphism

$$(1.5) \qquad S(\mathcal{J}/\mathcal{J}^2) \xrightarrow{\alpha} Gr(\mathcal{J}) \ .$$

Recently several authors have studied this homomorphism [8], [9]. Particularly criteria for α to be an isomorphism have been given. Clearly α is an isomorphism if W is CM and V a complete intersection. In this case, the conormal cone fibers locally trivial over V. In general, if only α is an isomorphism the fibers of the canonical morphism $\Lambda : \mathrm{Spec}(Gr(\mathcal{J})) \to V$ are all affine spaces. More precisely we have $\Lambda^{-1}(p) = \mathbb{C}^{\mu_p(\mathcal{J})}$, where $\mu_p(\mathcal{J})$ is the minimal number of generators of the $_{W,p}$-ideal $\mathcal{J}_p$. We consider some consequences of this in section 3.

Our last theme is the calculation of the cohomology of certain blow-up and their exceptional fibers. We first treat the blow-up of a d-dimensional point in an intersection of d hypersurfaces in general position. This applies to understand Macaulayfication and gives new characterizations of Buchsbaum singularities.

We then also consider some "symbolic" blow-up, which in most cases become (arithmetically) normal if the ground variety is, and which in general, are also CM if the ground variety V has only finitely many points which are not CM.

As we are only interested in local properties we always start with a noetherian local ring $(R,\mathfrak{m})$ and an ideal I. We then study the Rees ring $\mathcal{R}(I) = \bigoplus_{n\geq 0} I^n$ and the form ring $Gr(I) = \mathcal{R}(I)/I\mathcal{R}(I) = \bigoplus_{n\geq 0} I^n/I^{n+1}$. These are exactly the rings underlying the geometric objects we are interested in. By $\mathfrak{m}(I)$ we shall denote the homogeneous maximal ideal $\mathfrak{m} \oplus I \oplus I^2 \oplus \ldots$ of the Rees ring $\mathcal{R}(I)$. This is the ideal which corresponds to the vertex of the projecting cone over the blow-up $Bl_R(I) = \mathrm{Proj}(\mathcal{R}(I))$ of $\mathrm{Spec}(R)$ at $\mathrm{Spec}(R/I)$.

2. DEPTHS OF HIGHER CONORMAL SHEAVES

In this section we present some results of [3]. We consider a graded noetherian $R_o = R$ algebra $A = R_o \oplus R_1 \oplus R_2 \oplus \ldots$ which satisfies $A = R_o[R_1]$. We put $A_+ = R_1 \oplus R_2 \oplus \ldots$. A prime $\mathfrak{p}$ in A is called essential if $A_+ \not\subseteq \mathfrak{p}$. By $\mathrm{Spec}(A)^\bullet$ we denote the set of essential primes in A. If $\mathcal{M} \subseteq \mathrm{Spec}(A)$, $\mathcal{M}^\bullet$ stands for $\mathcal{M} \cap \mathrm{Spec}(A)^\bullet$. By $h\mathcal{M}$ we shall denote the set of homogeneous members of $\mathcal{M}$. Recall that $h\,\mathrm{Spec}(A)^\bullet$ is the topological space underlying to $\mathrm{Proj}(A)$. For simplicity we always will assume that $R/\mathfrak{m}$ is infinite. Now, let $M = \bigoplus_{n \geq n_o} M_n$ be a graded A-module of finite generation. A form $x \in R_i$ is called quasiregular with respect to M if $(0 : x)_M$ vanishes in almost all degrees or - equivalently - if the R-homomorphism $x : M_n \to M_{n+i}$ is injective for almost all n. It is easy to prove that

(2.1) $\quad x \in R_i$ is quasiregular with respect to M iff $x \notin \bigcup_{\mathfrak{p} \in \mathrm{Ass}(M)^\bullet} \mathfrak{p}$.

Now we have

(2.2) PROPOSITION: (i) $\bigcup_{n \geq n_o} \mathrm{Ass}_R(M_n) = \{\mathfrak{p} \cap R_o \mid \mathfrak{p} \in \mathrm{Ass}_A(M)\}$.

(ii) $\mathrm{Ass}_R(M_n)$ takes a constant value A^* for large n.

(iii) $A^* = \{\mathfrak{p} \cap R_o \mid \mathfrak{p} \in \mathrm{Ass}(M)^\bullet\}$.

Proof: "(i)" : We choose a member $\mathfrak{s}$ of $\mathrm{Ass}_R(M_n)$ and have to find a $\mathfrak{p} \in \mathrm{Ass}_A(M)$ which retracts to $\mathfrak{s}$. Localizing at $\mathfrak{s}$ we may assume that $\mathfrak{s} = \mathfrak{m}$. We find a $m \in M_n - (0)$ with $m = 0$. It follows $\mathfrak{m} \subseteq \bigcup R \cap \mathfrak{p}$ where $\mathfrak{p}$ runs through $\mathrm{Ass}_A(M)$. Now by the maximality of $\mathfrak{m}$ we find the requested $\mathfrak{p}$. If $\mathfrak{p} \in \mathrm{Ass}_A(M)$ there is some form $m \in M_n$ such that $(0 : m)_A = \mathfrak{p}$. It clearly follows that $(0 : m)_R = \mathfrak{p} \cap R \in \mathrm{Ass}_R(M_n)$.

"(ii)": By (i) $\bigcup_{n \geq n_o} \mathrm{Ass}_R(M_n)$ is finite. So it suffices to see that $\mathrm{Ass}_R(M_n) \subseteq \mathrm{Ass}_R(M_{n+1})$ for all large n. As $R/\mathfrak{m}$ is infinite we find a $x \in R_1 - \bigcup_{\mathfrak{p} \in \mathrm{Ass}_A(M)^\bullet} \mathfrak{p}$. By (2.1) x is quasiregular. So for all large n we have an injection $x : M_n \to M_{n+1}$, which induces the result.

"(iii)": Let $\mathfrak{s} \in A^*$. We want to construct an essential $\mathfrak{p} \in \mathrm{Ass}_A(M)$ which retracts to $\mathfrak{s}$. After localizing we may assume $\mathfrak{s} = \mathfrak{m}$. As $\mathfrak{m} \in A^*$

we clearly have that $(0 : \mathfrak{m})_{M_n} \neq 0$ for all large n. As $(0 : \mathfrak{m})_M$ is finitely generated we find some form $m \in (0 : \mathfrak{m})_{M_n}$ (for an appropriate n) such that Am is $\neq 0$ in all large degrees. It follows $(A_+)^s m \neq 0$ for all $s > 0$. So we have $(A_+)^s \not\subseteq (0 : m)_A$ for all such s. But now $(0 : m)_A$ is a homogeneous ideal with $\mathrm{Ass}(A/(0 : m)_A) \subseteq \mathrm{Ass}(M)$. By the previous statement $(0 : m)_A$ must have an essential minimal prime divisor $\mathfrak{q}$. It follows $\mathfrak{q} \in \mathrm{Ass}(M)$ and $\mathfrak{q} \cap R \supseteq (0 : m)_A \cap R \supseteq \mathfrak{m}$. Thus $\mathfrak{q} \cap R = \mathfrak{m}$.

Finally let $\mathfrak{q} \in \mathrm{Ass}(M)^\bullet$ and put $\mathfrak{s} = \mathfrak{q} \cap R$. We have to show that $\mathfrak{s} \in \mathrm{Ass}_R(M_n)$ for all large n. We localize to assume that $\mathfrak{s} = \mathfrak{m}$. We find some form $m \in M_n$ such that $\mathfrak{q} = (0 : m)_A$. As $\mathfrak{q}$ is essential we have $A_+ m \neq 0$. So for large n we have $Am \cap M_n \neq 0$, thus a m_n in $M_n - (0)$ with $\mathfrak{q} m_n = 0$ hence in particular with $\mathfrak{q} m_n = 0$. But this means that $\mathfrak{m} \in A^*$.

Introducing the morphisms $\Lambda : \mathrm{Spec}(A) \to \mathrm{Spec}(R)$ and $\lambda : \mathrm{Proj}(A) \to \mathrm{Spec}(R)$ our result may be formulated as follows.

(2.3) COROLLARY: In the previous notations we have

(i) $\bigcup_{n \geq n_o} \mathrm{Ass}_R(M_n) = \Lambda(\mathrm{Ass}(M))$.

(ii) $A^* = \lim_{n\to\infty} \mathrm{Ass}_R(M_n) = \lambda(\mathrm{Ass}(\tilde{M}))$, where $\tilde{M}$ stands for the $\mathrm{Proj}(A)$ - sheaf induced by M.

Now, based on (2.2), there is an inductive proof of the following result (for details see [3]).

(2.4) PROPOSITION: (i) $\min_n \mathrm{depth}(M_n) = \min \{\mathrm{depth}(M_{\mathfrak{q}}) \mid \Lambda^{-1}(\mathfrak{m})\}$.

(ii) $\mathrm{depth}(M_n)$ takes a constant value t^* for all large n.

(iii) $t^* = \min \{\mathrm{depth}(M_{\mathfrak{q}}) \mid \Lambda^{-1}(\mathfrak{m})^\bullet\} = \min \{\mathrm{depth}(M_p) \mid p \in \lambda^{-1}(\mathfrak{m})\}$.

(We made use of the convention $\min \phi = \infty$).

Now we choose an ideal I of R and introduce the following notations.

(2.5) $A(n) = \mathrm{Ass}(R/I^n)$, $B(n) = \mathrm{Ass}(I^{n-1}/I^n)$,

(2.6) $\quad t(n) = \mathrm{depth}(R/I^n)\ ,\ \bar{t}(n) = \mathrm{depth}(I^{n-1}/I^n)\ .$

We write again Λ, λ for the canonical morphisms $\mathrm{Spec}(Gr(I)) \to \mathrm{Spec}(R)$ resp. $\mathrm{Proj}(Gr(I)) \to \mathrm{Spec}(R)$.

Now, applying the precious results to the pair $A = \mathcal{R}(I)$ and $M = Gr(I)$ we get (using also the sequences $0 \to I^{n-1}/I^n \to R/I^n \to R/I^{n-1} \to 0$) the following result, which is an improvement of the main result of [2].

(2.7) PROPOSITION: (i) $\bigcup_{n\leq k} A(n) = \bigcup_{n\leq k} B(n)\ ,\ \bigcup A(n) = \bigcup B(n) = \Lambda(\mathrm{Ass}(Gr(I)))$.

(ii) $\lim_{n\to\infty} A(n) \supseteq \lim_{n\to\infty} B(n) = \{\mathfrak{p} \cap R \mid \mathfrak{p} \in \mathrm{Ass}(Gr(I))^{\bullet}\} = \lambda(\mathrm{Ass}(\mathrm{Proj}(Gr(I))))$.

(iii) If I contains regular elements the above limits coincide.

(2.8) PROPOSITION: (i) $\min_{n\leq k} t(n) = \min_{n\leq k} \bar{t}(n)\ ,\ \min t(n) = \min t(n) = \min\{\mathrm{depth}(Gr(I)_{\mathfrak{p}}) \mid \mathfrak{p} \in \Lambda^{-1}(\mathfrak{m})\}$.

(ii) $\lim_{n\to\infty} t(n) \leq \lim_{n\to\infty} \bar{t}(n) = \min\{\mathrm{depth}(Gr(I)_{\mathfrak{p}}) \mid \mathfrak{p} \in \Lambda^{-1}(\mathfrak{m})^{\bullet}\} = \min\{\mathrm{depth}(_{\mathrm{Proj}(Gr(I)),p}) \mid p \in \lambda^{-1}(\mathfrak{m})\}$.

(iii) If I is of positive height h and contains a regular sequence of length h the above limits coincide.

Recall that the analytic spread of I is defined as $\dim(\Lambda^{-1}(\mathfrak{m}) - \mathcal{F})$. Assume now that I is not nilpotent. Then - by Nakayama - $R/\mathfrak{m} \otimes Gr(I)$ is of positive dimension. This means that $\Lambda^{-1}(\mathfrak{m})^{\bullet}$ is not empty, hence that all the generic points of $\mathcal{F}$ are essential. So we get

(2.9) $$\lim_{n\to\infty} \bar{t}(n) = \min\{\mathrm{depth}(Gr(I)_{\mathfrak{p}}) \mid \mathfrak{p}\ \Lambda^{-1}(\mathfrak{m})^{\bullet}\} \leq \mathrm{ht}(\mathfrak{m}Gr(I)) \leq$$
$$\leq \dim(Gr(I)) - \ell(I) = \dim(R) - \ell(I)\ ,$$

where $\ell(I)$ stands for the analytic spread of I . This is the announced inequality for the analytic spread [10], [12].

3. THE GEOMETRY OF THE CONORMAL CONE

In this section we assume that the canonical map $\alpha : S(I/I^2) \to Gr(I)$ is an isomorphism. Then, as remarked in the introduction we have the following property of the fiber rings of Λ.

$$(3.1) \qquad K(\mathfrak{p}) \otimes Gr(I) \cong K(\mathfrak{p})[X_1,\ldots,X_{\mu_{\mathfrak{p}}(I)}] \quad , \quad \forall\, \mathfrak{p} \in V(I) \quad .$$

Thereby $\mu_{\mathfrak{p}}(I)$ denotes the minimal number of generators of the $R_{\mathfrak{p}}$-ideal $I_{\mathfrak{p}}$. We are interested to determine the minimal primes of $Gr(I)$ or - equivalently - the irreducible components of the conormal cone $Spec(Gr(I))$. To do so put $\tilde{\mathfrak{p}} = Gr(I) \cap \mathfrak{p}\, Gr(I)_{\mathfrak{p}}$, where $\mathfrak{p} \in V(I)$. By (3.1) $\tilde{\mathfrak{p}}$ is prime. We claim

(3.2) PROPOSITION: Assume that R is of pure dimension d, universally catenary and satisfies (3.1). Then it holds

$$Min(Gr(I)) = \{\tilde{\mathfrak{p}} \mid \mathfrak{p} \in V(I), \mu_{\mathfrak{p}}(I) = ht(\mathfrak{p})\} \quad .$$

Thereby Min stands for the set of minimal primes. Note also that always $\mu_{\mathfrak{p}}(I) \leq ht(\mathfrak{p})$.

Proof: By our hypothesis $Gr(I)$ is of pure dimension d and catenary. Let $\mathfrak{p} \in V(I)$, $\mu_{\mathfrak{p}}(I) = ht(\mathfrak{p})$. We want to show that $\tilde{\mathfrak{p}} \in Min(Gr(I))$. After localizing at $\mathfrak{p}$ we may assume that $\mathfrak{p} = \mathfrak{m}$, thus that $\tilde{\mathfrak{p}} = \tilde{\mathfrak{m}} = \mathfrak{m}Gr(I)$. By (3.1) we have $\dim(Gr(I)/\mathfrak{m}Gr(I)) = \mu_{\mathfrak{m}}(I) = ht(\mathfrak{m}) = d = \dim(Gr(I))$. This shows that $\tilde{\mathfrak{m}} \in Min(Gr(I))$.

Now let $\mathfrak{p}' \in Min(Gr(I))$ and put $\mathfrak{p} = \mathfrak{p}' \cap R$. Clearly we have $\tilde{\mathfrak{p}} \subseteq \mathfrak{p}'$, thus $\tilde{\mathfrak{p}} = \mathfrak{p}'$. It remains to show that $ht(\mathfrak{p}) = \mu_{\mathfrak{p}}(I)$. To do so we again may assume $\mathfrak{p} = \mathfrak{m}$. So we obtain $\mu_{\mathfrak{m}}(I) = \dim(Gr(I)/\tilde{\mathfrak{m}}) = \dim(Gr(I)) = d = ht(\mathfrak{m})$.

Now choose V, W and $\mathcal{J} \subseteq \mathcal{O}_W$ as in the introduction. For a non-negative integer μ let $V_\mu \subseteq V$ be the closed subvariety of those points p for which $\mu_p(\mathcal{J}) \geq \mu$. Assume that the fibers of $\Lambda : Spec(Gr(\mathcal{J})) \to V$ are all affine spaces. If h is the codimension of V with respect to W, let $V_{\mu,1},\ldots,V_{\mu,N_\mu}$ those irreducible components of V_μ which satisfy the following conditions:

(a) $V_{\mu,i} \not\subseteq V_{\mu-1}$.

(b) $V_{\mu,i}$ is of codimension $\mu-h$ in V .

Then - by (3.2) - Λ establishes a 1-1-correspondence between the varieties $V_{\mu,i}$ $(\mu=h,\ldots,\dim(W), i=1,\ldots,N_\mu)$ and the irreducible components of the conormal cone $\mathrm{Spec}(\mathrm{Gr}(\mathfrak{J}))$. The same situation clearly holds for the morphism $\lambda : \mathrm{Proj}(\mathrm{Gr}(\mathfrak{J})) \to V$. By $\tilde{V}_{\mu,i}$ we denote the irreducible component of $\mathrm{Spec}(\mathrm{Gr}(\mathfrak{J}))$ which corresponds to $V_{\mu,i}$, by $\overline{V}_{\mu,i}$ the corresponding irreducible component of the exceptional fiber $\mathrm{Proj}(\mathrm{Gr}(\mathfrak{J}))$. $\overline{V}_{\mu,i}$ is the projecting cone of $\tilde{V}_{\mu,i}$. Now we clearly have

(3.3) COMPLEMENT: Let R be as in (3.2). Then for each pair $\mathfrak{p}, \mathfrak{q} \in V(I)$ it holds $\mathrm{Min}(\mathrm{Gr}(I)/(\tilde{\mathfrak{p}} + \tilde{\mathfrak{q}})) = \{\tilde{\mathfrak{s}} \mid \mathfrak{s} \in \mathrm{Min}(R/(\mathfrak{p}+\mathfrak{q}))\}$.

Thus, denoting - for an irreducible subvariety T of V - the irreducible subvarieties in $\mathrm{Spec}(\mathrm{Gr}(\mathfrak{J}))$ resp. $\mathrm{Proj}(\mathrm{Gr}(\mathfrak{J}))$ which - via $\mathfrak{p} \to \tilde{\mathfrak{p}}$ - correspond to T by $\tilde{T}$ resp. $\overline{T}$ we see: The irreducible components of $\tilde{T} \cap \tilde{S}$ are exactly the varieties $\tilde{X}$, where X runs through the irreducible components of $T \cap S$. The same statement is also true for the varieties $\overline{T}$ and $\overline{S}$.

We sketch this situation (for the blow-up $\mathrm{Bl}_W(V)$) in the case where W is a surface.

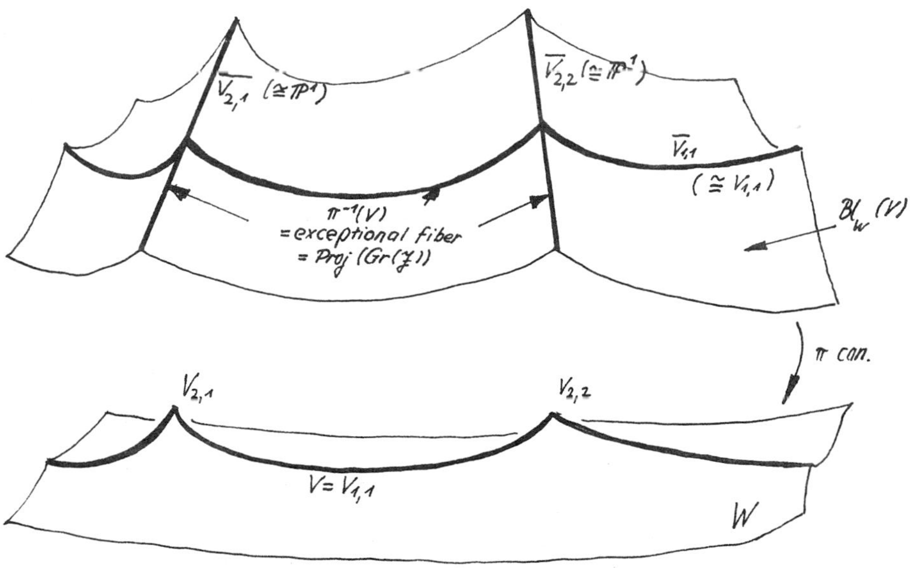

4. COHOMOLOGY OF CERTAIN BLOW-UP AND THEIR EXCEPTIONAL FIBER

In this section we present (without proofs) some of the results of [4] and [5]. Let $\mathcal{F}$ be a coherent sheaf over W and let $\mathcal{J}$ be any coherent ideal of the structure sheaf $\mathcal{O}_W$. We consider the blow-up $Bl_W(\mathcal{J}) = Proj(\mathcal{R}(\mathcal{J}))$ of W at $\mathcal{J}$ and the coherent sheaf $Bl_W(\mathcal{J},\mathcal{F})$ over $Bl_W(\mathcal{J})$ which is induced by the graded $\mathcal{R}(\mathcal{J})$-module $\mathcal{R}(\mathcal{J},\mathcal{F}) = \bigoplus_{n \geq 0} \mathcal{J}^n\mathcal{F}$. We call $Bl_W(\mathcal{J},\mathcal{F})$ the blow-up of $\mathcal{F}$ at $\mathcal{J}$. Let $V \subseteq W$ be the variety defined by $\mathcal{J}$. We now are interested in those global sections of $\mathcal{B} = Bl_W(\mathcal{J},\mathcal{F})$ which induce sections in $\Gamma(W-V,\mathcal{F})$. More precisely: if $\pi : Bl_W(V) \xrightarrow{\pi} W$ stands for the canonical projection, we are interested in the set $\Gamma_V(\mathcal{B}) = \{f \in \Gamma(\mathcal{B}) \mid f(p) = f(q), \forall p,q \in Bl_W(V) \text{ with } \pi(p) = \pi(q) \notin V\}$. $\mathcal{B} \to \Gamma_V(\mathcal{B})$ defines a functor, which gives rise to the functors $H^i_V(\cdot)$ of relative Serre cohomology (defined for all coherent sheaves over $Bl_W(V)$). If $\mathcal{L} \subseteq \mathcal{O}_W$ is an ideal we put $\mathcal{L}(\mathcal{J})$ for the ideal $\mathcal{L} \oplus \mathcal{J} \oplus \mathcal{J}^2 \oplus \ldots$ of the Rees algebra $\mathcal{R}(\mathcal{J})$. Then, as for the usual Serre cohomology over $\mathbb{P}^d$ we get the following relations

(4.1) (i) $\bigoplus_n \Gamma_V(\mathcal{B}(n)) = D_{\sqrt{\mathcal{J}}(\mathcal{J})}(\mathcal{R}(\mathcal{J},\mathcal{F}))$,

(ii) $\bigoplus_n H^i_V(\mathcal{B}(n)) = H^{i+1}_{\sqrt{\mathcal{J}}(\mathcal{J})}(\mathcal{R}(\mathcal{J},\mathcal{F}))$, $(i > 0)$.

Thereby D_J stands for the functor of ideal transform and H_J for the local cohomology with respect to the ideal J .

The following sketch shall visualize the meaning of $\Gamma_V(Bl(\mathcal{J},\mathcal{O}_V))$. We choose W as a surface, V as a point and $\mathcal{J}$ as the ideal induced by V . So the local cohomology modules $H^i_{\sqrt{\mathcal{J}}(\mathcal{J})}(\mathcal{R}(\mathcal{J}))$ are objects of a rather concrete geometric meaning. In the sequel we want to determine these modules (for all i in the range of finiteness) for a particular class of blow-up. We namely want to choose a point on W which has a punctured neighbourhood of CM points and we want to choose $\mathcal{J}$ as an ideal defined by d (=dim(W)) sections in general position in this point. The problem is in fact of local nature. So in view of Grothendieck's finiteness criterion for local cohomology modules all comes up to study the modules $H^i_{\mathfrak{m}(I)}(\mathcal{R}(I))$, where $(R,\mathfrak{m})$ is local of dimension d , where I is generated by a system of parameters (s.o.p.) and where $H^i_{\mathfrak{m}}(R)$ is finitely generated for all $i < d$. Easy examples show, that for an arbitrary s.o.p. the behaviour of the cohomology in question may be still very complex. So we have to subject I to another condition, which obviously should be mild enough to admit a suitable number of applications.

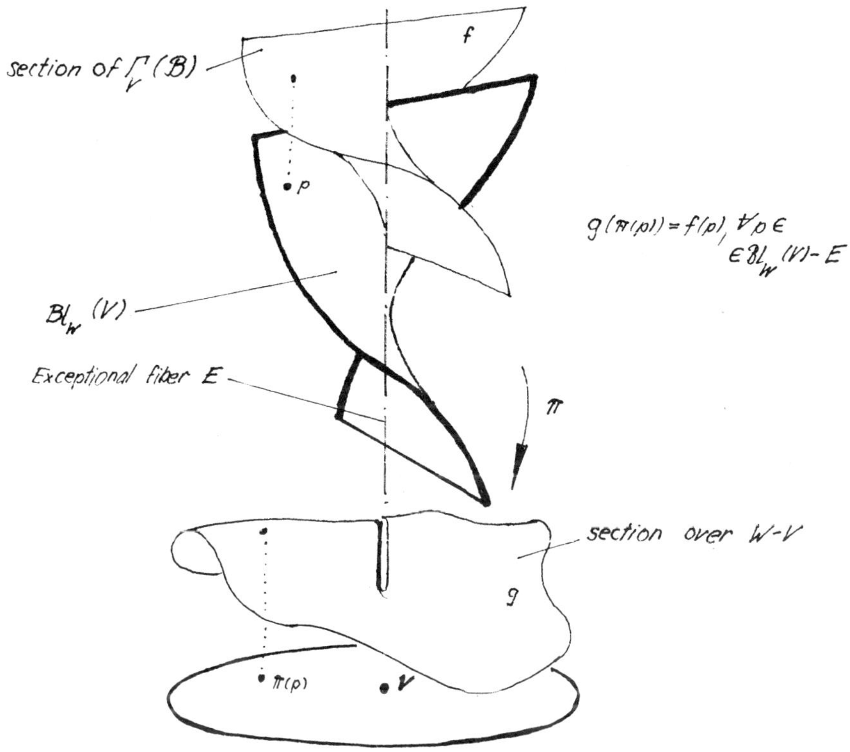

The condition we impose on the s.o.p. $x_1,\dots,x_d$ which generates I is the so-called (permutable) standard condition:

(4.2) For each permutation $\sigma \in \mathfrak{S}_d$ and each pair of non-negative integers i , j with $i + j < d$ it shall hold

$$I\, H^i_{\mathfrak{m}}(R/x_{\sigma(1)},\dots,x_{\sigma(d)})) = 0 \quad .$$

Under the above hypotheses on the modules $H^i_{\mathfrak{m}}(R)$ we always find an $\mathfrak{m}$-primary ideal $\mathfrak{a}$ such that each s.o.p. in $\mathfrak{a}$ satisfies (4.2). For a CM ring R each s.o.p. satisfies (4.2) as the occuring cohomology modules all vanish in this case. But there is another more general class of very interesting local rings for which each s.o.p. satisfies (4.2). We namely have:

(4.3) R is a Buchsbaum (=B)-ring iff (4.2) holds for each s.o.p. .

Systems satisfying (4.2) are permutable d-sequences [9], so we already know that $\mathcal{R}(I)$ and $Gr(I)$ are symmetric algebras [9]. Now we state a result on the shape of the cohomology we are interested in. Thereby we use the following notation: If M is a R-module, $[M]_j$ denotes the graded module over $\mathcal{R}(I)$ which has M in its j-th component and 0 elsewhere. Now we have

(4.4) PROPOSITION: Let I be generated by a s.o.p. which satisfies the condition (4.2). Then it holds:

$$\text{(i)} \quad H^i_{\mathfrak{m}(I)}(\mathcal{R}(I)) = \begin{cases} [H^0_{\mathfrak{m}}(R)]_0 , & \text{if } i = 0 \\ 0 , \text{ if } i \in \{1, \min(2,d) =: d'\} & \\ \bigoplus_{j=-i+i}^{-1} [H^{i-1}_{\mathfrak{m}}(R)]_j , & \text{if } d' < i \leq d . \end{cases}$$

$$\text{(ii)} \quad H^i_{\mathfrak{m}(I)}(Gr(I)) = [H^i_{\mathfrak{m}}(R)]_{-i} , \quad \text{for all } i < d .$$

We only mention a few applications of this result.

(4.5) COROLLARY: If $depth(R) > 0$, R is a B-ring iff $Proj(\mathcal{R}(I))$ is CM for each ideal I generated by a system of parameters.

Using the fact that local cohomology permutes with taking the Veronese transform $\mathcal{R}(I^h)$ of $\mathcal{R}(I)$ we then may deduce:

(4.6) COROLLARY: Let $h < d$. Then we have equivalence between:

(i) R is a B-ring with $H^i_{\mathfrak{m}}(R) = 0$ for all $i \quad \{1,\dots,h,d\}$.

(ii) $\mathcal{R}(I^h)$ is CM for each ideal I generated by a system of parameters.

(iii) $\mathcal{R}(I^t)$ is CM for each $t \geq h$ and each ideal I generated by a system of parameters.

For $h = 1$ this gives the main result of [7].

(4.4) clearly leads to the following processus of Macaulayfication [6]: Let W be a quasiprojective variety with only finitely many non-CM points $p_1,\dots,p_n$. Then there is an ideal $\mathcal{J} \subseteq \mathcal{O}_W$ satisfying $\mathcal{J}_p = \mathcal{O}_{W,p}$ for $p \neq p_1,\dots,p_n$ and such that $\mathcal{J}_{p_i} \subseteq \mathcal{O}_{W,p_i}$ is generated by s.o.p.

which satisfies (4.2). Then clearly $Bl_W(\mathcal{J}) = Proj(\mathcal{R}(\mathcal{J}))$ is CM. If the singularities p_i are all of Buchsbaum type we even easily may choose in such a way that $\mathcal{J}_{p_i}$ is a minimal reduction of the maximal ideal $\mathfrak{m}_{p_i} \subseteq \mathcal{O}_{W,p_i}$ for all $i \leq n$ (Use (4.3)!). If we do this, we know that $Bl_W(\mathcal{J})$ is homeomorphic to $Bl_W(\{p_1,\ldots,p_n\})$, if the reduced fibers of p_i in this latter are projective spaces [1]. So $Bl_W(\mathcal{J})$ is considered as a "exotic structure" on the blow-up $Bl_W(\{p_1,\ldots,p_n\})$ which has an improved cohomology-behaviour. To visualize this, we give the following example:

(4.7) Example: Let W be the affine surface whose coordinate ring is $\mathbb{C}[X,Y^2,XY,Y^3] \subseteq \mathbb{C}[X,Y]$. W is homeomorphic to the affine plane and has an isolated singularity p at the origin which is of Buchsbaum type. $W - \{p\}$ is isomorphic to the punctured plane. Then $\mathcal{J} = (X,Y^2)\mathcal{O}_W$ defines a minimal reduction in p and coincides with $\mathcal{O}_{W,q}$ for $p \neq q$. It turns out that $\mathcal{R}(\mathfrak{m}_p)$ is of depth 1, whereas $\mathcal{R}(\mathcal{J})$ becomes a CM-ring by (4.4). So the "exotic structure" is much better from the point of view of cohomology. On the other side this improvement goes to the cost of having more singularities. We namely get the following shapes (of the real parts) of both blow-up.

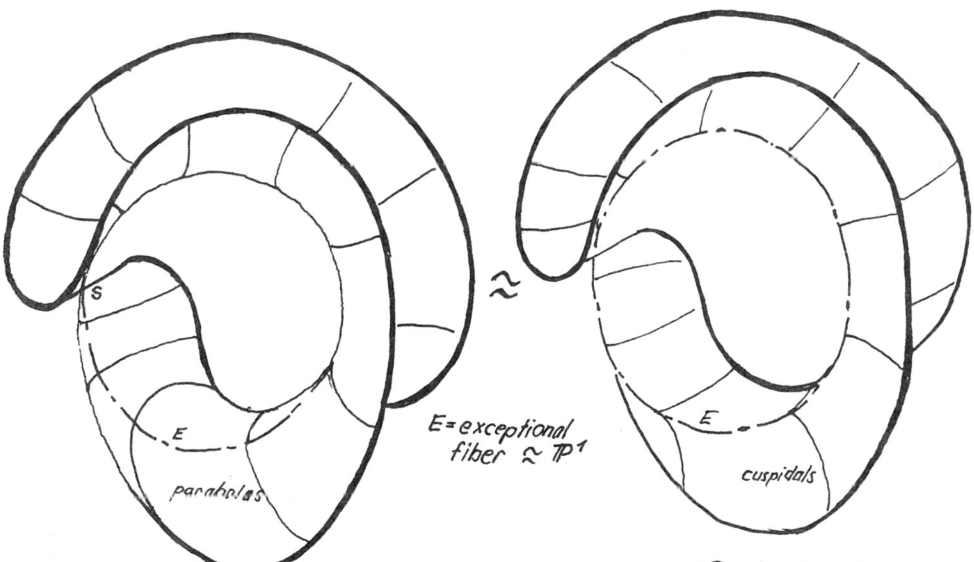

$T = Bl_W(p)$, $Sing(T) = \{s\}$

$S = Proj(\mathcal{R}(\mathcal{J}))$ $Sing(S) = E$
"Exotic structure"

Performing the proposed Macaulayfication with a normal singularity it shows up that in general we lose the property of normality. To eliminate this disadvantage there is another method of Macaulayfication, which bases on the following construction (for details see [5]). Let $x_1,\ldots,x_d$ be an s.o.p. which satisfies (4.2), put $L = (x_1,\ldots,x_{d-1})$ and consider the symbolic Rees ring $D_{\mathfrak{m}}(\mathcal{R}(L))$. It turns out that (if $d > 1$) this new ring is finite over $\mathcal{R}(L)$. It also turns out, that its local cohomology if of a type very close to the one of $\mathcal{R}(I)$ (with $I = (x_1,\ldots,x_d)$). So, as for cohomology we may replace $\mathcal{R}(I)$ by $D_{\mathfrak{m}}(\mathcal{R}(L))$, thus the blow-up $\mathrm{Proj}(\mathcal{R}(I))$ by the symbolic blow-up $\mathrm{Proj}(D_{\mathfrak{m}}(\mathcal{R}(L)))$. Now it turns out, that $D_{\mathfrak{m}}(\mathcal{R}(L))$ is in fact equal to the ordinary Rees ring $\mathcal{R}(\overline{L})$ of the ideal $\overline{L} := D_{\mathfrak{m}}(L)$ in the finite overring $D_{\mathfrak{m}}(R)$ of R . If R is normal, we have $R = D_{\mathfrak{m}}(R)$. So we get by the local Bertini theorems

(4.8) LEMMA: Let R be excellent and let $\mathfrak{a}$ be $\mathfrak{m}$-primary. Then for a general s.o.p. $x_1,\ldots,x_d$ in $\mathfrak{a}$, $\overline{L} = D_{\mathfrak{m}}(L)$ $(L = (x_1,\ldots,x_{d-1}))$ is reduced and generically a complete intersection.

Combining this with the previous remarks one then deduces the following result.

(4.9) PROPOSITION: Let R normal and excellent, $H^i_{\mathfrak{m}}(R)$ finitely generated for all $i < d = \dim(R)$. Then there is an $\mathfrak{m}$-primary ideal $\mathfrak{a}$ such that for a generic s.o.p. in $\mathfrak{a}$ $\mathrm{Proj}(\mathcal{R}(\overline{L}))$ is CM and arithmetically normal.

BIBLIOGRAPHY

1. Brodmann, M., Kohomologische Eigenschaften von Aufblasungen an lokal vollständigen Durchschnitten, Habilitationsschrift 1980

2. Brodmann, M., Asymptotic stability of $\mathrm{Ass}(M/I^nM)$, Proc. AMS 74 (1979) 16-18

3. Brodmann, M., Blow-up and asymptotic depth of higher conormal modules, preprint 1980

4. Brodmann, M., Local cohomology of certain Rees- and form-rings I, preprint 1980

5. Brodmann, M., Local cohomology of certain Rees- and form-rings II, preprint 1981

6. Faltings, G., Ueber Macaulayfizierung, Math. Ann. 238 (1978) 175-192

7. Goto, S., On the Macaulayfication of certain Buchsbaum-rings, preprint 1979

8. Herzog, J., Simis, M., Vasconcelos, V., Approximation complexes and blowing up rings, preprint 1981

9. Huneke, C., On the Symmetric and Rees-algebras of an ideal generated by a d-sequence, to appear

10. Burch, L., Codimension and analytic spread, Proc. Camb. phil. soc. 72 (1972) 369-373

11. Rees, D., On a problem of Zariski, Ill. J. of math. 2 (1968) 145-149

12. Brodmann, M., On the asymptotic nature of the analytic spread, Proc. Camb. phil. soc. 86 (1979) 35-39

ON FORM RING OF A ONE-DIMENSIONAL SEMIGROUP RING

Maria Pia Cavaliere - Gianfranco Niesi

Istituto di Matematica
Università di Genova
Genova (Italy)

1. Introduction

Let $n_0,\dots,n_p$ be positive integers generating a numerical semigroup and let k be a field. Our purpose is to study the "Cohen-Macaulayness" and the type of the form ring $G = \bigoplus_{i\geq 0} \mathfrak{m}^i/\mathfrak{m}^{i+1}$ of the semigroup ring $k[t^{n_0},\dots,t^{n_p}]$ relative to the homogeneous maximal ideal $\mathfrak{m} = (t^{n_0},\dots,t^{n_p})$.

On this subject we recall, in particular, some results of Robbiano - Valla and Herzog. The former authors determine all the rings $k[t^m, t^n, t^p]$ whose form rings are Cohen-Macaulay or complete intersection ([11]). The latter, using the notion of super-regular sequence, gives a necessary and sufficient condition in order that the form ring of $k[t^{n_0},\dots,t^{n_p}]$ is Cohen-Macaulay ([5]).

In our paper we first observe that the problem can be tackled studying the semigroup ring $k[S]$, where $S \subset \mathbb{N}^2$ is generated by $(n_0,0)$, $(n_1, n_1 - n_0)$,

---

This work was partially supported by C.N.R. (Consiglio Nazionale delle Ricerche)

..., $(n_p, n_p - n_o)$, $(0, n_o)$, so that we can use the semigroup ring theory of Goto-Suzuki-Watanabe ([4]). Then we introduce the notion of standard basis for S which enables us both to give a simple criterion (theorem 3.6) for the Cohen-Macaulayness of the rings $k[S]$ and G and also to compute their type (proposition 3.7). Our criterion can be easily traslated into a suitable algorithm for a computer.

2. Definitions and preliminary results

(2.1) Let $S \subset \mathbb{Z}^n$ be a finitely generated additive semigroup with identity and let L be the subgroup of $\mathbb{Z}^n$ generated by S . Let k be a field. The *semigroup ring* $k[S]$ of S over k is the subring of the polynomial ring $k[X_1, \ldots, X_n]$ generated by all the monomials $X^s = X_1^{s_1} \cdot \ldots \cdot X_n^{s_n}$ where $s = (s_1, \ldots, s_n)$ is in S .

It results $\dim k[S] = \dim k[L] = \mathrm{rank}_{\mathbb{Z}} L$.

(2.2) A semigroup S generated by natural numbers $n_o, \ldots, n_p$ is called *numerical semigroup* if $\mathrm{GCD}(n_o, \ldots, n_p) = 1$ (we will always assume that $n_o < n_1 < \ldots < n_p$). This is equivalent to say that $n + \mathbb{N} \subset S$ for some $n \in \mathbb{N}$. The smallest such integer c is called *conductor* of S . S is said to be *symmetric* if the number of elements in S that are less than c is $c/2$.

(2.3) Let S be a numerical semigroup and let $s \neq 0$ be an element of S ; the *s-standard basis* of S is the generating set $B = \{b_o, b_1, \ldots, b_{s-1}\}$ such that $b_o = s$ and , for $i > 0$, b_i is the least integer in S having s-residue distinct from those of $b_o, \ldots, b_{i-1}$.

Let B the s-standard basis of S . We put

$$\tilde{B} = \{ b \in B \;/\; b \neq b_o , \; b + b_i \notin B \text{ for every } i \}$$

(2.4) Let $(A, \mathfrak{m}, k)$ be a local d-dimensional Cohen-Macaulay ring. The *type* of A is the number $r(A) = \dim_k \mathrm{Ext}_A^d(k, A)$. If A is a Cohen-Macaulay ring (not necessarily local) the *(global) type* of A , still denoted by r(A) , is the supremum of the types of the local rings $A_{\mathfrak{m}}$ as $\mathfrak{m}$ ranges through the maximal ideals of A . If A is a graded ring it suffices to look at the localizations of A at the homogeneous maximal ideals ([1] ,theorem).

If r(A) = 1 , then A is said to be a *Gorenstein ring*.

REMARK. It is well-known that in the local case $r(A) = \dim_k \mathrm{Ann}_{A/I}(\mathfrak{m}/I)$, where I is an ideal generated by a maximal regular sequence.

We need the following graded version of this fact.

(2.5) LEMMA. *Let* $A = \bigoplus_{n\geq 0} A_n$ *be a graded Cohen-Macaulay ring such that* $A_0 = k$ *is a field. Then* $r(A) = \dim_k \mathrm{Ann}_{A/I}(\mathfrak{m}/I)$ *, where* $\mathfrak{m} = \bigoplus_{n>0} A_n$ *and* I *is an ideal generated by a homogeneous maximal regular sequence.*

Proof. By hypothesis $\mathfrak{m}$ is the only homogeneous maximal ideal of A , besides the homogeneous regular sequence generating I is a maximal regular sequence ([2], corollario 1.5). Then $r(A) = r(A_{\mathfrak{m}}) = \dim_k \mathrm{Ann}(\mathfrak{m}A_{\mathfrak{m}}/IA_{\mathfrak{m}}) =$ $= \dim_k(\mathrm{Ann}(\mathfrak{m}/I))_{\mathfrak{m}}$. Since A_0 is a field and I an $\mathfrak{m}$-primary ideal, every element of $(\mathrm{Ann}(\mathfrak{m}/I))_{\mathfrak{m}}$ can be written as f/1 with $f \in \mathfrak{m}/I$. It follows that $\dim_k(\mathrm{Ann}(\mathfrak{m}/I))_{\mathfrak{m}} = \dim_k \mathrm{Ann}(\mathfrak{m}/I)$.

(2.6) PROPOSITION. *Let* S *be a numerical semigroup, let* $s \in S$ *and let* $B = \{b_0,\dots,b_{s-1}\}$ *be the s-standard basis of* S *. Then the following conditions are equivalent:*

i) $k[S]$ *is Gorenstein*

ii) S *is symmetric*

iii) $b_{s-1} = b_i + b_{s-1-i}$ $\quad (1 \leq i \leq s-2)$

Proof. Condition (i) is equivalent to (ii) by [8], theorem and to (iii) by [10] , proposition 2.4 .

(2.7) PROPOSITION. *Let* S *and* B *be as in proposition 2.6 . Then:*

$$r(k[S]) = \mathrm{card}(\widetilde{B})$$

Proof. Since X^s is a non zero-divisor in $k[S]$, we have $r(k[S]) = \dim_k \operatorname{Ann}(\mathfrak{m}/(X^s))$ where $\mathfrak{m}$ is the homogeneous maximal ideal of $k[S]$ (lemma 2.5). Let $t \in S$. Then $t = ns + b$ for some $b \in B$. Thus a homogeneous element f of degree t in $k[S]/(X^s)$ is different from zero if and only if $t \in B \setminus \{s\}$, and $f \in \operatorname{Ann}(\mathfrak{m}/(X^s))$ if and only if $t + b_i \notin B$ for every $b_i \in B$. Hence $\operatorname{Ann}(\mathfrak{m}/(X^s)) = \bigoplus_{b \in \tilde{B}} X^b k$.

(2.8) EXAMPLE. Let S be the numerical semigroup generated by $s, s+1, \ldots, s+p$, where $p<s$. The s-standard basis of S consists of the element s, and of the elements $a_i s + i$ $(1 \leqslant i \leqslant s-1)$ where

$$a_i = \begin{cases} i/p & \text{if } i/p \in \mathbb{N} \\ [i/p]+1 & \text{otherwise} \end{cases}$$

($[i/p]$ denotes the greatest integer $\leqslant i/p$). Hence $\tilde{B} = \{a_i s + i \ / \ a_i = a_{s-1}\}$.

If $s-1 = pq + r$ $(r<p)$, then we have

$$\operatorname{card}(\tilde{B}) = \begin{cases} r & \text{if } r \neq 0 \\ p & \text{if } r = 0 \end{cases}$$

3. Cohen-Macaulayness and type of the form ring

From now on let $A = k[t^{n_0}, \ldots, t^{n_p}]$ where $n_0 < n_1 < \ldots < n_p$ generate minimally a numerical semigroup, let $\mathfrak{m} = (t^{n_0}, \ldots, t^{n_p})$ be the homogeneous maximal ideal of A, let $R = \bigoplus_{i \in \mathbb{Z}} \mathfrak{m}^i$ ($\mathfrak{m}^i = A$ for $i \leqslant 0$) be the Rees ring and $G = \bigoplus_{i \geqslant 0} \mathfrak{m}^i/\mathfrak{m}^{i+1}$ the form ring of A with respect to $\mathfrak{m}$.

Then $R = A[t^{n_0}X, \ldots, t^{n_p}X, X^{-1}] = k[t^{n_0}X, \ldots, t^{n_p}X, X^{-1}]$ is a semigroup ring associated to the semigroup S' generated by

$$(n_0, 1), \ldots, (n_p, 1), (0, -1).$$

As $G = R/(X^{-1})$, we have the following lemma:

(3.1) LEMMA. *i) G is Cohen-Macaulay if and only if so is R* ([7], theorem 4.11)

ii) $r(G) = r(R)$ ([3], theorem 3.7).

We note that the semigroup S' associated to R is contained in $\mathbb{Z}^2$ but not in $\mathbb{N}^2$.

(3.2) LEMMA. *The ring* R *is isomorphic to the semigroup ring* $k[S]$, *where* $S \subset \mathbb{N}$ *is the semigroup generated by:*

$$(n_0, 0), (n_1, n_1 - n_0), \ldots, (n_p, n_p - n_0), (0, n_0).$$

Proof. The group homomorphism $\varphi : \mathbb{Z}^2 \longrightarrow \mathbb{Z}^2$ defined by $\varphi(a, b) = (a, a - n_0 b)$ is injective and it induces a semigroup isomorphism between S' and $S = \varphi(S')$. It follows that the associated semigroup rings are isomorphic.

We observe that the rings R and $k[S]$ can be considered as graded rings over $\mathbb{Z}$ and also over $\mathbb{Z}^2$, but the ring isomorphism of lemma 3.2 is not a graded ring isomorphism.

(3.3) REMARKS. i) The group L generated by S is the subgroup of $\mathbb{Z}^2$ generated by $(n_0, 0)$, $(1,1)$, hence $L = \{(a, b) \in \mathbb{Z}^2 / a - b \in n_0\mathbb{Z}\}$.

ii) Let $\overline{S} = \{s \in L / as \in S$ for some positive integer $a\}$ be the *normalization* of S. Then $\overline{S} = L \cap \mathbb{N}^2$ and $\overline{S}$ is the semigroup generated by $(n_0, 0)$, $(1,1)$, $(0, n_0)$. From [6], §1 it follows that the normalization $\overline{R}$ of the ring R is isomorphic to $k[X^{n_0}, XY, Y^{n_0}]$.

Using the lemmas 3.1 and 3.2 we will investigate the Cohen-Macaulayness and the type of $k[S]$ instead of those of the form ring G.

Now we need the following lemma whose proof is a special case of theorem 2.2 of [4]. Nevertheless we prefer to give an easy indipendent proof of it.

(3.4) LEMMA. *The following conditions are equivalent:*

1) $k[S]$ *is Cohen-Macaulay*

2) X^{n_0}, Y^{n_0} *is a regular sequence*

3) $(X^{n_0}) \cap (Y^{n_0}) = (X^{n_0} Y^{n_0})$

4) $((n_0, 0) + S) \cap ((0, n_0) + S) = (n_0, n_0) + S$

Proof. First we remark that (1) is equivalent to (1') $k[S]_{\mathfrak{m}}$ is Cohen-Macaulay, where $\mathfrak{m}$ is the homogeneous maximal ideal of $k[S]$ ([9], theorem).

(1') ⇒ (2) $X^{n_o}/1$, $Y^{n_o}/1$ is a system of parameters for $k[S]_{\mathfrak{m}}$ and hence a $k[S]_{\mathfrak{m}}$-sequence. If f is a homogeneous element of $k[S]$ such that $fY^{n_o} \in (X^{n_o})$, then $f/1 \in (X^{n_o}/1)$, hence there exists $h \in k[S] \setminus \mathfrak{m}$ such that $hf \in (X^{n_o})$. Now all the homogeneous components of hf have to be in (X^{n_o}), in particular $h_o f \in (X^{n_o})$, but $h_o \in k$ hence $f \in (X^{n_o})$. Our thesis follows.

(2) ⇒ (3) It is an immediate consequence of the definition of regular sequence.

(3) ⇒ (1') It suffices to prove that $X^{n_o}/1$, $Y^{n_o}/1$ is a $k[S]_{\mathfrak{m}}$-sequence. Localizing at $\mathfrak{m}$ we have $(X^{n_o}/1) \cap (Y^{n_o}/1) = (X^{n_o}Y^{n_o}/1)$. Let $aY^{n_o}/1 \in (X^{n_o}/1)$. Then $aY^{n_o}/1 \in (X^{n_o}Y^{n_o}/1)$. Since $k[S]_{\mathfrak{m}}$ is an integral domain, $a \in (X^{n_o}/1)$.

(3) ⇔ (4) We consider $k[S]$ as a $\mathbb{Z}^2$-graded ring with degree in S. Then homogeneous ideals are equal if and only if so are the corresponding ideals of the semigroup S.

We put $S_i = pr_i(S)$ $(i=1,2)$, where pr_i is the natural projection to the i-th component. Then S_1 is the semigroup generated by $n_o, n_1, \ldots, n_p$ and S_2 is the semigroup generated by $n_o, n_1 - n_o, \ldots, n_p - n_o$. Thus $k[S_1]$ is the ring A and $k[S_2]$ is his quadratic tranform.

(3.5) DEFINITIONS. i) We call *n_o-standard basis* of S the generating set

$$B = \{(n_o,0), (n_1,n_1-n_o), \ldots, (n_p,n_p-n_o), (n_{p+1},m_{p+1}), \ldots, (n_{e-1},m_{e-1}), (0,n_o)\} =$$
$$= \{b_o, \ldots, b_e\}$$

where $\{n_o = e, n_1, \ldots, n_{e-1}\}$ (up to the order) is the n_o-standard basis of S_1 and m_i is the least element of S_2 such that $(n_i,m_i) \in S$ $(p+1 \leqslant i \leqslant e-1)$.

ii) We put $m_i = n_i - n_o$ $(1 \leqslant i \leqslant p)$. The n_o-standard basis B of S is said to be *good* if $\{n_o, m_1, \ldots, m_{e-1}\}$ (up to the order) is the n_o-standard basis of S_2 . This is equivalent to the condition $B = (B_1 \times B_2) \cap L$ where B_i is the n_o-standard basis of S_i $(i = 1,2)$.

(3.6) THEOREM. *The following conditions are equivalent:*

1) $k[S]$ is Cohen-Macaulay

2) The n_o-standard basis B of S is good

3) $S = (S_1 \times S_2) \cap L$

Proof. (1) ⇒ (2) Assume that B is not good and let $b_i = (n_i, m_i) \in B$ $(i > 0)$ be such that m_i does not belong to the n_o-standard basis of S_2 ; thus $c = m_i - n_o \in S_2$. Let a be the least element of S_1 such that $(a,c) \in S$. Then $a - n_i \in n_o\mathbb{Z}$. By definition of b_i we have $a > n_i$,hence $(a - n_o, m_i) \in S$. Now $(a,m_i) = (n_o,0) + (a - n_o,m_i) = (0,n_o) + (a,c) \in ((n_o,0) + S) \cap ((0,n_o) + S)$, but $(a,m_i) \notin (n_o,n_o) + S$ because $(a - n_o,c) \notin S$ by the choice of a . This is a contradiction (lemma 3.4).

(2) ⇒ (3) It suffices to show that $(S_1 \times S_2) \cap L \subset S$. Let $(s_1,s_2) \in (S_1 \times S_2) \cap L$. Then $s_1 = n_i + u\, n_o$, $s_2 = m_i + v\, n_o$ with n_i (resp. m_i) in the n_o-standard basis of S_1 (resp. S_2). Since s_1 and s_2 have the same n_o-residue and the n_o-standard basis B of S is good, $(n_i,m_i) \in B$ and hence $(s_1,s_2) = (n_i,m_i) + u\,(n_o,0) + v\,(0,n_o) \in S$.

(3) ⇒ (1) Because of lemma 3.4 it is enough to prove that $(n_o,n_o) + S \supset ((n_o,0) + S) \cap ((0,n_o) + S$. Let $(x,y) \in ((n_o,0) + S) \cap ((0,n_o) + S)$. Then $x - n_o \in S_1$ and $y - n_o \in S_2$. Hence $(x - n_o, y - n_o) \in S$ and our thesis follows.

As for the numerical semigroup rings (proposition 2.7), the type of $k[S]$ can be computed by means of the standard basis of S .

If $x \in \mathbb{N}$, $\bar{x}$ denotes the n_o-residue of x and for an element $s = (s_1,s_2)$ of S ,let $\bar{s}$ denote the common value of the n_o residue of s_1 and s_2 (remark 3.3 (i)).

(3.7) PROPOSITION. *Assume that $k[S]$ is Cohen-Macaulay. Let B be the*

n_o-standard basis of S and let $\tilde{B} = \{b \in B \,/\, \bar{b} \neq 0,\ b + b_i \notin B$ for all $b_i \in B\}$. Then $r(k[S]) = \text{card}(\tilde{B})$.

Proof. Let $\mathfrak{m}$ be the homogeneous maximal ideal of $k[S]$ and let $I = (X^{n_o}, Y^{n_o})$. Then $r(k[S]) = \dim_k \text{Ann}(\mathfrak{m}/I)$ (lemmas 2.5 and 3.4). Because of theorem 3.6, B is good, so for every element $s \in S$ there exists $b \in B$ such that $\bar{s} = \bar{b}$ and $s = b + u(n_o,0) + v(0,n_o)$ $(u, v \in \mathbb{N})$. Therefore a homogeneous element f of degree s in $k[S]/I$ is not zero if and only if $s = b \in B$ and $\bar{b} \neq 0$, moreover $f \cdot (\mathfrak{m}/I) = 0$ if and only if $s + b_i \notin B$ for all $b_i \in B$.

(3.8) COROLLARY. *$r(k[S_i]) \leq r(k[S])$ $(i = 1, 2)$.*

The corollary 3.8 for $i = 1$ is a special case of [3], theorem 3.7.

(3.9) PROPOSITION. *Assume that $k[S]$ is Cohen-Macaulay and let $B = \{b_i = (n_i, m_i) \,/\, 0 \leq i \leq e\}$ be the n_o-standard basis of S ordered so that $b_o = (n_o, 0)$, $b_e = (0, n_o)$ and $n_1 < n_2 < \ldots < n_{e-1}$. Then $k[S]$ is Gorenstein if and only if $b_{e-1} = b_i + b_{e-1-i}$ for all $i = 1, \ldots, e-2$.*

Proof. Suppose that $k[S]$ is Gorenstein. By corollary 3.8 and proposition 2.6, the semigroups S_1 and S_2 are symmetrics. Moreover if m_i is the greatest element of the n_o-standard basis of S_2 different from n_o, then $i = e-1$ otherwise $\tilde{B} \supset \{b_{e-1}, b_i\}$. By proposition 2.6 we have $n_{e-1} = n_i + n_{e-1-i}$ for all i and there exists j such that $m_i + m_j = m_{e-1}$. But $\bar{m}_j = \bar{m}_{e-1-i}$ hence $j = e-1-i$.

The converse follows from proposition 3.7.

(3.10) REMARKS. i) In general even if $k[S]$ is Gorenstein is not true that $m_i < m_j$ if $n_i < n_j$. For example let S_1 be the semigroup generated by 10, 12, 33. Then we have

$B = \{(10,0), (12,2), (24,4), (33,23), (36,6), (45,25), (48,8), (57,27), (69,29), (81,31), (0,10)\}$.

ii) In proposition 3.9 the assumption "$k[S]$ is Cohen-Macaulay" is necessary.

In fact if S_1 is the semigroup generated by 6 , 7 , 15 , the 6-standard basis of S satisfies to the condition of proposition 3.9 , but k [S] is not Cohen-Macaulay.

(3.11) EXAMPLE. Let s be a positive integer and let S_1 be the semigroup minimally generated by $n_o , n_1 , \ldots , n_p$, where $n_o = s$ and $n_1 = s+1$. Thus $S_2 = \mathbb{N}$. If $n_p > 2s$, then $n_p - n_o$ is not in the n_o-standard basis of S_2 , so k [S] is not Cohen-Macaulay.

Even if $n_p < 2s$, k [S] may not be Cohen-Macaulay (look at S_1 generated by 7 , 8 , 13). However in the following special case k [S] is always Cohen-Macaulay.

Let S_1 be the semigroup generated by $s , s+1 , \ldots , s+p$ $(p<s)$ and let $B_1 = \{ n_o , \ldots , n_{s-1} \}$ be the s-standard basis of S_1 (example 2.8). Then the s-standard basis of S is $B = \{ (n_o,0) , (n_1,1) , \ldots , (n_{s-1},s-1) , (0,n_o) \}$. Clearly B is good and we have $r(k[S]) = r(k[S_1])$.

This last example shows a class of rings A such that $r(A) = r(G)$; in general we have only $r(A) \leq r(G)$ ([3] ,theorem 3.7) .

Added in proof:

Recently we have known that the question whether the form ring of a one-dimensional semigroup ring is Cohen-Macaulay has been dealed also by Garcia in: "Cohen-Macaulayness of the associated graded of a semigroup ring", Comm. Algebra 10 (1982).

REFERENCES

[1] Y.Aoyama - S.Goto, On the type of graded Cohen-Macaulay rings, J.Math. Kyoto Univ. 15 (1975) 19-23.

[2] R.Barattero, Alcune caratterizzazioni degli anelli H-Macaulay ed H-Gorenstein, Le Matematiche 29 (1974).

[3] M.P.Cavaliere - G.Niesi, On Serre's conditions in the form ring of an ideal, J.Math.Kyoto Univ. 21 (1981).

[4] S.Goto - N.Suzuki - K.Watanabe, On affine semigroup rings, Japan J. Math. 2 (1976) 1-12.

[5] J.Herzog, When is a regular sequence super regular?,Nagoya Math.J. (to appear).

[6] M.Hochster, Rings of invariants of tori, Cohen-Macaulay rings generated by monomials, and polytopes, Ann. of Math. 96 (1972) 318-337.

[7] M.Hochster - L.J.Ratliff,Jr. , Five theorems on Macaulay rings, Pacific J. of Math. 44 (1973) 147-172.

[8] E.Kunz, The value- semigroup of a one-dimensional Gorenstein ring, Proc.Amer.Math.Soc. 25 (1970) 748-751.

[9] J.Matijevic - P.Roberts, A conjecture of Nagata on a graded Cohen-Macaulay ring, J.Math.Kyoto Univ. 14 (1974) 125-128.

[10] D.S.Rim - M.A.Vitulli, Weierstrass points and monomial curves, J.Algebra 48 (1977) 454-476.

[11] L.Robbiano -G.Valla, On the equations defining tangent cones, Math. Proc. Camb. Phil.Soc. 88 (1980) 281-297.

ON THE ORDER OF BRANCHES

Caterina Cumino

Politecnico di Torino
Torino (Italy)

INTRODUCTION

There exists a lot of classical literature about the concept of branch at a closed point of an algebraic variety, in particular it is well known the notion of order of a branch; however, what is deepely treated is only the case of branches at a point of an algebraic curve; already for the case of surfaces, indeed, we cannot find too much material (recall, for example, the works of F.Enriques, especially [4], where he speaks of "falde" of a surface); for instance, the behaviour of branches along a singular subvariety is nowhere closely examined, mostly in the case of nonlinear branches (in this sense some idea is implicit in Jung's work on local uniformization, see [12]).

The aim of the present paper is to study the algebraic notion of order of an analytic branch and to apply it to examine the behaviour of the branches of an algebraic variety X along a singular irreducible subvariety Y.

We deal with this problem using the theory of henselian and strict henselian rings. From this point of view the branches (resp. the geometric branches) of X at $x \in X$ are the minimal primes of the henselization ${}^{h}\mathcal{O}_{X,x}$ (resp. of a strict henselization ${}^{hs}\mathcal{O}_{X,x}$). Correspondingly we have the notion of ring of a (geometric) branch.

The geometric branches at the generic point y of Y are the main tool to study the branches at the closed points of Y; it must be remembered, however, that the strict henselization ${}^{hs}\mathcal{O}_{X,y}$ is not canonical, hence the notion of geometric branch at y is without meaning, unless one has fixed in advance a separable closure K of k(y).

In order to avoid this difficulty we prove, in section 1, that any two geometric branches over a fixed branch are isomorphic, no matter in which strict henselization they are viewed; from here we can deal without trouble with any notion which depends only on the isomorphism classes.

Such a notion is the order of a (geometric) branch, which is introduced in section 2. The main result in this section is proposition 2.7, which says that if X is equidimensional the multiplicity of a point $z \in X$ is equal to the sum of the orders of the (geometric) branches at z.

In section 3 we construct, for every point $x \in Y$, a map ψ_x from the set of branches at x to the set of geometric branches at y; the above result on multiplicity allows us to prove that ψ_x is bijective for almost all $x \in Y$ and that, for such an x, ψ_x induces a surjective canonical map from the branches at x to the branches at y; therefore one can divide the branches into families consisting of branches with the same order.

It is reasonable to think that this result applies to investigate the relations betweeen tangent cones and branches along a singular subvariety of codimension 1, hence we intend to come back to this last subject in a subsequent paper.

1. *ISOMORPHIC BRANCHES*

In this section, after some preliminaries, we prove that the geometric branches over a fixed branch are all isomorphic.

We assume that all rings are noetherian and commutative with identity and all schemes are locally noetherian. We use notations and basic facts contained in [8], [13] and [20].

Later on we shall use the following definitions:

DEFINITION 1.1 If X is a scheme over a field k, a *geometric point* of X (relatively to k) is a point of the K-scheme $X \otimes_k K$, where K is an algebraic closure of k.

DEFINITION 1.2 If X is a scheme, denote by X' the normalization of X_{red} and by $p: X' \to X$ the canonical morphism; let $x \in X$, a *branch* (resp. *geometric branch*) of X at x is a point (resp. a geometric point) of the k(x)-scheme $p^{-1}(x)$.

Moreover, if γ is the branch corresponding to the point x' of $p^{-1}(x)$, we call *geometric branches over* γ the geometric points of $p^{-1}(x)$ lying over x'.

We denote by b(x) (resp. g(x)) the number of branches (resp. geometric branches) of X at x.

If A is a local ring, the *branches* (resp. *geometric branches*) of A are the branches (resp. geometric branches) of spec(A) at its closed point; the number of branches (resp. geometric branches) of A is denoted by b(A) (resp. g(A)).

PROPOSITION 1.3 *With the same notations as 1.2, let* x' *be a point of* $p^{-1}(x)$ *and* γ *the corresponding branch. Then the number of the geometric branches of* X *over* γ *is* $[k(x'):k(x)]_s$.

PROOF. It follows by [9] 4.5.10.

REMARK 1.4 We point out that the definition of geometric branch at a point x presupposes that one has fixed in advance either an algebraic closure K or (by [9] 4.5.21) a separable closure $K' \subseteq K$ of k(x). The number of geometric branches, however, is clearly independent of the choice of K or K' (see 1.3).

If A is a ring, we denote by c(A) the number of minimal primes of A; if X is a scheme and $x \in X$, we put $c(x) = c(\mathcal{O}_{X,x})$.

All over the paper, ${}^{h}A$ (resp. ${}^{hs}A$) shall indicate a henselization of the local ring A (resp. a fixed strict henselization which contains ${}^{h}A$); see for reference [9] § 18 or [17]. Note that, denoting by k the residue field of A, to fix ${}^{hs}A$ is the same as to fix a separable closure of k.

More details about the theory of (geometric) branches of a local ring or of an algebraic scheme can be found in [6].

THEOREM 1.5 (see [6] 2.1) *Let* A *be a local ring. Then we have the following equalities, each induced by canonical bijections:*

(i) $b(A) = c({}^{h}A)$;

(ii) $g(A) = c({}^{hs}A)$.

Moreover if A *is excellent:*

(iii) $b(A) = c(\hat{A})$;

(iv) $g(A) = c({}^{hs}(\hat{A})) = c(({}^{hs}A)\hat{})$.

COROLLARY 1.6 *Let* A *be a local ring and let* p *be a minimal prime of* ${}^{h}A$. *Then the geometric branches over the branch corresponding to* p *can be canonically identified with the minimal primes of* ${}^{hs}A$ *lying over* p.

PROOF. It follows by 1.5 and by the canonical isomorphism ${}^{hs}({}^{h}A/p) \cong {}^{hs}A/p\,{}^{hs}A$.

DEFINITION 1.7 Let A be a local ring; if p is a minimal prime of ${}^{h}A$ (resp. of ${}^{hs}A$) we call *ring of the branch* (resp. *ring of the geometric branch*) corresponding to p, the ring ${}^{h}A/p$ (resp. ${}^{hs}A/p$). We will say that two *branches* (resp. *geometric branches*) of A are *isomorphic* if their rings are such as A-algebras.

PROPOSITION 1.8 *Let* (A,m) *be a local henselian domain. Let* B, B' *be two strict henselizations of* A *and let* p, p' *be minimal primes of* B, B' *respectively. Then there exists an* A*-isomorphism* B $\to$ B' *, which induces an isomorphism* B/p $\cong$ B'/p' *(in particular the isomorphism class of* B/p *depends only on* A*).*

PROOF. Since B and B' are A-isomorphic (see [9] 18.8.8 (vi)) we may assume B = B'. $\bar{A}$ is a local ring (see [17] ch.IX, def.1); let K denote its residue field, k_{sep} the separable closure of k in K and h a primitive element of k_{sep} over k; let k' be the normal closure of k with respect to h, in an algebraic closure of k.

k' is a Galois extension of k and we have k' = k[X]/(f), where f $\in$ $\in$ k[X] is a monic polynomial; let F $\in$ A[X] be a monic polynomial which lifts f; put A' = A[X]/(F) = A[x] and let f' be the derived polynomial of f; f'(x) is not zero in k' and hence F'(x) is invertible in A[x], since mA[x] $\subseteq$ rad A[x]; therefore A' is a local A-algebra, which is finite and étale over A and henselian (see [17] ch.I §3).

Let G be the Galois group of k'; G is finite and acts also on A' (see [17] ch.VIII, prop.1). Denote by A'^G the set of G-invariants of A'; A'^G is a local ring having k as residue field; let m' be the maximal ideal of A', then for each $\sigma \in G$ (σ different from the identity) there exists x $\in$ A' such that, putting $\bar{x}$ = x mod m', one has $\sigma(\bar{x}) \neq \bar{x}$, that is $(\sigma(x) - x) \notin m'$; in fact the set $\{\bar{x} \in k' \mid \sigma(\bar{x}) = \bar{x}\}$ is different from k'; it follows by [3] ch.III, prop.1.2, that A' is étale over A'^G; moreover A' is étale over A, hence A'^G is étale over A (see [3] ch.III, th.2.4); then we have in particular that mA'^G is the maximal ideal of A'^G, so $A'^G = A$.

We can now say (see [1] ch.V §2, th.2.4) that G acts transitively on the primes of A' which lie over a fixed prime of A; in particular, as A' is flat over A, G acts transitively on the set of the minimal primes of A'.

As A' is henselian, b(A') = c(A') (see 1.5(i)) and, by construction,

$g(A) = g(A')$ (see [6] 2.2); moreover, by [9] 18.8.6, there exists an inductive system $(A_j)_{j \in J}$ of local finite étale A-algebras such that A' is isomorphic to some A_j and $\varinjlim A_j = B$; hence B is also a strict henselization of A'.

Let $\overline{A'}$ be the integral closure of A'; by [9] 18.6.8 $\overline{A'} = \overline{A} \otimes_A A'$, hence $\overline{A'} \otimes_{A'} k' = \overline{A} \otimes_A k'$ and so $b(A') = g(A')$, as k' is a Galois extension of k (see [9] 4.5.10). But by 1.5(ii) $g(A') = c(B)$, therefore $c(A') = c(B)$.

Let $p_1, \ldots, p_n$ be the minimal primes of B; the primes $q_i = p_i \cap A'$ are the only minimal primes of A', because $A' \to B$ is faithfully flat; moreover they are all distinct, as $c(A') = c(B)$; if q_i and q_j are two such primes there exists $\sigma \in G$ such that $\sigma(q_i) = q_j$; now, by the universal property of strict henselization, σ extends to an A-automorphism τ of B; but $q_i B = p_i$ and $q_j B = p_j$, therefore τ induces an A-isomorphism $B/p_i \cong B/p_j$.

COROLLARY 1.9 *Let* (A,m) *be a local reduced ring. Let* γ *be a branch of* A. *Then the geometric branches of* A *over* γ *(with respect to any strict henselization) are pairwise isomorphic.*

PROOF. γ corresponds to a minimal prime p of ${}^hA$. Hence the conclusion follows by 1.7, 1.8 and [9] 18.7.5.

REMARK 1.10 The main point of the preceeding proof is the existence of a finite étale Galois A-algebra A', such that $c(A') = c(B)$.

The referee has suggested that the existence of such an algebra can be proved, in a slightly shorter way, as follows: since $K = \varinjlim k_i$, where k_i/k is finite and Galois, one has $B = \varinjlim B_i$, where B_i is finite étale and local with residue field k_i, hence Galois (same argument used for A'); choose then, for each minimal prime p_i of B, an x_i in p_i and not in $\bigcap_{i \neq j} p_j$; let C be a subring of B containing the x_i's and such that $A \to C$ is étale and Galois; the conclusion follows by the same argument applied above to A'. The only advantage of our original proof is that A' is explicitly constructed.

2. *GEOMETRIC BRANCHES*

An algebraic definition of order of a branch is well known for branches of an algebraic curve. In this section we introduce a definition of order of a (geometric) branch, in terms of henselization and strict henselization, for a local ring of any dimension; we verify that it coincides with the classical one, whenever this makes sense, and that it fulfilles analogous elementary properties.

All over this section (A,m,k) shall denote a local reduced ring; by $m_1, \ldots, m_n$ and $k_1, \ldots, k_n$ we shall denote respectively the maximal ideals of $\bar{A}$ and the residue fields of $\bar{A}_{m_1}, \ldots, \bar{A}_{m_n}$. Moreover we fix, once for all, a separable closure K of k; we denote by ${}^{hs}A$ the corresponding strict henselization of A.

DEFINITION 2.1 (i) Let p be a minimal prime of ${}^{h}A$; we call *order of the branch* γ of A corresponding to p, the multiplicity of the ring of γ, i.e. $e(\gamma) = e({}^{h}A/p)$.

(ii) Let P be a minimal prime of ${}^{hs}A$; we call *order of the geometric branch* Γ of A corresponding to P, the multiplicity of the ring of Γ, i.e. $e(\Gamma) = e({}^{hs}A/P)$.

Note that by 1.9 also (ii) is a good definition, because it is independent of the choice of K.

The following proposition gives the order of a branch in terms of the normalization.

PROPOSITION 2.2 (i) *Let* $p_1, \ldots, p_n$ *be the minimal primes of* ${}^{h}A$, *then, for each* i = 1, ..., n, *we have:*

$$e({}^{h}A/p_i) = e(m\bar{A}_{m_i})[k_i:k]$$

(ii) *If moreover* A *is excellent, let* $p_1, \ldots, p_n$ *be the minimal primes of* $\hat{A}$; *then, for each* i = 1, ..., n, *we have :*

$$e(\hat{A}/p_i) = e(m\bar{A}_{m_i})[k_i:k]$$

PROOF. (i) Since A is reduced, ${}^hA$ is reduced; let $\overline{{}^hA}$ be the integral closure of ${}^hA$; then

$$\overline{{}^hA} = \oplus_{i=1}^n \overline{({}^hA/p_i)}$$

moreover (see [9] §18) we have the following isomorphisms :

$$\overline{{}^hA} \cong {}^hA \otimes_A \bar{A}; \qquad {}^h(\bar{A}) \cong {}^hA \otimes_A \bar{A} \cong \oplus_{i=1}^n {}^h(\bar{A}_{m_i});$$

it follows that, for each i, ${}^h(\bar{A}_{m_i}) \cong \overline{({}^hA/p_i)}$ and hence, using [20] vol.II, th.24:

$$e({}^hA/p_i) = e(m\,{}^h(\bar{A}_{m_i}))[k_i:k]$$

where $e(m\,{}^h(\bar{A}_{m_i})) = e(m\bar{A}_{m_i})$, since, for each $n \in \mathbb{N}$,

$${}^h(\bar{A}_{m_i}/m^n\bar{A}_{m_i}) \cong (\bar{A}_{m_i} \otimes_A A/m^n) \otimes_A {}^hA \cong {}^h(\bar{A}_{m_i})/m^n\,{}^h(\bar{A}_{m_i}).$$

(ii) follows by [9] 7.6.2 and 1.5(iii), as above.

REMARK 2.3 Using 2.2(ii) it is easy to verify that the notion of order we have introduced coincides with the classical one, whenever this makes sense.

In fact, suppose that A is the local ring at a point P of a curve (i.e. of a 1-dimensional algebraic scheme over an algebraically closed field of arbitrary characteristic) and let (B,n) be the local ring at a place Q having center P; then it is well known that the order of the branch joining P with Q is the integer $e(mB) = \ell_B(B/mB) = v(mB)$, where v is the valuation corresponding to Q; this fact, by 2.2, agrees with our definition, being the concept of place nothing else than the algebraic counterpart of that of branch; (as reference see e.g., for the case of characteristic o, [19] ch.IV §5.3 or the classical approach of [18] ch.II, whereas for the case of arbitrary characteristic one can see [20] vol.II ch.VI and VIII).

More generally, let us consider, for example, an irreducible hypersurface V of $\mathbb{A}^n(\mathbb{C})$ passing through the origin o; we have then: $\hat{\mathcal{O}}_o$ =

$= \mathbb{C}[[X_1, \ldots, X_n]]/(f)$; let now $f = f_1 \ldots f_m$ be the factoring of f in irreducible power series factors, then each f_i represents a branch of V in o and one usually assumes $\mathrm{ord}(f_i)$ as its order; note that in this case too $\mathrm{ord}(f_i) = e(\mathbb{C}[[X_1, \ldots, X_n]]/(f_i))$, according to our definition.

PROPOSITION 2.4 *Let* A *be a local henselian domain; denote by* P_i, i = 1, ..., n, *the minimal primes of* ${}^{hs}A$. *Then:*

$$e(A) = \Sigma_{i=1}^{n} e({}^{hs}A/P_i)$$

PROOF. Let P be a minimal prime of ${}^{hs}A$, B the integral closure of ${}^{hs}A$ and n the unique maximal ideal of B containing P, that is $B_n = (\overline{{}^{hs}A/P})$. Being A an henselian domain, $\bar{A}$ is local (see [17] ch.IX, prop.1) and as B = $= \bar{A} \otimes_A {}^{hs}A$, the morphism $\bar{A} \to B$ is absolutely flat (in the sense of [7] 5.1); let $\bar{m}$ be the maximal ideal of $\bar{A}$; we have then $B_n/\bar{m}B_n = k(n)$, hence, by flatness, $\dim(\bar{A}) = \dim(B_n)$; but $\dim(A) = \dim(\bar{A})$, $\dim(B_n) = \dim({}^{hs}A/P)$; it follows that $\dim({}^{hs}A/P) = \dim(A)$, for each minimal prime P. The conclusion follows then by [14] 23.5, since A and ${}^{hs}A$ have the same multiplicity.

COROLLARY 2.5 *Let* γ *be a branch of* A, *then if* Γ *is any geometric branch of* A *over* γ, *its order depends only on* γ *and precisely we have* $e(\gamma) =$ $= r\, e(\Gamma)$, (where r is the number of geometric branches over γ).

PROOF. By 1.3, 1.9 and 2.4.

The next proposition, which can be considered as an analogue of 2.2, gives the order of a geometric branch in terms of the normalization.

PROPOSITION 2.6 *Let* B *be the integral closure of* ${}^{hs}A$ *and* n *a maximal ideal of* B *lying over a fixed maximal ideal* m_* *of* $\bar{A}$; *let* k_* *be the residue field of* $\bar{A}_{m_*}$ *and* P *the minimal prime of* ${}^{hs}A$ *corresponding to* n. *Then:*

(i) $e({}^{hs}A/P) = e(mB_n)[k(n):K]$;

(ii) $e(\overline{{}^{hs}A/P}) = e(\bar{m}A_{m_*})[k_*:k]_i$.

PROOF. (i) follows by [20] vol.II th.24, being $B_n = (\overline{{}^{hs}A/P})$. Put now $p = P \cap {}^hA$ and $C = {}^hA/p$; then by 2.2 we have $e(C) = e(m\bar{A}_{m_*})[k_*:k]$, while by 2.5 $e(C) = [k_*:k]_s\, e({}^{hs}A/P)$. Thus, being $[k_*:k] = [k_*:k]_s [k_*:k]_i$, we have (ii).

PROPOSITION 2.7 *If* A *is universally catenary and equidimensional, denoting by* γ_i, $i = 1, \ldots, n$, *(resp.* Γ_j, $j = 1, \ldots, m$), *the branches (resp. geometric branches) of* A, *we have:*

(i) $e(A) = \Sigma_{i=1}^{n} e(\gamma_i)$;

(ii) $e(A) = \Sigma_{j=1}^{m} e(\Gamma_j)$.

PROOF. $\bar{A}$ is the direct product of the integral closures of A/p, as p ranges in the set of minimal primes of A; let $p_1, \ldots, p_n$ be the minimal primes of ${}^hA$; we have then $\dim(A) = \dim({}^hA)$ and, for each i, $\dim({}^hA/p_i) = \dim(\overline{{}^hA/p_i})$; moreover $(\overline{{}^hA/p_i}) \cong ({}^h\bar{A}/p_i)$; since for each p, A/p is universally catenary and $\dim(A/p) = \dim(A)$, then, by [9] 5.6.10, ${}^hA$ is equidimensional. The conclusion follows then by [14] 23.5.

(ii) follows by (i) and 2.4.

REMARKS 2.8 (i) As a consequence of 2.6 we get $e(m\bar{A}_{m_*}) = e(mB_n)$; in fact, the local homomorphism $\bar{A}_{m_*} \to B_n$ is absolutely flat and we can calculate multiplicities as at the end of the proof of 2.2. Consequently, by 2.6, we obtain in particular that $[k(n):K] = [k_*:k]_i$. More generally one has the following statement: let k be a field and $\bar{k}$ an algebraic closure of k, let $K \subseteq \bar{k}$ be a finite algebraic extension of k and let k' and K' denote respectively the separable closures of k and K in $\bar{k}$, then $[K':k'] = [K:k]_i$.

(ii) In [6] §2 it is stated that $g(A) \leq e(A)$; 2.7 gives some more details.

(iii) 2.7 generalizes a classical result about the branches of an algebraic curve (see for instance [18] pag.89).

(iv) Nothing we can say for the moment about the other characters of the branches (see for example [18]); it might be interesting to characterize them as it was done for the order.

We examine now the case of linear branches. The next statement make precise and generalize some results which are well known in the analytic case (see for example [15] §3 and §5).

DEFINITION 2.9 We say that a *branch* (resp. a *geometric branch*) of A, associated to the prime p of ${}^{h}A$ (resp. P of ${}^{hs}A$) is *linear* if ${}^{h}A/p$ (resp. ${}^{hs}A/P$) is regular. By 2.2(ii) if A is excellent, the branch associated to the prime p is *linear* if $\hat{A}/p$ is regular.

PROPOSITION 2.10 *When* A *is* 1-*dimensional or excellent, a branch (resp. a geometric branch) of* A *is linear if and only if its order is* 1.

PROOF. If dim(A) = 1 the conclusion is clear. Assume A is excellent, then by [9] 18.7.6 ${}^{h}A$ is excellent. If p is a minimal prime of ${}^{h}A$, $({}^{h}A/p)\hat{}$ is equidimensional (see [9] 7.8.3(x)); hence the conclusion follows by [14] 40.6; the same argument we can apply to a minimal prime P of ${}^{hs}A$, since by [7] 5.6(iii) ${}^{hs}A$ is an excellent ring too.

3. *BRANCH BEHAVIOR*

In this section, applying the results of the preceding ones, we reach aim of the paper: we study the behaviour of the branches along a singular irreducible subvariety of an algebraic variety in terms of order of the branches; we conclude with some illustrative examples.

We use the same notations as in §2; moreover X will denote an algebraic variety over a field k (i.e. a k-scheme of finite type) and Y will be a closed subvariety of X, reduced and irreducible.

LEMMA 3.1 *Let* p *be a prime of* A *and* P *a prime of* ${}^{h}A$ *such that* $P \cap A = p$; *fix a separable closure* L *of* k(P) , *then the strict henselizations of* A_p *and* $({}^{h}A)_P$ *with respect to* L *are canonically isomorphic.*

PROOF. The canonical homomorphism $A \to {}^{h}A$ is absolutely flat (see [6] 5.1), hence induces an absolutely flat homomorphism $A_p \to ({}^{h}A)_P$. Then the conclusion follows by [16] §3, proof of th.3.

THEOREM 3.2 *Let* X *be an equidimensional reduced variety over an algebraically closed field* k, *let* Y *be a reduced irreducible subvariety of* X *and* y *the generic point of* Y. *Then, by using* 3.1, *one can construct for each closed point* $x \in Y$ *a map*

$$\psi_x: \{\text{geometric branches at } y\} \to \{\text{branches at } x\}$$

such that:

(i) *for almost all* $x \in Y$, ψ_x *is bijective and preserves the order of branches;*

(ii) *for such an* x, ψ_x *induces a surjective canonical map*

$$\phi: \{\text{branches at } x\} \to \{\text{branches at } y\}$$

such that if γ, γ' *are branches at* x, $\phi(\gamma) = \phi(\gamma')$ *implies* $e(\gamma) = e(\gamma')$.

PROOF. Let x be any closed point of Y, put $A = \mathcal{O}_x$, let p be the prime ideal of A corresponding to Y and P a prime ideal of ${}^{h}A$, whose contraction to A is p.

As the homomorphism $A_p \to ({}^{h}A)_P$ is flat and local, the homomorphism $u: A_p \to ({}^{h}A)_P \to {}^{h}(({}^{h}A)_P)$ is flat and local; let $i: A_p \to {}^{h}(A_p)$ be the canonical homomorphism, then, as ${}^{h}(({}^{h}A)_P)$ is henselian, there exists a unique homomorphism $\tilde{u}: {}^{h}(A_p) \to {}^{h}(({}^{h}A)_P)$ such that $u = \tilde{u} \circ i$ (see [17] def.1, ch.VIII).

Moreover, ${}^{h}(A_p)$ and ${}^{h}(({}^{h}A)_P)$ are ind-étale algebras over A_p, hence, by [3] ch.III, th.2.4, ${}^{h}(({}^{h}A)_P)$ is ind-étale over ${}^{h}(A_p)$. Hence, by 3.1, we have the following commutative diagram of canonical homomorphisms:

$$\begin{array}{ccccccc} {}^{h}A & \to & ({}^{h}A)_{P} & \to & {}^{h}(({}^{h}A)_{P}) & \searrow & \\ \uparrow & & \uparrow & & \uparrow & & {}^{hs}(A_{p}) \\ A & \to & A_{p} & \to & {}^{h}(A_{p}) & \nearrow & \end{array}$$

and, by contracting minimal primes, we get a map

$$\psi_x : \mathrm{Min}({}^{hs}(A_p)) \to \mathrm{Min}({}^{h}A)$$

By [9] 9.7.8 there exists a non empty open set $V \subseteq Y$ such that $g(x) = g(y)$ for all closed points $x \in V$; moreover by Krull-Seidenberg theorem (see [10] th.1 pag.188) we can choose a non empty open subset W of Y such that X is normally flat along Y in each point $x \in W$ and each point of W is regular for Y; as a consequence we have (see [10] cor.2 pag.186) that $e(\mathcal{O}_x) = e(\mathcal{O}_y)$, for all $x \in W$.

Choose an open non empty subset $U \subseteq V \cap W$. Now, let x be any closed point of U and let A, p and P be as above, (note that P is the only prime of ${}^{h}A$ over p, because A/p is regular). Put $B = ({}^{h}A)_P$; as A is equidimensional, by 2.7(ii) $e({}^{h}A) = \Sigma_{i=1}^{n} e({}^{h}A/p_i)$, where p_i are the minimal primes of ${}^{h}A$.

The minimal primes of B are the ideals p_iB such that $p_i \subseteq P$; thus we renumber the p_i's in such a way that $p_i \subseteq P$ if and only if $i \leq m$; then, since B too is equidimensional, we get $e(B) = \Sigma_{i=1}^{m} e(B/p_iB)$, with $m \leq n$; on the other hand, being ${}^{h}A$ excellent, we are by [9] 7.6.7(ii)(a) in the hypothesis of [14] 40.1; hence for each $i = 1, \ldots, m$, $e({}^{h}A/p_i) \geq e(B/p_iB)$; moreover, by the choice of U, $e({}^{h}A) = e(B)$; therefore $m = n$ and $e({}^{h}A/p_i) = e(B/p_iB)$, for each i.

Now, as $b(x) = g(x) = g(y)$ and $g(y) = g(B)$, in any strict henselization of B there exists only one minimal prime lying over each minimal prime of B. Then, by applying 2.4 to B/p_iB, for each i, we obtain (i). It follows also: $\mathrm{Min}({}^{h}A) \cong \mathrm{Min}(({}^{h}A)_P) \cong \mathrm{Min}({}^{h}(({}^{h}A)_P))$; hence we can define a map $\phi: \mathrm{Min}({}^{h}A) \to \mathrm{Min}({}^{h}(A_p))$, by extending a minimal prime of ${}^{h}A$ to ${}^{h}(({}^{h}A)_P)$ and contracting it then to ${}^{h}(A_p)$. Therefore by 1.9 we get (ii).

REMARKS 3.3 (i) Note that the main point of the proof of 3.2 is to substitute to the ring $(\mathcal{O}_x)_p$ the ring $B = ({}^h\mathcal{O}_x)_P$, whose minimal primes are in 1-1 correspondence, which preserves the orders of branches, both with the branches of $\mathcal{O}_x$ and the geometric branches of $(\mathcal{O}_x)_p$.

(ii) With regard to the geometric meaning of 3.2, it might be interesting to go deeper into some arguments of Enriques (see [4] libro IV, §39).

(iii) Statement 3.2(i) may be false outside the open set U; see for example 3.4(i) below, when char k $\neq$ 2.

(iv) In 3.2 the map ψ_x depends in particular on the separable closure of k(P) (see 3.1), while the map ϕ is canonical; this means: fix a branch γ at y, take another separable closure of k(P) and get a map ψ'_x, then ψ_x and ψ'_x may be different, but anyway they map the geometric branches over γ to the same set of branches at x (i.e. to $\phi^{-1}(\gamma)$).

EXAMPLES 3.4 (i) Consider in k^3 the surface V having equation $XZ^2 - Y^2 = 0$ (k algebraically closed); let y be the generic point of the double line $Z = Y = 0$, let $A = k[U^2, UV, V]$ be the coordinate ring of V and $p = (UV, V)$ the prime ideal of A corresponding to the double line; then $A_p = k[U^2][UV,V]_{(UV,V)}$ is the local ring of V at y and $\bar{A}_p = k(U)\cdot[V]_{(V)}$ is its integral closure.

Now we have at y one branch of order 2, since $[k(U):k(U^2)] = 2$ (see 2.2). If char k $\neq$ 2, at each point of the double line (except the origin) there are two linear branches. If char k = 2, y has only one geometric branch of order 2 and at each point of the double line there is only one branch of order 2.

(ii) Let $k_\circ$ be an algebraically closed field of characteristic 3. Put $A = k_\circ[X,Y]$, $f_1 = Y^3 - X$, $f_2 = Y^3 - X - 1$, $f_3 = Y^2 - 2X^3Y - X^3$, $k = k_\circ(X)$, $V = \operatorname{spec}(A)$, $W_i = \operatorname{spec}(A/(f_i))$ and $k(W_i) = K_i$ $(i = 1, 2, 3)$.

Clearly for i = 1, 2 K_i/k is a purely inseparable extension and $[K_i:k] = [K_i:k]_i = 3$; while K_3/k is a separable extension and $[K_3:k] = [K_3:k]_s = 2$.

Consider the variety obtained from V by gluing together the W_i's, that is the variety **spec**$(A \times_{K_1 \times K_2 \times K_3} \bar{K})$ (fiber product, see [5]) and let W be its subvariety corresponding to the W_i's.

Then codim(W) = 1 and at its generic point we have branches $\gamma_1, \gamma_2, \gamma_3$ (corresponding respectively to K_1, K_2, K_3) of order 3; over γ_1, γ_2 we have respectively geometric branches Γ_1, Γ_2 of order 3 and over γ_3 geometric branches Γ_3^1, Γ_3^2 of order 1.

Using the same method one can construct other analogous examples.

APPENDIX

We report here a remark of the referee, which justifies the way we have used for the study of branches: if A is the local ring of a point of a curve, there exists a factorization $A \subseteq B \subseteq \bar{A}$ where B is unramified over A and $B \to \bar{A}$ is radicial (see D.Ferrand, "Set theoretical complete intersection in characteristic p > o", Proceedings Summer Meeting Copenhagen 1978, L.N.Math. 322, 82-89 (1979) Springer-Verlag); in other words, each branch ${}^{hs}A/p$ is the strict henselization of the corresponding local ring of B; therefore one can work out the analysis of the branches directly from the ring B, which is very close to A. In dimension ≥ 2, such a factorization does not exist in general; e.g. look at the local ring at the origin of the surface in 3.4(i) above.

REFERENCES

[1] N.BOURBAKI. *Algèbre Commutative*, ch.V, Hermann, 1964.

[2] C.CUMINO. *Sulle sottovarietà singolari delle varietà algebriche*, Boll. U.M.I. (5) 15-B (1978), 906-914.

[3] F.DEMEYER, E.INGRAHAM. *Separable algebras over commutative rings*, Lecture Notes Math., 181, Springer-Verlag, 1971.

[4] F.ENRIQUES, O.CHISINI. *Sulla teoria geometrica delle equazioni e delle funzioni algebriche*, vol.III, Zanichelli, 1915.

[5] P.GIANNI. *Morfismi interi delle varietà algebriche*, Boll. U.M.I. 15-A (1978), 187-196.

[6] S.GRECO. *On the theory of branches*, Proc. Symp. Alg. Geom. Kyoto (1977), 477-493.

[7] S.GRECO. *Two theorems on excellent rings*, Nagoya Math. J., vol.60 (1976), 139-149.

[8] A.GROTHENDIECK, J.DIEUDONNE! *Eléments de Géométrie Algébrique*, ch.I, Erg. der Math. b. 166, Springer-Verlag, 1971.

[9] A.GROTHENDIECK, J.DIEUDONNE! *Eléments de Géométrie Algébrique*, ch.IV, Pubbl. Math. I.H.E.S., 20, 24, 28, 32, 1964/67.

[10] H.HIRONAKA. *Resolution of singularities of an algebraic variety over a field of characteristic zero*, Ann. of Math. 79 (1964), 109-326.

[11] N.JACOBSON. *Lectures in abstract algebra*, vol.III, Van Nostrand, 1964.

[12] H.W.E.JUNG. *Darstellung der Funktionen eines algebraischen Körpers zweier unabhängigen Veränderlichen x, y in der Umgebung einer Stelle $x=a$, $y=b$*, J. f. Math. 142 (1912), 289-314.

[13] H.MATSUMURA. *Commutative algebra*, Benjamin, 1970.

[14] M.NAGATA. *Local rings*, Interscience, 1962.

[15] D.G.NORTHCOTT. *Some results concerning the local analytic branches of an algebraic variety*, Proc. of Camb. Phil. Soc. 49 (1953), 386-396.

[16] G.PAXIA. *Ascent of properties for absolutely flat morphisms*, Comm. in Algebra, vol.6 n.4 (1978), 393-407.

[17] M.RAYNAUD. *Anneaux locaux henséliens*, Lecture Notes Math. 169, Springer-Verlag, 1969.

[18] F.SEVERI. *Trattato di geometria algebrica*, vol.I, Zanichelli, 1929.

[19] R.J.WALKER. *Algebraic curves*, Princeton University Press, 1950.

[20] O.ZARISKI, P.SAMUEL. *Commutative algebra*, vol.I, vol.II, Van Nostrand, 1960.

ON PSEUDO-VALUATION DOMAINS AND THEIR GLOBALIZATIONS (1)

David E. Dobbs (2) and Marco Fontana (3)

Abstract. The present paper gives a unified approach to the study of pseudo-valuation domains and, through the analysis of a Noetherian number-theoretic example, leads to a very satisfactory global counterpart of the notion of pseudo-valuation domain.

The concept of a pseudo-valuation domain (for short, PVD) was introduced a few years ago by J.R. Hedstrom and E.G. Houston (cf. [17], [18]). Considerable interest in this class of quasi-local, seminormal integral domains has been witnessed by several recent papers devoted to the topic (cf. [3], [4], [5], [7], [8], [10], [11], [14]). We recall that a *pseudo-valuation domain* is an integral domain A, with quotient field K, satisfying (one of) the following equivalent conditions:

(i) For every $P \in \mathrm{Spec}(A)$, $xy \in P$, with $x, y \in K$, implies that $x \in P$ or $y \in P$ (such a prime P is called a *strongly prime ideal*);

(ii) For each $P \in \mathrm{Spec}(A)$ and $x \in K \setminus A$, $x^{-1} P \subseteq P$;

(iii) For each $P \in \mathrm{Spec}(A)$ and $x \in K$, the ideals xA and P are comparable;

(iv) The ring A is quasi-local and, denoting by M its maximal ideal, $x^{-1} M \subseteq M$ for every $x \in K \setminus A$;

---

(1) The lecture given at the meeting by the second-named author dealt with some of the arguments discussed in the present paper and with some of those contained in [11].

(2) University of Tennessee, Knoxville, TN 37916, USA; partially supported by a University of Tennessee Faculty Development Grant.

(3) Istituto Matematico "G. Castelnuovo", Università di Roma, 00185 Roma, Italy; work performed under the auspices of CNR (Gruppo Nazionale Strutture Algebriche Geometriche e loro Applicazioni).

(v) For each element g of the group of divisibility G of A, either $g \geq 0$ or $g < h$ for all $h \in G$ such that $h > 0$.
Cf. [17; Prop. 1.2, Th. 1.4] , [3; Prop. 4.1, Prop. 5.1].

The name of "pseudo-valuation" domain is justified by the fact that these rings share many ideal-theoretic properties with the valuation rings to which they are closely related. The motivations for studying such a generalization of a valuation ring are various and meaningful. For instance, when one considers the localization at any maximal ideal of the (proper) subring $\mathbb{Z}[\sqrt{d}]$ of the ring of the integers of the quadratic field $Q(\sqrt{d})$, d being a square-free integer such that $d \equiv 5 \pmod 8$, one has a 1-dimensional local domain, which is not necessarily a valuation ring (because it need not be integrally closed), but which turns out to be naturally a PVD (cfr. Example 4). On the other hand, it happens that every PVD is a divided domain, a type of quasi-local going-down domain introduced in [6] (cf. also [1]). Hence, by means of the following theorem displaying an easy constructive method for obtaining every PVD, one describes the (probably) most significant class of quasi-local going-down domains of arbitrary Krull-dimension, which need not be integrally closed. Theorem 1 also describes how each PVD admits an associated valuation overring with the *same* set of prime ideals. Parts of Theorem 1 are already in [4]; the present proof outlines a unified point of view for various important properties of a pseudo-valuation domain.

THEOREM 1. *Let* A *be an integral domain and let* K *be its quotient field. Then* A *is a* PVD *if, and only if, there exists a valuation ring* $(V, M, k(V))$, *a subfield* k *of* $k(V)$ *and a Cartesian diagram of the following type:*

$$(\square)\qquad \begin{array}{ccc} A & \xrightarrow{\ v'\ } & k \\ {\scriptstyle u'}\downarrow & & \downarrow{\scriptstyle u} \\ V & \xrightarrow[\ v\ ]{} & k(V) \end{array}$$

where v *is the canonical projection and* u *is the inclusion.*

In this situation, after identifying A *with* $u'(A)$, *we have:*

(a) $M \cap A = M$ *is the conductor of* $A \overset{u'}{\hookrightarrow} V$;

(b) *the canonical map* u'^{*}: Spec $(V) \longrightarrow$ Spec (A) *is the identity homeomorphism; it is the identity scheme-theoretic isomorphism outside of* $\{M\}$;

(c) A *and* V *have the same quotient field;*

(d) *if* A *is not a valuation ring, then the valuation ring* V *in* ($\square$) *is uniquely determined by* A; *more precisely:*

$$V = (M :_K M) = (A :_K M) = M^{-1}.$$

PROOF. (Sketch). It suffices to consider the case $M \neq (0)$. For the "if" part, we know that a Cartesian square of the form ($\square$) yields conditions (a), (b) and (c) (cf. [12]), hence (A,M) is a quasi-local domain with residue field $k(A) \cong k$. It is easy to show that $x^{-1} M \subseteq M$, for every $x \in K\setminus A$, hence A is a PVD. Finally, from (a) and from the fact that (V,M) is a valuation ring, it follows that $V = (M:M) \subseteq (A:M)$. If $A \subsetneq V$, one shows easily that $(M:M) \subsetneq (A:M)$ implies that M is principal in A and hence, that $(M:M) = A$, contradicting the assumption $A \subsetneq V$.

For the "only if" part, we set $V = (M:M)$, where M is the maximal ideal of A. From the statement (iv) of the definition of PVD, one can deduce that V is a valuation ring. The conclusion follows from the fact that M is a (maximal) ideal both in A and in V, taking $k = A/M$ in the diagram ($\square$).$\square$

From the previous theorem and from the general machinery

concerning pullbacks of commutative rings as in [12], it is easy to show the close "resemblance" of the ideal-theoretic structure between a PVD and its canonically associated valuation overring. (Further informations on this subject can be found in [17, Sec. 2] and [14, Cor. 1.11]). By way of contrast, we note, in passing to polynomials, that the behavior of $A[X]$ (where A is a PVD) and $V[X]$ (where V is the valuation overring canonically associated to A) can be very different. For instance, it is wellknown that if $\dim(V) = n$, then $\dim(V[X]) = n+1$ and $V[X]$ is catenarian. However, even though $\dim(A) = n$, then it can happen that $\dim(A[X]) = n+2$ and $A[X]$ can fail to be catenarian (cf. [18, Th. 2. 5] and [15]).

Among the several interesting consequences of the Theorem 1, we limit ourself to point out the following two.

COROLLARY 2. *Let* $(A, M, k(A))$ *be a* PVD *and* $(V=(M:M), M, k(V))$ *the valuation overring associated to* A *(cf.* Th.1*). We denote by* k' *the algebraic closure of* $k(A)$ *into* $k(V)$*. Then*

(a) The integral closure A' *of* A *coincides with the pullback of* k' *into* V*, i.e.* $A' \cong k' \times_{k(V)} V$;

(b) A *is integrally closed if, and only if,* $k(A)$ *is algebraically closed in* $k(V)$.

(c) The following statements are equivalent:

(c_1) $A' = V$;

(c_2) A' *is a valuation ring;*

(c_3) $k(A) \hookrightarrow k(V)$ *is an algebraic extension;*

(c_4) *For every overring* T *of* A *such that* $A \subseteq T \subseteq V$ *the canonical map* $\mathrm{Spec}(T) \longrightarrow \mathrm{Spec}(A)$ *is the identity;*

(c_5) A *is an* i-*domain (in the sense of* [20]*);*

(c_6) *Every overring of* A *is a* PVD.□

We take this opportunity to emphasize that every PVD (though not necessarily integrally closed) is seminormal, essentially because it can be obtained from a valuation ring by a glueing-type process [14, Prop. 1.10].

COROLLARY 3. *With the same notations as in* Corollary 2, *the following statements are equivalent:*

(i) A *is Noetherian;*

(ii) M *is finitely generated over* A *and* V *is a discrete valuation ring (for short,* DVR*);*

(iii) $[k(V):k(A)] < \infty$ *and* V *is a* DVR;

(iv) *Every overring of* A *is Noetherian.*

In particular, every Noetherian PVD A *is such that* $A' = V$ *and* $\dim(A) \leqslant 1$. □

Several other consequences of the Theorem 1 (giving a rather satisfactory description of the properties of pseudo-valuation domains), in addition to the proofs of the Corollary 1 and 2, are collected in [14].

It is natural to ask whether PVD's admit a global counterpart, forming a class of seminormal domains intermediate between Prüfer domains and locally divided domains (cf. [9], [13]), possibly such that each of these new domains has an associated unibranched Prüfer overring from which its structure may be recovered by a pullback. We start our search for such a globalization with one simple but clearly suggestive example ([4]).

EXAMPLE 4. Let d be a square-free integer such that $d \equiv 1 \pmod 4$. Set $A = \mathbb{Z}[\sqrt{d}]$ and $K = \mathbb{Q}(\sqrt{d})$. Let B be the integral closure of $\mathbb{Z}$ (or, equivalently, of A) in K. Then, if $d \equiv 5 \pmod 8$, $A \hookrightarrow B$ is unibranched and A_P is a PVD, having B_Q as canonically associated valuation overring where $P = Q \cap A$, for

([4]) cf. also [17, Example 3.6] for the local case.

every Q in Spec (B); if $d \equiv 1 \pmod 8$, for some prime P in Spec (A) (lying over $2\mathbb{Z}$), A_P cannot be a PVD (although A_P is a discrete valuation ring for every P in Spec (A) such that $P \cap \mathbb{Z} \neq 2\mathbb{Z}$).

It is wellknown that precisely when $d \equiv 1 \pmod 4$, $B = \mathbb{Z}[\omega_d]$, where $\omega_d = (1+\sqrt{d})/2$ (cf. for instance [21; Ch. X, Sec. 6]). Furthermore, B is a Dedekind domain [19, Th. 98] and A is a 1-dimensional Noetherian integral domain [19; Th. 48 and Th. 69]. It is easy to show that, for every *odd* prime p, the canonical homomorphism $\mathbb{Z}[\sqrt{d}] \subset \mathbb{Z}[\omega_d] \twoheadrightarrow \mathbb{Z}[\omega_d]/p\mathbb{Z}[\omega_d]$ is surjective and its kernel is, exactly, $p\,\mathbb{Z}[\sqrt{d}]$, that is, $A/pA \cong B/pB$. Hence, the fiber above $p\mathbb{Z}$ in Spec(A) is isomorphic to the corresponding fiber above $p\mathbb{Z}$ in Spec(B). By a standard homomorphism theorem, if P in Spec(A) satisfies $P \cap \mathbb{Z} = p\mathbb{Z}$ and if Q denotes the unique prime ideal of B such that $Q \cap A = P$, then the canonical map $k_A(P) \hookrightarrow k_B(Q)$ is an isomorphism. Furthermore, it is classically known that:

$$pB = \begin{cases} (p\cdot\mathbb{Z} + \frac{a+\sqrt{d}}{2}\cdot\mathbb{Z})\cdot(p.\mathbb{Z} + \frac{a-\sqrt{d}}{2}\mathbb{Z}), & \text{if } (\frac{\Delta}{p}) = 1; \\ (p\cdot\mathbb{Z} + \frac{p+p\sqrt{d}}{2}\cdot\mathbb{Z}), & \text{if } (\frac{\Delta}{p}) = -1; \\ (p\cdot\mathbb{Z} + \frac{p+\sqrt{d}}{2}\cdot\mathbb{Z})^2, & \text{if } (\frac{\Delta}{p}) = 0; \end{cases}$$

where $\Delta(= d)$ is the discriminant of K, $(\frac{\Delta}{p})$ is the Legendre symbol and a is an (odd) integer such that $a^2 \equiv d \pmod p$, [21, p. 348]. Consequently, we see easily that $QB_Q \subset A_P$ and, hence, $PA_P = QB_Q$, using the fact that 2 is invertible in A_P. In conclusion, we have for every $P \in \text{Spec}(A)$ such that $P \cap \mathbb{Z} \neq 2\mathbb{Z}$, that $A_P \cong B_Q$ is a discrete valuation ring.

Now, we turn our attention to the fibers above the prime ideal $2\mathbb{Z}$. We have (cf. [21, p. 347]):

$$2B = \begin{cases} (2\cdot\mathbb{Z} + \frac{1+\sqrt{d}}{2}\cdot\mathbb{Z})\cdot(2\cdot\mathbb{Z} + \frac{1-\sqrt{d}}{2}\cdot\mathbb{Z}), & \text{if } d \equiv 1 \pmod 8; \\ (2\cdot\mathbb{Z} + (1+\sqrt{d})\cdot\mathbb{Z}), & \text{if } d \equiv 5 \pmod 8; \end{cases}$$

In particular when $d \equiv 5 \pmod 8$, $2B \subset A$ (remark that $2A \subsetneq 2B$!) and $2B = P$ is a prime ideal of A *and* B; thus it is easy (cf.[17]) to see in this case that A_P is a (Noetherian) PVD (but not a valuation ring), having B_P as its associated (discrete) valuation overring. On the other hand, if $d\equiv 1 \pmod 8$ and Q is one of the (two) prime ideal(s) of B lying over $2\mathbb{Z}$, then $P = Q \cap A$ satisfies $\mathbb{Z}/2\mathbb{Z} = k_{\mathbb{Z}}(2\mathbb{Z}) \hookrightarrow k_A(P) \hookrightarrow k_B(Q) \cong \mathbb{Z}/2\mathbb{Z}$, hence $k_A(P) \cong k_B(Q)$. However, at least one of the two possible values of Q satisfies $A_P \subsetneq B_Q$ (otherwise, $A = B$, by globalization, using the previous results on localizations), and so $PA_P \neq QB_Q$. In conclusion, if $d\equiv 1 \pmod 8$, Theorem 1 implies that A_P cannot be a PVD, for some P in Spec (A) lying over $2\mathbb{Z}$, since B_Q is the unique (discrete) valuation overring dominating A_P.

Naïvely, the natural global counterpart to the concept of PVD is the notion of *locally pseudo-valuation domain* (for short LPVD), which can be defined, given an integral domain A with quotient field K, by one of the following equivalent conditions:

(i) A_M is a PVD, for every M in Max(A);

(ii) A_P is a PVD, for every P in Spec(A);

(iii) For every P in Spec (A) and $x \in K\setminus A_P$, $x^{-1}PA_P \subseteq PA_P$.

The integral domains of this type provide a large class of locally divided (cf. [9]) and seminormal (cf.[14, Prop. 1.10] , [5, Lemma 2.1 (b)]) domains, not necessarily integrally closed, of arbitrary Krull-dimension.

Clearly, every Prüfer domain is an integrally closed LPVD. We emphasize that the converse is not true in general, even in the local case: cf. Corollary 2 (b). In any case, it can be shown that each overring of an integral domain A is a LPVD if, and only if, A is an LPVD and its integral closure is a Prüfer domain (cf. [10, Th. 2.9]). The Example 4 supplies an LPVD which is not integrally closed, but which has a Prüfer (indeed, Dedekind) domain as integral closure. We remark also that the techniques of Example 3.2 of [13, p. 68] can lead to a quasi-semi-local, but not quasi-local, LPVD which though integrally closed is not Prüfer (cf. also [10, Ex. 2.5]).

One of the reasons which makes the theory of PVD's so rich and tractable is the presence of a valuation overring sharing the prime spectrum of the given PVD : cf. Theorem 1. Though a non-quasi-local LPVD cannot have a proper (Prüfer) overring with the *same* spectrum (cf. [4, Prop. 3.3]), there is however present in nature a well-behaved class of LPVD's admitting *uni-branched* Prüfer overring (cf. Example 4). These considerations lead to the following more cogent generalization of the notion of PVD.

Let A be a subring of a Prüfer domain B, then A is called a *global pseudo-valuation domain* (for short, GPVD) if one of the following (two) equivalent conditions holds:

(j) (a) $A \hookrightarrow B$ is a unibranched extension;

(b) There exists a non-zero radical ideal I, common to A and B such that each prime ideal of A (resp., B) containing I is maximal in A (resp., B);

(jj) There exists a non-zero radical ideal J, common to A and B, such that $\bar{A} = A/J$ and $\bar{B} = B/J$ satisfy the following:

($\bar{a}$) $\bar{A} \hookrightarrow \bar{B}$ is a unibranched extension;

($\bar{b}$) $\dim(\bar{A}) = \dim(\bar{B}) = 0$.

It is easy to see that, if (j) or (jj) holds then $J(A) =$

$J(B)$ and

$$A \cong B \times_{\Pi k_B(N)} \Pi k_A(N \cap A)$$

where the index N runs over the set of all the maximal ideals of B (or, equivalently, the set of all the maximal ideals of B containing I, in case (j), or J, in case (jj)). Furthermore, it is straightforward to see that, for every N in $\mathrm{Max}(B)$, the following square:

$$\begin{array}{ccc} A_{N\cap A} & \twoheadrightarrow & k_A(N\cap A) \\ \downarrow & & \downarrow \\ B_N & \twoheadrightarrow & k_B(N) \end{array}$$

is a pullback diagram. In this situation, B will be called the *Prüfer domain associated to the* GPVD A. (For the equivalence between (j) and (jj), cf. [10, Th. 3.1]).

It is straightforward to see that each GPVD in an LPVD and that, in the quasi-local case, the notions of GPVD, LPVD and PVD coincide. Evidently, every Prüfer domain is a GPVD. The ring $\mathbb{Z}[\sqrt{d}]$ is a GPVD (but not a Prüfer domain), having $B = \mathbb{Z}[\omega_d]$ as its associated Prüfer domain, whenever $d \equiv 5 \pmod 8$ (take $I = 2B$): cf. Example 4. We remark that there exists: (a) a quasi-semi-local GPVD such that its localization at *every* maximal ideal are PVD's but not valuation domains; (b) an LPVD *not* GPVD, such that its integral closure is a Prüfer domain; and (c) a *non* quasi-semi-local GPVD which is not a Prüfer domain. Examples with these properties are given explicitly in [10].

The global counterpart of Corollary 3 is given next.

PROPOSITION 5. *Let* A *be a* GPVD , *with associated Prüfer domain* B *and integral closure* A'. *Then, the following conditions are equivalent:*

(i) A *is Noetherian;*

(ii) B *is a Dedekind domain and* $[k_B(N):k_A(N\cap A)]<\infty$, *for each* N *in* $\mathrm{Max}(B)$;

(iii) B *is a Dedekind domain and* NB_N *is a finitely generated* $A_{N\cap A}$*- module, for each* N *in* $\mathrm{Max}(B)$.

Moreover, if one of the above conditions holds, then $\dim(A) \leq 1$ *and* $A' = B$.

PROOF. (Sketch). Although the Noetherian property is not a local property, one can nevertheless reduce the problem to a local one here, by using a standard argument which assures that a locally Noetherian ring such that every non-zero element lies in only finitely many maximal ideals is necessarily Noetherian. In the local case, one applies Corollary 3. □

The main theorem concerning the GPVD's in the context of the classical Prüfer's ascent theorem [16, Th. 22.3] is the following:

THEOREM 6. *Let* A *be a* GPVD, *with associated Prüfer domain* B *and quotient field* K. *Let* K^* *be an extension field of* K *and let* A^* *(resp.,* B^**) denote the integral closure of* A *(resp.,* B *) into* K^*. *Then* A^* *is a* GPVD, *with associated Prüfer domain* B^*.

PROOF. (Sketch) (5). The result can be obtained after several reductions.

(5) Cf. also [10, Theorem 3.7].

1^ Reduction. We may suppose A (and B) quasi-local. This is possible in part because the integral closure behaves "well" in regard to localizations [16, Prop. 10.2].

2^ Reduction. We may suppose A is integrally closed and $A \subsetneq B$. As a matter of fact, A and its integral closure in K have the same integral closure in K^*, moreover, if $A = B$ the statement is wellknown [16, Th. 20.1].

3^ Reduction. We may assume that K^* is algebraic over K. Otherwise, replace K^* with the algebraic closure of K into K^*.

4^ Reduction. It is sufficient to show the result when K^* is a finite extension of K. In fact, failure of condition (jj) ($\bar{a}$) implies a corresponding failure when K^* is replaced by a suitable finite extension of K.

After the reductions, one can attack the remaining problem with reasonable finite-type techniques (similar to those used in [13, E. 3.2]), being then concerned only with domains obtained by intersection of a finite family of valuation [16, Cor. 20.3] or pseudo-valuation domains. □

In conclusion, we mention one problem, connected with some results obtained recently by K. Uchida [22] (cf. also T. Albu [2]), whose solution would have important number theoretic consequences: Characterize the simple finite extensions of Dedekind domains which are (Noetherian) GPVD's. We remark that a partial answer to this problem can be deduced from Example 4: The simple quadratic extension $\mathbb{Z}[\delta]$, where $\delta = \sqrt{d}$ with d square-free integer, is a GPVD if, and only if, $d \equiv 2, 3 \pmod 4$ or $d \equiv 5 \pmod 8$.

REFERENCES

[1] T. Akiba, *A note on AV-domains.* Bull. Kyoto Ed. Ser. B 31 (1967), 1-3.

[2] T. Albu, *On a paper of Uchida concerning simple finite extensions of Dedekind domains.* Osaka J. Math. 16 (1979), 65-69.

[3] D.F. Anderson, *Comparability of ideals and valuation overrings.* Houston J. Math. 5 (1979), 451-463.

[4] D.F. Anderson & D.E.Dobbs, *Pair of rings with the same prime ideals.* Canad. J. Math. 32 (1980), 362-384.

[5] D.F. Anderson,D.E.Dobbs & J.A. Huckaba, *On seminormal overrings.* Comm. Algebra (to appear).

[6] D.E. Dobbs, *Divided rings and going-down.* Pac. J. Math. 67 (1976), 353-363.

[7] D.E. Dobbs, *Coherence, ascent of going-down and pseudo-valuation domains.* Houston J. Math. 4 (1978), 551-567.

[8] D.E. Dobbs, *On the weak global dimension of pseudo-valuation domains.* Canad. Math. Bull. 21 (1978), 159-164.

[9] D.E. Dobbs. *On locally divided integral domains and CPI-overrings.* Internat. J. Math. & Math. Sciences 4 (1981), 119-135.

[10] D.E. Dobbs & M. Fontana, *Locally pseudo-valuation domains.* In preparation.

[11] D.E. Dobbs, M. Fontana, J.A. Huckaba & I.J. Papick, *Strong ring extensions and pseudo-valuation domains.* Houston J. Math. (to appear).

[12] M. Fontana, *Topologically defined classes of commutative rings.* Annali Mat. Pura Appl. 123 (1980), 331-355.

[13] M. Fontana, *Carrés cartésiens, anneaux divisés et anneaux localement divisés.* Pre-Publ. Math. Université Paris-Nord Fasc. 21, 1980.

[14] M. Fontana, *Carrés cartésiens et anneaux de pseudo-valuation.* Publ. Math. Univ. Lyon 17 (1980), 57-95.

[15] M. Fontana, *Sur quelques classes d'anneaux divisés.* Rend. Sem. Mat. Milano (à paraître).

[16] R. Gilmer, *Multiplicative Ideal Theory.* Dekker, New York 1972.

[17] J.R. Hedstrom & E.G. Houston, *Pseudo-valuation domains.* Pac. J. Math. 75 (1978), 137-147.

[18] J.R. Hedstrom & E.G. Houston, *Pseudo-valuation domains, II* Houston J. Math. 4 (1978), 199-207.

[19] I. Kaplansky, *Commutative rings.* Allyn & Bacon, Boston 1970.

[20] I.J. Papick, *Topologically defined classes of going-down rings.* Trans. Amer. Math. Soc. 219 (1976), 1-37.

[21] L.W. Reid, *The elements of the theory of algebraic numbers (with an introduction by* D. Hilbert*).* MacMillan, New York, 1910.

[22] K. Uchida, *When is* $\mathbb{Z}[\alpha]$ *the ring of integers?* Osaka J. Math. 14 (1977), 155-157.

KOSZUL HOMOLOGY AND BLOWING-UP RINGS

J. Herzog, A. Simis and W. V. Vasconcelos

Universitat Essen Gesamthochschule
Essen, West Germany

Departamento de Matematica
Universidade Federal de Pernambuco
Recife, Brazil

Department of Mathematics
Rutgers University
New Brunswick, New Jersey

These notes are a combined and expanded account of the individual talks given by the authors at the Trento conference. They aim at providing a framework for studying arithmetical properties of Rees algebras, associated graded rings, symmetric algebras and the theory of linear resolutions of ideals and modules.

We avail ourselves of the opportunity to make a fuller exposition by drawing liberally - and likely with considerable personal biases in view of the as yet unsettled state of the subject - from various other sources.

For all of these sins, we are much grateful to the organizers of the conference, Professors Silvio Greco and Guiseppe Valla.

1. Introduction.

Given a Noetherian ring R and an ideal I, there are several algebras that conveniently code information, of an analytic or algebraic nature, on the variety $V(I)$ that I defines. Because these algebras are defined in terms of various conormal bundles it is not surprising that their properties mirror how $V(I)$ sits inside $Spec(R)$.

A brief catalog of these algebras, which will be loosely designated as 'blowing-up' rings, is the following:

(a) $\mathcal{R}(I) = R[IT] = \Sigma\, I^s T^s$ = Rees algebra of I.

(b) $\mathcal{R}(I)^* = R[IT, T^{-1}]$ = Rees* algebra of I.

(c) $gr_I(R) = \Sigma\, I^s/I^{s+1}$ = graded algebra associated to the I-adic filtration.

(d) Sym(I) = symmetric algebra of I.

(e) $Sym(I/I^2)$ = symmetric algebra of the first conormal bundle I/I^2.

If, besides, I is a prime ideal and $I^{(s)}$ denotes its s-th symbolic power - a procedure that can be adequately extended to any ideal - the following algebras may be added:

(f) $Symb(I) = \Sigma\, I^{(s)} T^s$ = symbolic power algebra of I.

(g) $Symb(I)^* = Symb(I)[T^{-1}]$ = symbolic* power algebra of I.

(h) $gr_I^!(R) = \Sigma\, I^{(s)}/I^{(s+1)}$ = graded algebra associated to the I-symbolic filtration.

Each of the listed algebras packs considerable information about the ideal I. Noteworthy are (a) for its role in the process of blowing-up a variety along a subvariety, and (f) in view of Zariski's main lemma on holomorphic functions. The last three algebras may fail to be Noetherian, thus making their usefulness less approachable.

There are several relationships among these algebras, expressed by algebra homomorphisms. For instance, $gr_I(R) = \mathcal{R}(I) \otimes_R (R/I) = \mathcal{R}(I)^*/(T^{-1})$, or $\mathrm{Sym}(I/I^2) = \mathrm{Sym}(I) \otimes_R (R/I)$, and several other less explicit ones.

One of our main themes will be the study of the comparisons involved in the canonical diagram:

$$\begin{array}{ccccc} \mathrm{Sym}(I) & \xrightarrow{\alpha} & \mathcal{R}(I) & \longrightarrow & \mathrm{Symb}(I) \\ \downarrow & & \downarrow & & \downarrow \\ \mathrm{Sym}(I/I^2) & \xrightarrow{\beta} & gr_I(R) & \longrightarrow & gr_I^!(R) \end{array}$$

The approach to be used will be that of expressing the comparison in the first square by certain differential graded algebras - labelled approximation complexes - that, among other properties, reflect the obstructions for the morphisms α and β to be isomorphisms. These complexes are constructed over Sym(I) and $\mathrm{Sym}(I/I^2)$, and are intimately connected to the homology modules $H_i(\underline{x};R)$ of the ordinary Koszul complex $K(\underline{x})$, built from a sequence $\underline{x}$ generating the ideal I. This is a rather natural approach since Sym(I) and $\mathrm{Sym}(I/I^2)$ are more directly accessible than the other rings, but places considerable emphasis on retrieving properties of the Koszul homology modules from the specifics of the presentation of the ideal. When successful this method allows for describing many arithmetical properties of the blowing-up rings $\mathcal{R}(I)$ and $gr_I(R)$, and permits certain comparisons between $\mathcal{R}(I)$ and Symb(I) as well.

The main approximation complexes associated to the sequence $\underline{x}$, $\mathcal{M}(\underline{x})$ and $\mathcal{Z}(\underline{x})$, display different sensitivities vis-a-vis $K(\underline{x})$, being acyclic in situations much broader than the usual context of regular sequences. It has been found that certain generalizations thereof, d-sequences and proper sequences, play here the role of 'acyclic sequence'. This turns out to be rather fruitful, for its connections to the theory of linear and algebra resolutions, multiplicities, coarse Cohen-Macaulayfication of local rings, and for spawning classes of Cohen-Macaulay algebras arising from symmetric algebras.

Before we describe, in some detail, the contents of these notes, we point out that although various of the results could be stated in greater generality, the context will be that of Noetherian rings. For notation, terminology, and basic references - especially when dealing with Cohen-Macaulay rings - we shall use [Kap] and [Mat]. In addition, [Mat] along with [GL] will be our sources for the more readily accessible properties of Koszul complexes.

An early focus of this study is the canonical homomorphism that exists between the symmetric algebra of the ideal I, Sym(I), and the corresponding Rees ring $\mathcal{R}(I)$:

$$\alpha : \mathrm{Sym}(I) \longrightarrow \mathcal{R}(I),$$

induced by the identification $\mathrm{Sym}_1(I) = \mathcal{R}(I)_1 = I$. α is always a surjection, and the case when it is also injective occurs often enough to deserve a designation. Thus we shall call I an ideal of linear type if α is an isomorphism.

Initial examinations of this mapping were carried out by various authors ([MSS],[Mi],[Sal]), who pointed out several classes of ideals of linear type. In §2 we begin listing some of these ideals and introduce the first measure of comparison. That will simply be $\delta_1(I) = \mathrm{kernel}(\mathrm{Sym}_2(I) \longrightarrow I^2)$. It is easy to express for certain determinantal ideals, and later it will be interpreted cohomologically.

In §3 we discuss lifting theorems. These are of two kinds. The first revolves around a surprising discovery of Valla ([Va]), to wit, if the reduction of

$$\beta : \mathrm{Sym}_{R/I}(I/I^2) = \mathrm{Sym}(I/I^2) \longrightarrow \mathcal{gr}_I(R)$$

is an isomorphism, then α is an isomorphism also.

The other results compare the homomorphism α for the ideals I, $I/(f)$ and (I,f). As an application one obtains a result of Huneke ([Hu_2]) that $\mathrm{Sym}(I) \cong \mathcal{R}(I)$ for ideals generated by d-sequences. This notion belongs to a family of extensions of the notion of regular sequence and will play an ubiquitous role here.

Given a set of generators $\underline{x} = \{x_1,\ldots,x_n\}$ of the ideal I and an R-module M, we construct in §4 the associated approximation complexes. These are derived from a double Koszul complex and are closely linked to $\mathcal{K}(\underline{x};M)$, the ordinary Koszul complex associated to $\underline{x}$ and M. Two of them will be at the center of interest : the so-called $\mathcal{M}$ and $\mathcal{Z}$ complexes. They are complexes of graded modules over the polynomial ring $S = R[T_1,\ldots,T_n]$:

$$\mathcal{M}(\underline{x};M) : 0 \longrightarrow H_n \otimes S[-n] \longrightarrow \ldots \longrightarrow H_1 \otimes S[-1] \longrightarrow H_0 \otimes S \longrightarrow 0$$

and

$$\mathcal{Z}(\underline{x};M) : 0 \longrightarrow Z_n \otimes S[-n] \longrightarrow \ldots \longrightarrow Z_1 \otimes S[-1] \longrightarrow Z_0 \otimes S \longrightarrow 0,$$

where H_i and Z_i denote the homology and cycles of $\mathcal{K}(\underline{x};M)$. In contrast to $\mathcal{K}(\underline{x}:M)$, whose homology depends on the generating set $\underline{x}$, the homology of either $\mathcal{M}$ or $\mathcal{Z}$ depends solely on I.

When $M = R$, we have $H_0(\mathcal{Z}) = \mathrm{Sym}(I)$ and $H_0(\mathcal{M}) = \mathrm{Sym}(I/I^2)$. There are several relationships between $H(\mathcal{M})$ and $H(\mathcal{Z})$: It allows for instance to conclude that if $H_1(\mathcal{M}) = 0$, then

Sym(I) $\cong$ $\mathcal{R}$(I). Thus these complexes serve as an obtruction theory for the homomorphism α.

In §5 we consider conditions for the acyclicity of these complexes (see also § 12). Those depend on having sufficiently high depth on the Koszul homology modules H_i. It will prepare the ground for one of our main applications in §10. Here, as illustration, we consider instances of ideals that are nearly enough to complete intersections.

Various extensions of the notion of regular sequence, e.g. relative regular sequence, d-sequence and proper sequence, are taken up in §6. They exist in more profusion than regular sequences. For instance, we prove that any ideal of height n contains a d-sequence of height n. This will be used to provide a coarse form of Cohen-Macaulayfication of local rings. The relationships among these sequences will be looked at and new measures of syzygeticity introduced.

In §7 we discuss methods for determining the depth of the Koszul homology modules of an ideal. The main result here is a theorem of Huneke ([Hu_5]), stating how that depth may change along the linkage class of the ideal. Around this result one can organize several earlier ones of ideals with Cohen-Macaulay Koszul homology.

The next section develops a tool for reading the Krull dimension of a module M from the details of a free presentation. It works often enough for the uses in §§9 and 10.

In §10 we present some of the main applications of the approximation complexes so far. It is proved that, if R is Cohen-Macaulay and the Koszul homology of the ideal I is

Cohen-Macaulay and a natural local bound on the number of generators hold, then $\mathrm{Sym}(I) \cong \mathcal{R}(I)$ and $\mathrm{Sym}(I/I^2) \cong gr_I(R)$. Furthermore, all of these algebras are Cohen-Macaulay, and if R is Gorenstein then $gr_I(R)$ is Gorenstein also.

By weakening some of the hypotheses of §9 - that reflect requirements for the acyclicity of the $\mathcal{M}$-complex - one is able in §10 and using mostly the $\mathcal{Z}$-complex to obtain new classes of Cohen-Macaulay symmetric algebras and develop a theory of their specializations.

In §11 we introduce a new array of syzygetic groups associated to a sequence $\underline{x}$ and an R-module M. They extend the δ's of §6 and reflect the properties of the Koszul homology not of the ideal $I = (\underline{x})$ alone but of the irrelevant ideals of the graded algebras $\mathcal{R}(I)$ and $gr_I(R)$. It is used to provide the precise conditions for an ideal to be of linear type ([Kü]) and to prepare the ground for §12. This section contains proofs of the following structural aspects of the approximation complexes. Essentially (i.e. locally and modulo a faithfully flat extension) : (a) The complex $\mathcal{M}(I;M)$ is acyclic if and only if I is generated by a d-sequence relative to M; (b) (cf. [Kü]) The complex $\mathcal{Z}(I;M)$ is acyclic if and only if I is generated by a proper sequence relative to M.

Finally, in §13, we study the $\mathcal{M}$-complex relative to the maximal ideal of a regular local ring and its connections to the theory of linear resolutions. It allows for finding algebra resolutions of certain ideals, determination of Hilbert-Poincaré series of resolutions and display the lack of rigidity of the approximation complexes themselves.

§2. Ideals of linear type.

In this section we begin to introduce measures of comparison between the symmetric and Rees algebras of ideals. Let R be a commutative ring and let I be an ideal of R. Denote by α the canonical surjection of R-algebras

$$0 \longrightarrow \mathcal{A} = \ker(\alpha) \longrightarrow \mathrm{Sym}(I) \xrightarrow{\alpha} \mathcal{R}(I).$$

If $\underline{a} = \{a_1,\dots,a_n\}$ is a generating set for I, the morphism α can also be described in the following manner. Present I as

$$0 \longrightarrow Z_1 \longrightarrow R^n \xrightarrow{\phi} I \longrightarrow 0.$$

ϕ induces a surjection of the polynomial ring $R[T_1,\dots,T_n]$ onto $\mathrm{Sym}(I)$, with T_i mapping into $a_i \in \mathrm{Sym}_1(I)$. Its kernel is the ideal Q generated by the elements in Z_1, that is, Q is generated by all linear forms $\Sigma\, c_iT_i$ such that $\Sigma\, c_ia_i = 0$. $\mathcal{R}(I)$ may also be viewed as a homomorphic image of $R[T_1,\dots,T_n]$, with T_i mapping into a_iT. The kernel, Q_∞, is generated by all forms $F(T_1,\dots,T_n)$ such that $F(a_1,\dots,a_n) = 0$. We thus have a diagram

$$\begin{array}{ccccc} Q_\infty & & & & Q \\ & \searrow & & \swarrow & \\ & & R[T_1,\dots,T_n] & & \\ & \swarrow & & \searrow & \\ \mathrm{Sym}(I) & & \xrightarrow{\quad\alpha\quad} & & \mathcal{R}(I) \end{array}$$

with $\mathcal{A} = \ker(\alpha) = Q_\infty/Q$. It is clear, from this discussion, that alternative ways to present this comparison are needed. This difficulty will be somewhat dealt with in the next sections, especially with the introduction of the approximation complexes.

The case when α is an isomorphism occurs often enough to deserve its own designation. The following terminology was suggested by Robbiano and Valla.

Definition 2.1. The ideal I is said to be of linear type if the corresponding morphism α is an isomorphism.

The module-theoretic properties of Sym(I) play a dominant role in the study of the comparison morphism α, as witnesses the following

Proposition 2.2. Let I be an ideal such that $I \cap (0:I) = 0$. The following conditions are equivalent:

(a) I is of linear type.

(b) Sym(I) has no I-torsion (i.e. 0 is the only element of Sym(I) annihilated by I).

Proof. That (a) => (b) is clear since, by hypothesis, $\mathcal{R}(I)$ has no I-torsion. Conversely, suppose $w \in \mathrm{Sym}_t(I)$; since

$$I^t w = \alpha(w)\mathrm{Sym}_t(I),$$

we easily get the desired assertion.

Note that, as a consequence, if $I \cap (0:I) = 0$ and $\mathrm{Sym}(I) \cong \mathcal{R}(I)$ under some other R-morphism, then α is an isomorphism as well.

The first large class of ideals of linear type is provided by the following construction of Salmon ([Sal]; see also [HSV_1]).

Example 2.3. Let E be a module over the ring A and let $R = \mathrm{Sym}_A(E)$. Pick for I the augmentation ideal of R,

$$I = R_+ = \bigoplus_{t>0} \mathrm{Sym}_t(E).$$

We have the natural R-morphism

$$E \otimes_A \mathrm{Sym}_A(E) \longrightarrow \mathrm{Sym}_A(E)_+ \longrightarrow 0,$$

whose kernel is generated by the elements of the form $e\otimes f\cdot w - f\otimes e\cdot w$, $w, f \in E = \mathrm{Sym}_1(E)$, $w \in \mathrm{Sym}_A(E)$. Taking the n-th symmetric power of this homomorphism - and using the universal property of such powers with respect to the change of rings

$A \longrightarrow R$ - we get

$$\mathrm{Sym}_n(E \otimes_A R) = \mathrm{Sym}_n(E) \otimes_A R \longrightarrow \mathrm{Sym}_n(I) \longrightarrow 0,$$

from which we obtain $I^n \cong \mathrm{Sym}_n(I)$ when the relations above are taken into account.

An important module-theoretic obstruction to "$\mathrm{Sym}(I) = \mathcal{R}(I)$" is provided by

Proposition 2.4. Let I be an ideal of the Noetherian ring R. If $\mathrm{Sym}(I) \cong \mathcal{R}(I)$, then for each prime ideal $P \supset I$, I_P can be generated by height(P) elements.

Proof. We may assume that (R,P) is a local ring of dimension d. We work with the derived homomorphism $\mathrm{Sym}(I/I^2) = gr_I(R)$.

Let $\mathcal{R}(I)^*$ be the 'large' Rees ring, that is $\mathcal{R}(I)^* = R[IT,u]$, $u = T^{-1}$; then $gr_I(R) = \mathcal{R}(I)^*/(u)$. Since $\dim \mathcal{R}(I)^* \leq d+1$ ([Ra_1]) and u is a regular element, we have $\dim \mathrm{Sym}(I/I^2) \leq d$. In this case its homomorphic image $\mathrm{Sym}_{R/P}(I/PI)$ has dimension at most d, as desired.

To compare, in general, the algebras Sym(I) and $\mathcal{R}(I)$, it is often useful to look at the graded components of the morphism α. Thus, denote

$$\alpha_t\colon\ \mathrm{Sym}_t(I) = S_t(I) \longrightarrow I^t = \mathcal{R}(I)_t.$$

Since α_o and α_1 are the ordinary identifications, the first place these algebras may differ is for $t = 2$.

In order to spell out $A_2 = \ker(\alpha_2)$ we bring in the Koszul complex on a set of generators of the ideal I. This idea will play an ever increasing role as we go on. We shall refer to [Mat] and [GL] for basic properties of the various Koszul complexes.

Tensoring by R/I the implied presentation of I yields an

exact sequence of R/I-modules

$$0 \longrightarrow Z_1 \cap IR^n/IZ_1 \longrightarrow Z_1/IZ_1 \longrightarrow R^n/IR^n \longrightarrow I/I^2 \longrightarrow 0.$$

Thus $Z_1/Z_1 \cap IR^n$ is the first-kind syzygy module of the conormal module I/I^2. It fits into the exact sequence

$$0 \longrightarrow Z_1 \cap IR^n/B_1 \longrightarrow H_1 \longrightarrow Z_1/Z_1 \cap IR^n \longrightarrow 0$$

where H_1 (resp. B_1) is the first homology module (resp. 1-boundaries) of the Koszul complex on the chosen set of generators of I. We set $H_1^* = Z_1/Z_1 \cap IR^n$, and call it the <u>syzygy part</u> of H_1 (cf. [Si]; see also §4).

Although H_1 and H_1^* both depend on the set of generators for I, $Z_1 \cap IR^n/B_1$ depends only on I. To put this in evidence, we denote $Z_1 \cap IR^n/B_1$ by $\delta_1(I)$.

<u>Proposition 2.5</u>. There exists a natural homomorphism of R/I-modules $\sigma\colon \Lambda^2 I \longrightarrow \mathrm{Tor}_1^R(I,R/I)$ ('antisymmetrization') such that $\delta_1(I) = \mathrm{Tor}_1^R(I,R/I)/\sigma(\Lambda^2 I)$.

<u>Proof</u>. Since I annihilates H_1, $IZ_1 \subset B_1$. We thus see that $\delta_1(I)$ fits naturally into the exact sequence

$$0 \longrightarrow B_1/IZ_1 \longrightarrow Z_1 \cap IR^n/IZ_1 \longrightarrow \delta_1(I) \longrightarrow 0.$$

By the long exact sequence of Tor, as applied to the initial presentation of I, we have $\mathrm{Tor}_1^R(I,R/I) = Z_1 \cap IR^n/IZ_1$. But $\mathrm{Tor}_1^R(I,R/I)$ can also be viewed as the kernel of the multiplication mapping $I \otimes I \longrightarrow I^2$, a fact we shall now use. Let then $\tilde{\sigma}\colon \Lambda^2 I \longrightarrow I \otimes I$ be the antisymmetrization mapping, that is, $\tilde{\sigma}(u \wedge v) = u \otimes v - v \otimes u$, for $u,v \in I$. The composite $\Lambda^2 I \longrightarrow I \otimes I \longrightarrow I^2$ is zero, so that we actually have a homomorphism $\sigma\colon\Lambda^2 I \longrightarrow \mathrm{Tor}_1^R(I,R/I)$.

Let $I = (a_1,\ldots,a_n)$ and let $\{e_1,\ldots,e_n\}$ be a basis of R^n such that $\phi(e_i) = a_i$. The homomorphism

$$\Lambda^2 I \longrightarrow \mathrm{Tor}_1^R(I,R/I) = Z_1 \cap IR^n/IZ_1$$

as applied to the element $a_i \wedge a_j$ gives the element $a_j e_i - a_i e_j$ (mod. IZ_1), which belongs to B_1/IZ_1. The composite is evidently onto, so $\sigma(\Lambda^2 I) = B_1/IZ_1$.

The claim then follows from the exact sequence above.

Remark. Since the image of σ is killed by I, σ factors through $\Lambda^2 I \otimes (R/I) = \Lambda^2(I/I^2)$. On the other hand, if 2 is invertible, then σ is injective and gives, actually, the splitting $\mathrm{Tor}_1^R(I,R/I) = \delta_1(I) \oplus \Lambda^2(I/I^2)$.

Corollary 2.6. $\delta_1(I) = A_2 = \ker(S_2(I) \longrightarrow I^2)$.

Proof. We show that $\ker(S_2(I) \longrightarrow I^2) = \mathrm{Tor}_1^R(I,R/I)/\sigma(\Lambda^2 I)$. To see this note that the sequence

$$\Lambda^2 I \longrightarrow I \otimes I \longrightarrow S_2(I) \longrightarrow 0$$

is exact, where the last mapping is the canonical one. But $\sigma(\Lambda^2 I) \subset \mathrm{Tor}_1^R(I,R/I)$, and therefore we get the commutative diagram

$$\begin{array}{ccccccccc} 0 & \longrightarrow & \mathrm{Tor}_1^R(I,R/I) & \longrightarrow & I \otimes I & \longrightarrow & I^2 & \longrightarrow & 0 \\ & & \uparrow & & \uparrow & & \uparrow & & \\ & & \Lambda^2 I & \longrightarrow & I \otimes I & \longrightarrow & S_2(I) & \longrightarrow & 0 \end{array}$$

from which the assertion follows.

To consider the relative strength of the vanishing of $\delta_1(I)$ we call on two examples.

Proposition 2.7. (cf. [SV_1]) Let (R,m) be a local ring of depth two and let I be an ideal of height two and projective dimension one. The following are equivalent:

(a) $\delta_1(I) = 0$.

(b) I is generated by a regular sequence of two elements.

Proof. Let

$$0 \longrightarrow F_1 \longrightarrow F_o \longrightarrow I \longrightarrow 0$$

be a minimal projective resolution of I. Assume $\delta_1(I) = 0$; we must show that F_1 has rank one. Associated to this resolution there is a canonical complex [Wey]:

$$0 \longrightarrow \Lambda^2 F_1 \longrightarrow F_1 \otimes F_o \longrightarrow S_2(F_o) \longrightarrow S_2(I) \longrightarrow 0.$$

From the minimality hypothesis and the construction of this complex, it is a minimal projective resolution of $S_2(I)$. Since $\delta_1(I)$ is the torsion-submodule of $S_2(I)$ (cf. (2.6)), and the depth two condition, it follows that $\delta_1(I) = 0$ implies $\Lambda^2 F_1 = 0$, as desired.

The converse is obvious.

Remark. The proof also shows the following. Let R be a Noetherian ring and let I be an ideal of height two and projective dimension one. Then $\delta_1(I) = 0$ if and only if I is generically a complete intersection. In particular, making use of Macaulay's examples, we conclude that if $R = k[x,y,z]$, there is no bound for the minimal number of generators of ideals with $\delta_1(I) = 0$.

Proposition 2.8. Let R be a regular local ring and let I be a Gorenstein ideal of height 3. If $1/2 \in R$, then $\delta_1(I) = 0$.

Proof. It was proved in [He$_1$] (without the hypothesis $1/2 \in R$) that for such ideals I/I^2 is a Cohen-Macaulay module and that if $\underline{x}$ is a system of parameters for $S = R/I$, then $\lambda(I/I^2 \otimes S/(\underline{x})) = 3\lambda(S')$, where $S' = S/(\underline{x})$ and λ = length. On the other hand, it follows from [Le$_1$] that $\Lambda^2 I$ has projective dimension 3. Since $1/2 \in R$, $\Lambda^2 I = \Lambda^2(I/I^2)$ and thus $\Lambda^2(I/I^2)$ is a Cohen-Macaulay module. Furthermore, since it is a specialization of the generic case, it follows from the methods of [He$_1$] that

$$\lambda(\Lambda^2(I/I^2) \otimes S') = 3\Lambda(S').$$

Now let

$$0 \longrightarrow R \longrightarrow R^n \longrightarrow R^n \longrightarrow I \longrightarrow 0$$

be a minimal resolution of I. Tensoring with S and taken into account that the entries of the last 'matrix' lie in I, we get the exact sequence

$$0 \longrightarrow \mathrm{Tor}_1^R(I,S) \longrightarrow S^n \longrightarrow S^n \longrightarrow I/I^2 \longrightarrow 0.$$

Since I/I^2 is Cohen-Macaulay, $\mathrm{Tor}_1^R(I,S)$ will also be Cohen-Macaulay. Now to show $\delta_1(I) = 0$ note that we have the decomposition

$$\mathrm{Tor}_1^R(I,S) = \Lambda^2(I/I^2) \oplus \delta_1(I).$$

Tensoring it with S' and comparing lengths, we get $\delta_1(I) \otimes S' = 0$, and thus $\delta_1(I) = 0$ by Nakayama lemma.

Later a whole array of measures of comparison will be introduced and conveniently interpreted cohomologically.

§3. Ascent and descent.

The comparison of Sym(I) and $\mathcal{R}(I)$ is enhanced by the existence of certain natural mappings defined on Sym(I). One of these - later to be explained as a homology connecting homomorphism - is the downgrading homomorphism λ,

$$\lambda_t\colon\ S_{t+1}(I) \longrightarrow S_t(I)$$

that sends $b_1 \cdot b_2 \cdot \ldots \cdot b_{t+1} \in S_{t+1}(I)$ into the element $b_1(b_2 \cdot \ldots \cdot b_{t+1}) \in S_t(I)$. It is clear that it does not matter which element is pulled out of the pack. The image of λ_t is the submodule $I(S_t(I))$ of $S_t(I)$, that is, the kernel of the canonical homomorphism $S_t(I) \longrightarrow S_t(I/I^2)$.

We thus have the commutative exact diagram

$$\begin{array}{ccccccccc}
 & & 0 & & 0 & & 0 & & \\
 & & \downarrow & & \downarrow & & \downarrow & & \\
 & & A_{t+1} & \longrightarrow & A_t & \longrightarrow & A_t/\lambda(A_{t+1}) & \longrightarrow & 0 \\
 & & \downarrow & & \downarrow & & \downarrow & & \\
 & & S_{t+1}(I) & \xrightarrow{\lambda_t} & S_t(I) & \longrightarrow & S_t(I/I^2) & \longrightarrow & 0 \\
 & & \downarrow{\scriptstyle \alpha_{t+1}} & & \downarrow{\scriptstyle \alpha_t} & & \downarrow{\scriptstyle \beta_t} & & \\
0 & \longrightarrow & I^{t+1} & \longrightarrow & I^t & \longrightarrow & I^t/I^{t+1} & \longrightarrow & 0 \\
 & & \downarrow & & \downarrow & & \downarrow & & \\
 & & 0 & & 0 & & 0 & &
\end{array}$$

The following surprising lifting property was discovered by Valla ([Va]; see also [HSV_1]).

Theorem 3.1. Let I be an ideal of the Noetherian ring R. Then α: Sym(I) $\longrightarrow \mathcal{R}(I)$ is an isomorphism if and only if the reduction β: Sym(I/I^2) $\longrightarrow gr_I(R)$ is an isomorphism.

Proof. We only have to show that β lifts to an isomorphism.

The preceding diagram and the snake lemma leads to the exact sequence:

$$0 \longrightarrow A_t/\lambda(A_{t+1}) \longrightarrow S_t(I/I^2) \xrightarrow{\beta_t} I^t/I^{t+1} \longrightarrow 0.$$

By hypothesis we have $A_t = \lambda(A_{t+1})$ for $t \geq 2$. Since A is a finitely generated homogeneous ideal of Sym(I), there exists an integer $s \geq 2$ such that

$$A_{t+1} = S_1(I)\cdot A_t, \quad t \geq s.$$

Applying λ to this equality and taking into account the hypothesis above, we get

$$A_t = \lambda(A_{t+1}) = \lambda(S_1(I)\cdot A_t) = I(A_t)$$

and thus, by localization and the Nakayama lemma, $A_t = 0$ for $t \geq s$.

By an obvious descending induction we get $A_t = 0$ for all t.

Similar is the following (cf. [HSV_1]).

Theorem 3.2. $A = \ker(\alpha)$ is nilpotent if and only if $A^* = \ker(\beta)$ is nilpotent.

Proof. Since A is finitely generated, the equation of (2.2) implies that there exists an integer r such that $I^r A = 0$.

We only have to prove that A is nilpotent along with A^*. For this we again refer to the earlier diagram. As $A^* = \ker(\beta)$ is finitely generated, there exists an integer s such that $[A_t/\lambda(A_{t+1})]^s = 0$, that is, $A_t^s \subset \lambda(A_{st+1}) \subset IA_{st}$. Using the integer r, we now get $A_t^{rs} \subset I^r A_{st}^r = 0$, as desired.

The next result (cf. [Va]) is useful in showing that certain ideals generated by special sequences are of linear type.

Proposition 3.3. Let I be an ideal of the ring R. Let f be an element of I such that for each integer t, $I^t \cap (0:f) = (0:f)I^t$. If $\mathrm{Sym}_{R/(f)}(I/(f)) \cong R(I/(f))$, then $\mathrm{Sym}(I) \cong R(I)$.

Remark. The condition is satisfied if $I \cap (0:f) = 0$, in particular if f is a regular element.

Proof. From the inclusion $(f) \subset I$ consider the exact sequence of symmetric algebras with its natural mappings into the corresponding Rees rings:

$$\begin{array}{ccccccccc} 0 & \longrightarrow & (f)\cdot\mathrm{Sym}(I) & \longrightarrow & \mathrm{Sym}(I)_+ & \longrightarrow & \mathrm{Sym}(I/(f))_+ & \longrightarrow & 0 \\ & & \downarrow & & \downarrow \alpha & & \downarrow & & \\ 0 & \longrightarrow & fTR(I) & \longrightarrow & R(I)_+ & \longrightarrow & R(I/(f))_+ & \longrightarrow & 0, \end{array}$$

where all the vertical mappings are induced by α. Since we are assuming the isomorphism on the right, the bottom row is exact as well. To show that α is an isomorphism, assume $0 \neq w$ is a homogeneous element of least degree in $A = \ker(\alpha)$. From the exactness of the top row, we have $w = f\cdot w'$, $w' \in S_{t-1}(I)$. Since $\alpha(w) = 0 = f\alpha(w')$, $\alpha(w') \in I^{t-1} \cap (0:f) = (0:f)I^{t-1}$, and thus $w' = u + v$, with $u \in (0:f)S_{t-1}$ and $v \in \ker(\alpha_{t-1})$ In this case, $w = f\cdot v = 0$ by the minimality of t.

For an application, we recall the notion of a <u>d-sequence</u>. It is a broad generalization of 'regular sequence' and has been developed into a theory by Huneke ([Hu_1]). Along with several other notions of sequence it will be given a close scrutiny in later sections.

<u>Definition 3.4</u>. A sequence $\underline{x} = \{x_1,\ldots,x_n\}$ of elements in the ring R is called a d-sequence if:

(a_1) $\underline{x}$ is a minimal generating set of elements for the ideal $I = (\underline{x})$.

(a_2) $(x_1,\ldots,x_i):x_{i+1}x_k = (x_1,\ldots,x_i):x_k$, for $i = 0,\ldots,n-1$, $k \geq i+1$.

Note that regular sequences are prime examples of d-sequences. For $n = 1$, $\{x\}$ is a d-sequence if and only if $0:x = 0:x^2$, that is $(0:x)\cap(x) = 0$.

<u>Theorem 3.5</u>. (cf. [Hu_2]; see also [Va]) Let I be an ideal generated by a d-sequence. Then $\mathrm{Sym}(I) \cong \mathcal{R}(I)$.

To prove this, via (3.3) and induction, we only need

<u>Proposition 3.6</u>. (cf. [Hu_1]) If $\underline{x} = \{x_1,\ldots,x_n\}$ is a d-sequence, then

(a) The images of $x_2,\ldots,x_n$ in $R/(x_1)$ form a d-sequence.

(b) $(0:x_1)\cap(x_1,\ldots,x_n) = 0$.

<u>Proof</u>. (a) is immediate from the definition. We prove (b) by induction on n. By the remark above we assume $n > 1$. Let

$\Sigma r_i x_i \in 0:x_1$. As $0:x_1 \subset 0:x_1 x_n = 0:x_n$, we have $r_n x_n^2 \in (x_1,\ldots,x_{n-1})$ and hence $r_n x_n \in (x_1,\ldots,x_{n-1})$ as required.

Remark. There exist ideals of linear type which are not generated by a d-sequence (cf. Example 6.5).

Corollary 3.7. (cf. [Mi]) Let I be an ideal generated by a regular sequence. Then $\mathrm{Sym}(I) \cong \mathcal{R}(I)$.

Corollary 3.8. (cf. [Mi]) Let (R,m) be a Noetherian local ring. Then $\mathrm{Sym}(m) \cong \mathcal{R}(m)$ if and only if R is regular.

Proof. If R is regular, m is generated by a regular sequence and thus it is of linear type by (3.7).

For the converse, note that by (2.5), $\delta_1(m) = 0$. This means that every element in Z_1 with coefficients in m is a boundary. But if we consider a minimal presentation of m in (2.5), $Z_1 \subset m.R^n$, and therefore $H_1 = 0$. Thus, by [Mat], m is generated by a regular sequence.

Proposition 3.9. (cf. [Va]) Let $J = (I,f)$ be an ideal of linear type. If f lies in the Jacobson radical of R and $I : f = I$, then I is also of linear type.

Proof. Consider the presentation

$$0 \longrightarrow L \longrightarrow I \oplus R \xrightarrow{\phi} J \longrightarrow 0$$

where $\phi(a,r) = a+rf$. Since $I:f = I$, $L = \{(-bf,b),\ b \in I\}$.

The associated diagram of symmetric algebras and corresponding Rees rings is the following:

$$\begin{array}{ccccccccc} 0 & \longrightarrow & G & \longrightarrow & \mathrm{Sym}(I)[U] & \longrightarrow & \mathrm{Sym}(J) & \longrightarrow & 0 \\ & & \downarrow & & \downarrow \alpha[U] & & \downarrow & & \\ 0 & \longrightarrow & H & \longrightarrow & \mathcal{R}(I)[U] & \longrightarrow & \mathcal{R}(J) & \longrightarrow & 0 \end{array}$$

where G is generated by the elements $bU - f\underline{b} \in \mathrm{Sym}(I)[U]$, $b \in I$, $\underline{b}$ = corresponding 1-form in $S_1(I)$, and H by the polynomials of bi-degree 1 $bU - fbT \in \mathcal{R}(I)[U]$.

As in (3.3), from the isomorphism at the right we have the exactness of the bottom row. Let w be a homogeneous element in $A_t = \ker(\alpha_t)$. Then $w \in G$ and thus

$$w = \sum_{i=1}^{n} (b_iU - f\underline{b}_i)\left(\sum_{j=0}^{t-1} c_{ij}U^j\right), \quad c_{ij} \in S_{t-1-j}(I).$$

As $w \in S_t(I)$, we must have the equations:

$$w = \sum f\underline{b}_i \cdot c_{io}$$

$$0 = \sum b_i c_{io} - f\underline{b}_i \cdot c_{i1}$$

$$\vdots$$

$$0 = \sum b_i c_{i,t-1}.$$

They show, successively, that $\sum b_i \cdot c_{ij}$ lies in A, and consequently $A_t \subset f A_t$.

<u>Remark</u>. A converse of this result is proved in [Va], under the condition $I^t : f = I^t$ for all t.

Another aspect of ideals generated by d-sequences is that they satisfy the Artin-Rees lemma 'on the nose' (cf. [Hu_1]). Let us point this out in the context of ideals of linear type.

<u>Proposition 3.10</u>. Let $J \subset I$ be ideals of the ring R. Assume that I and I/J are ideals of linear type (with respect to R and R/J respectively). Then $J \cap I^m = J.I^{m-1}$ for $m > 0$.

<u>Proof</u>. Consider the diagram of symmetric algebras and Rees rings associated to the ideals J and I:

$$\begin{array}{ccccccccc} 0 & \longrightarrow & J\cdot\mathrm{Sym}(I) & \longrightarrow & \mathrm{Sym}(I)_+ & \longrightarrow & \mathrm{Sym}(I/J)_+ & \longrightarrow & 0 \\ & & \downarrow & & \downarrow & & \downarrow & & \\ 0 & \longrightarrow & JT\mathcal{R}(I) & \longrightarrow & \mathcal{R}(I)_+ & \longrightarrow & \mathcal{R}(I/J)_+ & \longrightarrow & 0. \end{array}$$

From the assumption, the lower sequence is exact. But this means that for each m, $(I/J)^m = I^m + J/J \cong I^m/JI^{m-1}$, and the assertion follows.

§4. Approximation complexes.

In this section we give the construction of a double complex associated to a sequence of elements in a ring R. It bears a strong similarity to the ordinary Koszul complex but its added usefulness lies on the points where the two complexes differ.

Let us recall the definition of the graded Koszul complex associated to a map of R-modules $\theta : G \longrightarrow B$. For a module M over the commutative ring R, we shall denote the exterior and symmetric algebras of M respectively by ΛM and $\mathrm{Sym}(M)$. Their components of degree t will be indicated by $\Lambda^t M$, and, in the case of the symmetric algebra, by either $\mathrm{Sym}_t(M)$ or $S_t(M)$.

Let

$$\begin{array}{ccc} G & \xrightarrow{\psi} & B \\ {\scriptstyle\phi}\downarrow & & \\ R & & \end{array}$$

be a diagram of R-modules and homomorphisms. We shall make the algebra $\Lambda G \otimes_R \mathrm{Sym}(M)$ into a double complex.

$$d_\phi = \partial : \Lambda^r G \otimes S_t(B) \longrightarrow \Lambda^{r-1} G \otimes S_t(B)$$

is the usual graded Koszul complex with coefficients in $\mathrm{Sym}(B)$:

$$\partial(e_1\wedge\ldots\wedge e_r\otimes w) = \Sigma(-1)^{i-1} e_1\wedge\ldots\wedge\hat{e}_i\wedge\ldots\wedge e_r \otimes \phi(e_i)w.$$

The other differential is:

$$d_\psi = \partial' : \Lambda^r G \otimes S_t(B) \longrightarrow \Lambda^{r-1} G \otimes S_{t+1}(B)$$

$$\partial'(e_1\wedge\ldots\wedge e_r \otimes w) = \Sigma(-1)^{i-1} e_1\wedge\ldots\wedge\hat{e}_i\wedge\ldots\wedge e_r \otimes \psi(e_i)\cdot w,$$

where $\psi(e_i)\cdot w$ indicates the product of $\psi(e_i)$ and w in $\mathrm{Sym}(B)$.

Straightforward verification shows that ∂ and ∂' (skew) commute. The resulting differential graded algebra will denoted by $L = L(\phi,\psi)$.

Let $\underline{x} = \{x_1,\ldots,x_n\}$ be a generating set of the ideal I, and let $\phi\colon G = R^n \longrightarrow R$ be the mapping defined by the matrix $[x_1,\ldots,x_n]$.

Definition. $L = L(\phi,\text{identity})$ will be called the double Koszul complex associated to the sequence $\underline{x}$.

To make the next observations apparent, we display a portion of the complex L :

$$\begin{array}{ccccc}
\Lambda^{r+2}(R^n) \otimes S_{t-1}(R^n) & \longrightarrow & \Lambda^{r+1}(R^n) \otimes S_t(R^n) & \longrightarrow & \Lambda^{r}(R^n) \otimes S_{t+1}(R^n) \\
\downarrow & & \downarrow & & \downarrow \\
\Lambda^{r+1}(R^n) \otimes S_{t-1}(R^n) & \longrightarrow & \Lambda^{r}(R^n) \otimes S_t(R^n) & \xrightarrow{\partial'} & \Lambda^{r-1}(R^n) \otimes S_{t+1}(R^n) \\
\downarrow & & \partial\downarrow & & \downarrow \\
\Lambda^{r}(R^n) \otimes S_{t-1}(R^n) & \longrightarrow & \Lambda^{r-1}(R^n) \otimes S_t(R^n) & \longrightarrow & \Lambda^{r-2}(R^n) \otimes S_{t+1}(R^n)
\end{array}$$

We note that the single complex $L(\partial)$ is simply the ordinary Koszul complex associated to the sequence $\underline{x}$ in the polynomial ring $\mathcal{S} = \mathrm{Sym}(R^n) \cong R[T_1,\ldots,T_n]$. As for the 'horizontal' complex $L(\partial')$, it is also an ordinary Koszul complex, but constructed over the sequence $\underline{T} = \{T_1,\ldots,T_n\}$. Thus we have a grading of $L(\partial')$ by the subcomplexes of R-modules, $L = \Sigma L_t$

$$L_t = \sum_{r+s=t} \Lambda^r(R^n) \otimes \mathrm{Sym}_s(R^n)$$

and the L_t are exact for $t > 0$. This observation plays a key role in the sequel.

This construction may be extended by attaching coefficients from an R-module. Thus to the sequence $\underline{x}$ and the R-module M there is the associated double complex $L(\underline{x};M) = M \otimes \Lambda(R^n) \otimes \mathrm{Sym}(R^n)$, obtained by tensoring the previous complex by M. The differentials ∂ and ∂' are, respectively, obtained by viewing $L(\underline{x};M)$ as one of the Koszul complexes, $K^S(\underline{x};M \otimes_R S)$ and $K^S(\underline{T};M \otimes_R S)$, respectively.

If we denote by $Z = Z(\underline{x};M)$ the submodule of cycles of $K(\underline{x};M)$, by $B = B(\underline{x};M)$ its boundaries and by $H = H(\underline{x};M)$ its homology, the commutativity of ∂ and ∂' allows to derive from the complex $L(\underline{x};M)$ several other complexes:

$$\mathcal{Z} = \mathcal{Z}(\underline{x};M) = \{Z \otimes S, \partial'\}$$

$$\mathcal{B} = \mathcal{B}(\underline{x};M) = \{B \otimes S, \partial'\}$$

and

$$\mathcal{M} = \mathcal{M}(\underline{x};M) = \{H \otimes S, \partial'\}.$$

These complexes are called the *approximation complexes* associated with $\underline{x}$ and M. We shall focus on the complexes $\mathcal{M}$ and $\mathcal{Z}$.

The $\mathcal{M}$-complex, as well as the $\mathcal{Z}$-complex, are graded complexes over the polynomial ring S. The t-th homogeneous part $\mathcal{M}_t$ of $\mathcal{M}$ is a complex of finitely generated R-modules of the following form:

$$\mathcal{M}_t: 0 \longrightarrow H_n \otimes S_{t-n} \xrightarrow{\partial'} \cdots \xrightarrow{\partial'} H_1 \otimes S_{t-1} \xrightarrow{\partial'} H_o \otimes S_t \longrightarrow 0.$$

In a similar manner $\mathcal{Z}_t$ is a complex of finitely generated R-modules. For certain uses however we must view $\mathcal{M}$ and $\mathcal{Z}$ as complexes over the polynomial ring S. Thus, for instance, the $\mathcal{Z}$-complex may be written:

$$\mathcal{Z}: 0 \longrightarrow Z_n \otimes S[-n] \xrightarrow{\partial'} \cdots \xrightarrow{\partial'} Z_1 \otimes S[-1] \xrightarrow{\partial'} Z_o \otimes S \longrightarrow 0,$$

where, as usual, $S[k]$ denotes the graded S-module with S_{k+m} for its component in degree m. In this fashion all the maps in these complexes have degree 0.

There are still other complexes associated to $L(\underline{x};M)$. Thus if we denote by $Z^*(\underline{x};M)$ the cycles of $K(\underline{x};M)$ with coefficients in $(\underline{x})M$, we see that $Z^*(\underline{x};M)$ is a subcomplex of $Z(\underline{x};M)$

and $B(\underline{x};M) \subset Z^*(\underline{x};M)$. The module $H^* = H^*(\underline{x};M) = Z(\underline{x};M)/Z^*(\underline{x};M)$ will be referred to as the syzygy part of H (cf. [Si]). Finally, the quotient $Z^*(\underline{x};M)/B(\underline{x};M)$ will be denoted by $\delta(\underline{x};M)$.

$\delta(\underline{x};M)$ will be one of our most explicit syzygietic measures. In degree 1 and for $M = R$, note that $\delta(\underline{x};R)_1$ is just $\delta_1(\underline{x})$ defined in §2.

The corresponding complexes induced by ∂' will be denoted:

$$M^* = M^*(\underline{x};M) = \{Z^* \otimes S, \partial'\}, \text{ and}$$

$$\Delta = \Delta(\underline{x};M) = \{\delta(\underline{x};M) \otimes S, \partial'\}.$$

These complexes made their appearance in [SV_1, SV_2, HSV_1 and HSV_2], to where we shall occasionally refer for some of their properties, especially those that mimic the properties of the ordinary Koszul complexes. We shall now remark on their difference.

While the complexes defined above depend on the sequence $\underline{x}$ and the module M, whenever the context allows we shall adopt the simpler notation of M, Z, etc. Also, we whall write several proofs for $M = R$, since the general case is similar.

We first consider the dependence of M and Z on the generating set $\{x_1,\ldots,x_n\}$ of the ideal $(\underline{x}) = I$. For this purpose we recall two elementary properties of the graded Koszul complexes associated to a map $\phi: G \longrightarrow B$.

Let $\phi': G' \longrightarrow B'$, $G = G' \oplus R$, $B = B' \oplus R$ and put $\phi = \phi' \oplus \mathrm{id}: G \longrightarrow B$. Denote by L' and L the corresponding complexes. Note that L' is a subcomplex of L.

Proposition 4.1. $H_\bullet(L') = H_\bullet(L)$.

This is the "cancelling by R" property of L. It is the consequence of an underlying mapping cylinder construction in L.

Let now $\phi': G' \longrightarrow B$ and define $\phi : G = G' \oplus R \longrightarrow B$, $\phi = \phi' + 0$. Denote again by L' and L the corresponding complexes.

Proposition 4.2. $H_\bullet(L) = H(L') \otimes R[z]$, where degree(z) = 1 and $z^2 = 0$.

This is immediate from the definition of L.

Assume now $G = G' \oplus Re = R^{n-1} \oplus Re$, $\phi(e) = 0$, $Re \cong R$. Let L' be the the (sub-)complex determined by G' and the restriction of ϕ.

Proposition 4.3. $H_\bullet(L') = H_\bullet(L)$.

Proof. According to the definition of M, its component $M_{r,t}$ is given by

$$H_r(K) \otimes S_t(B) = (H_r(K') \oplus H_{r-1}(K') \otimes Re) \otimes (\sum_{j=0}^{t} S_j(B')e^{t-j})$$

which we rearrange as the sum of three subcomplexes:

$$\underbrace{[H_r(K') \otimes S_t(B')]}_{M'_{r,t}} \oplus \underbrace{[H_r(K') \otimes (\sum_{j=0}^{t-1} S_j(B')e^{t-j}]}_{A_{r,t}} \oplus \underbrace{[H_{r-1}(K') \otimes Re \otimes (\sum_{j=0}^{t} S_j(B')e^{t-j}]}_{B_{r,t}}.$$

Note that $A_{r,t} \simeq B_{r+1,t-1}$ induced by multiplication by e, on the first factor. It is clear that the complex

$$C_d = \{A_{r,t} \oplus B_{r,t} \mid r+t = d\}$$

is a mapping cylinder of the same kind that underlies (4.1). As a consequence we have the desired isomorphism.

A corollary is that the $H_r(M)$'s do not depend on ϕ. Indeed, if $\phi: G = R^{n-1} \oplus Re$ and $\phi(e) \in \phi(R^{n-1})$, we may easily arrange a change of basis so that $G = R^{n-1} \oplus Re'$ but now $\phi(e') = 0$. (This shows the role of ψ = identity, and we did not have to take it into account.) Since any two presentations of I, $\phi: R^n \longrightarrow R$ and $\theta: R^m \longrightarrow R$, may be compared, in the sense above, to the sum map $\sigma = \phi + \theta: R^{n+m} \longrightarrow R$, the invariance of the $H_r(\mathcal{M})$'s results.

Similarly the homology of the complex $\mathcal{Z}$ is independent of the generating set $\underline{x}$.

Note that while the homology of the ordinary Koszul complexes depends on the generating sets of the ideal it does have certain properties - rigidity, among others - not shared by the present complexes (cf. §13).

We now begin pointing out the reasons for the usefulness of the complexes $\mathcal{Z}$ and $\mathcal{M}$.

First, we claim that $H_o(\mathcal{Z}(I;R)) = \mathrm{Sym}_R(I)$. Indeed, looking at the presentation of I provided by

$$0 \longrightarrow Z_1(K) \longrightarrow R^n \longrightarrow I \longrightarrow 0$$

we obtain

$$Z_1(K) \otimes \mathrm{Sym}(R^n) \longrightarrow Z_o(K) \otimes \mathrm{Sym}(R^n) = \mathrm{Sym}(R^n) \longrightarrow H_o(\mathcal{Z}(I;R)) \longrightarrow 0,$$

the assertion follows from the universal properties of symmetric algebras.

Similarly, one sees that $H_o(\mathcal{M}(I;R)) = \mathrm{Sym}_A(I/I^2)$, $A = R/I$. (As a notation, we shall write $\mathrm{Sym}_A(I/I^2) = \mathrm{Sym}(I/I^2)$.) It is also easy to see that $H_\bullet(\mathcal{Z}(I;R))$ (resp. $H_\bullet(\mathcal{M}(I;R))$) is a finitely generated $\mathrm{Sym}(I)$ (resp. $\mathrm{Sym}(I/I^2)$)-module.

For $M \neq R$, the meaning of $H_o(\mathcal{Z}(I;M))$ and $H_o(\mathcal{M}(I;M))$ is not so clear-cut. There are however natural surjections

$$\alpha: H_o(\mathcal{Z}(I;M)) \longrightarrow \mathcal{R}(I;M) = \bigoplus_j I^j M = \text{associated Rees module,}$$

and

$$\beta: H_o(\mathcal{M}(I;M)) \longrightarrow gr_I(M) = \bigoplus_j I^j M/I^{j+1}M = \text{associated}$$

graded module. α and β are natural identifications in degree 0.

We now point out a basic relationship between the complexes $\mathcal{Z} = \mathcal{Z}(I;M)$ and $\mathcal{M} = \mathcal{M}(I;M)$.

Proposition 4.4. For each positive integer t there exists an exact sequence of R-modules:

$$\cdots \longrightarrow H_r(\mathcal{Z}_{t+1}) \longrightarrow H_r(\mathcal{Z}_t) \longrightarrow H_r(\mathcal{M}_t) \longrightarrow H_{r-1}(\mathcal{Z}_{t+1}) \longrightarrow \cdots$$

Proof. Consider the defining exact sequences

$$0 \longrightarrow \mathcal{Z}_t \longrightarrow \mathcal{L}_t \longrightarrow \mathcal{B}_{t-1}[-1] \longrightarrow 0$$

and

$$0 \longrightarrow \mathcal{B}_t \longrightarrow \mathcal{Z}_t \longrightarrow \mathcal{M}_t \longrightarrow 0.$$

As observed earlier, $\mathcal{L}(\partial')$ is the Koszul complex associated to the indeterminates of the ring $R[T_1,\ldots,T_n]$ and the module $M \otimes R[T_1,\ldots,T_n]$. Thus

$$H_r(\mathcal{L}_t) = \begin{cases} M & \text{for } r = t = 0 \\ 0 & \text{otherwise.} \end{cases}$$

When this is taken into the two long homology exact sequences, the desired exactness ensues.

Of interest here will be the tail of this exact sequence, that is

$$H_1(\mathcal{M}_t) \overset{\sigma}{\dashrightarrow} H_o(\mathcal{Z}_{t+1}) = S_{t+1}(I) \overset{\lambda}{\longrightarrow} H_o(\mathcal{Z}_t) = S_t(I) \longrightarrow H_o(\mathcal{M}_t) = S_t(I/I^2) \longrightarrow 0,$$

where we denote by σ (or σ_t) and λ (or λ_t) the connecting homomorphism and the downgrading homomorphism, respectively. Tracing through the meaning of the complex $\mathcal{L}_t$, it is easy to verify that λ_t is just the downgrading mapping from $S_{t+1}(I)$ to $S_t(I)$ that was condidered in §3.

Corollary 4.5. The following are equivalent for the complexes $\mathcal{M}$ and $\mathcal{Z}$.

(a) $\mathcal{M}$ is acyclic.

(b) $\mathcal{Z}$ is acyclic and λ is injective.

Proof. Since $H_r(\mathcal{Z}_o) = 0$ for $r > 0$, the assertion follows by induction and the proposition above.

Taking into account the exact sequence of (4.4), one has the following statement about the mappings σ, λ and α. It will be our key technical device in checking whether an ideal is of linear type.

Theorem 4.6. Let I be an ideal and M an R-module. Let $\mathcal{Z}$ and $\mathcal{M}$ be the complexes constructed on a generating set for I. The following are equivalent:

(a) σ is the zero mapping.

(b) λ is injective

(c) α is an isomorphism.

Proof. Consider the commutative exact diagram

$$\begin{array}{ccccccccc}
H_1(\mathcal{M}_t) & \xrightarrow{\sigma} & H_o(\mathcal{Z}_{t+1}) & \xrightarrow{\lambda} & H_o(\mathcal{Z}_t) & \longrightarrow & H_o(\mathcal{M}_t) & \longrightarrow & 0 \\
 & & \downarrow{\scriptstyle \alpha_{t+1}} & & \downarrow{\scriptstyle \alpha_t} & & \downarrow{\scriptstyle \beta_t} & & \\
0 & \longrightarrow & I^{t+1}M & \longrightarrow & I^tM & \longrightarrow & I^tM/I^{t+1}M & \longrightarrow & 0.
\end{array}$$

Since α and β are isomorphisms in degree 0, the assertion follows by induction and the exactness of the diagram.

Corollary 4.7. Let I be an ideal of the ring R. If the complex $\mathcal{M}(I;R)$ is acyclic, then I is of linear type.

One may extend the definition of ideals of linear type (relative to R) to coefficients in modules as well.

Let I be an ideal of the ring R and let M be an R-module.

Definition 4.8. I is of linear type with respect to M if the mapping $\alpha : H_o(\mathcal{Z}(I;M)) \longrightarrow \mathcal{R}(I;M)$ is an isomorphism.

The argument of (3.1) carries over to this setting, and we have

Proposition 4.9. Let R be a Noetherian ring and let M be a finitely generated module. Then α is an isomorphism if and only if β is an isomorphism.

§5. Depth and acyclicity.

In this section we consider exactness in the complexes $\mathcal{M}$ and $\mathcal{Z}$. For simplicity we shall assume from now on that R is a Cohen-Macaulay ring. It allows us to state the next results with 'heights', which is more convenient.

In the sequel we shall denote the minimal number of generators of a module E by $\nu(E)$, and the height of an ideal I by ht(I).

Theorem 5.1. Let R be a Cohen-Macaulay ring and let I be an ideal of height ℓ generated by $n = \ell + s$ elements. Assume:

(a) For each prime ideal $P \supset I$, $\nu(I_P) \leq ht(P)$.

(b) For each integer r and each prime $P \supset I$

$$\operatorname{depth}(H_r)_P \geq \inf\{ ht(P/I), r \},$$

where the H_r denote the homology modules of the Koszul complex on the chosen n generators of I.

Then the complex $\mathcal{M}(I;R)$ is acyclic.

The proof (cf. [HSV_1]) uses the following combination of the acyclicity lemma of [PS_1] and [Kap, (p.103)].

Proposition 5.2. Let R be a local ring and let

$$C: \quad 0 \longrightarrow C_s \longrightarrow \cdots \longrightarrow C_1 \longrightarrow C_o \longrightarrow 0$$

be a complex of finitely generated R-modules.

(a) If (i) depth $C_j \geq j$, and (ii) depth $H_j(\mathcal{C}) = 0$ or $H_j(\mathcal{C}) = 0$ for $j \geq 1$, then $\mathcal{C}$ is acyclic.

(b) Moreover, if for some positive integer t, depth $C_j \geq t+j$ for each $j \geq 0$, then depth $H_0(\mathcal{C}) \geq t$.

Remark. Note that the condition (a) in (5.1) (cf. (4.6) and (2.4)) is necessary for the acyclicity of $\mathcal{M}(I;R)$. The formulation of (b) is recognizably cumbersome. Nevertheless, it is, due to (4.1) and (4.2), independent of the chosen generating set of n generators. Later we consider hypotheses that blanket the number of generators and the local behaviour of the depths in the Koszul homology.

Proof of (5.1). (cf. [SV_2]) We may assume that R is a local ring and that I is minimally generated by $\ell + s$ elements, $\ell =$ grade(I), $s \geq 1$. (Recall that grade may increase under localization, but this just improves the hypotheses.)

Consider the complex

$$\mathcal{M}_t: \quad 0 \longrightarrow H_n \otimes S_{t-n} \longrightarrow \cdots \longrightarrow H_1 \otimes S_{t-1} \longrightarrow H_0 \otimes S_t \longrightarrow 0.$$

These complexes have length at most s, since $H_r = 0$ for $r > s$. We thus have complexes of type

$$0 \longrightarrow L_s \longrightarrow \cdots \longrightarrow L_1 \longrightarrow L_0 \longrightarrow 0$$

where, for a given r, $L_r = H_r \otimes S_{t-r}$.

By (b) depth $(L_r) \geq \inf\{\dim(R/I), r\} \geq r$, and we may appeal to (5.2) to complete the proof.

Because of (4.5), the $\mathcal{Z}$-complex is also acyclic under the conditions of (5.1). The following is however a slightly improved version.

Theorem 5.3. Let R be a Cohen-Macaulay ring and let I be an ideal of height $\ell \geq 1$ generated by $n = \ell + s$ elements. Assume:

(a) For each prime $P \supset I$, $v(I_P) \leq 1+ht(P)$.

(b) For each integr r and each prime $P \supset I$

$depth(H_r)_P \geq \inf\{ ht(P/I), r\} - 1$,

where the H_r denote the homology modules of the Koszul complex on the chosen n generators of I.

Then the complex $Z(I;R)$ is acyclic.

The proof is similar, with the hypothesis $\ell \geq 1$ making for $Z_n = 0$, which will shorten the complexes to which (5.2) will apply.

We shall now illustrate the uses of these complexes for ideals that are sufficiently close to complete intersections. Although not strictly necessary, we shall assume that R is a Cohen-Macaulay local ring. Let I be an ideal of height ℓ minimally generated by $v(I) = \ell + s$ elements; we discuss the cases $s \leq 3$.

(i) $s = 0$: In this case both complexes are exact, and $gr_I(R) = Sym(I/I^2)$ is isomorphic to the polynomial ring $(R/I)[T_1,\ldots,T_\ell]$. On the other hand, the dimension and depth of Sym(I) can readily be determined from the acyclicity of the Z-complex - in fact, Sym(I) is Cohen-Macaulay along with R. These facts will however be taken up, with greater generality, in §9.

(ii) $s = 1$: By (5.3) the Z-complex is always acyclic. The M-complex will be exact if for each minimal prime $P \supset I$, I_P is generated by a regular sequence. Indeed, it suffices, by (5.1), to show that if R is local and $s = 1$, then $depth(H_1) \geq 1$. More generally:

Proposition 5.4. ([Ao], [Mats]) Let R be a Cohen-Macaulay

local ring and let I be an ideal generated by $1 + ht(I)$ elements. Then

$$\operatorname{depth}(H_1) \geq \inf\{\dim(R/I),\ 2+\operatorname{depth}(R/I)\},$$

with equality if R is Gorenstein.

This situation provides examples of non-Cohen-Macaulay ideals I, for which both $\operatorname{Sym}(I/I^2)$ and $\operatorname{Sym}(I)$ are Cohen-Macaulay (cf. [SV_2] and [HSV_2]).

(iii) $s = 2$: Assume now that R is a Gorenstein local ring and that I is a Cohen-Macaulay ideal. The depth of the Koszul homology modules of such ideals was determined in [AH]:

Proposition 5.5. Let R be a Gorenstein local ring and let I be a Cohen-Macaulay ideal minimally generated by $2+\text{height}(I)$ elements. Then the Koszul homology modules H_i are Cohen-Macaulay (R/I)-modules.

Proof. Let $K_\bullet$ be the Koszul complex on a minimal generating set of I. Since R is assumed to be a Gorenstein ring, the last non-vanishing homology module of $K_\bullet$, H_2, is the canonical module of R/I, and therefore a Cohen-Macaulay module. We thus only have to show that H_1 is Cohen-Macaulay. Let $x \in R/I$ be a regular element and let $y \in R$ be an R-regular element which is a lifting of x. From the short exact sequence

$$0 \longrightarrow K_\bullet \xrightarrow{y} K_\bullet \longrightarrow K'_\bullet \longrightarrow 0$$

we obtain the long exact sequence

$$0 \longrightarrow H_2 \xrightarrow{x} H_2 \longrightarrow H_2(K'_\bullet) \longrightarrow H_1 \xrightarrow{x} H_1 \longrightarrow H_1(K'_\bullet) \longrightarrow$$
$$\longrightarrow H_0 \xrightarrow{x} H_0 \longrightarrow H_0(K'_\bullet) \longrightarrow 0.$$

In the above sequence the mapping $H_2 \longrightarrow H_2(K'_\bullet)$ is surjective, and the mapping $H_0 \xrightarrow{x} H_0$ is injective. Therefore x is H_1-regular and $H_1/xH_1 = H_1(K'_\bullet)$. By induction on the dimension of R/I, the assertion follows.

In this case (s = 2) we have (cf. (5.1) and (5.3)):

(iii$_a$) If $v(I_P) \leq ht(P)$ for each prime $P \supset I$, then the $\mathcal{M}$-complex is acyclic.

(iii$_b$) If $v(I_P) \leq 1+ht(P)$ for each prime $P \supset I$, then the $\mathcal{Z}$-complex is acyclic.

(iv) s = 3: Let R be a Gorenstein local ring and let I be a Cohen-Macaulay ideal generated by $v(I) = 3+ht(I)$ elements. As in (5.5), considering the Koszul complex associated to a minimal generating set of I, we conclude that H_3 is Cohen-Macaulay. Here however the depths of H_1 and H_2 are not known - much less whether they are Cohen-Macaulay.

(iv$_a$) If, in the situation above, for every prime ideal $P \supset I$, $v(I_P) \leq ht(P)$, then the $\mathcal{Z}$-complex is acyclic.

Proof. Let us apply (5.2) directly to the $\mathcal{Z}$-complex ($n=\ell+3$):

$$\mathcal{Z}_t\colon 0 \longrightarrow Z_n \otimes S_{t-n} \longrightarrow \cdots \longrightarrow Z_2 \otimes S_{t-2} \longrightarrow Z_1 \otimes S_{t-1} \longrightarrow Z_o \otimes S_t \longrightarrow 0.$$

Since we may assume $\ell \geq 1$, $Z_n = 0$ and the complex $\mathcal{Z}_t$ has length at most $\ell+2$. If K denotes the corresponding Koszul complex, for $r \geq 4$ the truncated complex

$$0 \longrightarrow K_{\ell+3} \longrightarrow \cdots \longrightarrow K_{r+1} \longrightarrow B_r \longrightarrow 0$$

provides a minimal projective resolution of B_r $(=Z_r)$ and thus, by the Auslander- Buchsbaum equality,

$$\operatorname{depth}(Z_r) = \dim R - (\ell+3-r-1) = (\dim R - \ell - 2) + r \geq r+1$$

since $v(I) = \ell+3 \leq \dim R$.

For $r = 3$: From the exact sequence

$$0 \longrightarrow B_r \longrightarrow Z_3 \longrightarrow H_3 \longrightarrow 0$$

we get depth $Z_3 \geq 3$, as H_3 is Cohen-Macaulay of depth equal to $\dim(R/I) \geq 3$, and depth $B_3 \geq 3$, by the truncated Koszul complex (which it is still exact for $r = 3$).

On the other hand, Z_1 and Z_2 are second-syzygy modules, and thus depth $Z_2 \geq 2$ and depth $Z_1 \geq 2$. (In fact, Z_1 has a much higher depth.)

The acyclicity of the $\mathcal{Z}$-complex now follows from a direct application of (5.3).

It is easy to see that the depth of the irrelevant prime ideal of Sym(I) is at least $\ell+3$.

(iv_b) As for the $\mathcal{M}$-complex, its exactness is a bit harder to deal with since the use of (5.1) would require depth $H_2 \geq 2$, and depth $H_1 \geq 1$. An alternative would be to show directly the isomorphism of the blowing-up rings Sym(I) and $\mathcal{R}$(I), and then make use of the preceding and (4.5).

§6. Sequences.

The aim of this section is a discussion of certain generalizations of the notion of a regular sequence that play, in the setting of the approximation complexes, a role comparable to that of regular sequences for ordinary Koszul complexes.

A key notion is that of a d-sequence - already briefly discussed in §2 in the context of ideals of linear type - which will assume a dominant role in §§12 and 13. We show that the existence of 'straight-up' d-sequences are ubiquitous in Noetherian rings and use it to prove a coarse one-step Cohen-Macaulayfication.

(a) Generalizations of regular sequences.

We begin with a listing of the definitions of the various kinds of sequences.

Definition 6.1. Suppose $\underline{x} = \{x_1, \ldots, x_n\}$ is a sequence of elements in a ring R and M is an R-module. The sequence $\underline{x}$ is called a:

(a) d-sequence with respect to M, if

(a_1) $\underline{x}$ is a minimal generating set of the ideal $I = (\underline{x}) = (x_1, \ldots, x_n)$.

(a_2) $(x_1, \ldots, x_i)M :_M x_{i+1}x_k = (x_1, \ldots, x_i)M :_M x_k$, for $i = 0, \ldots, n-1$, and $k \geq i+1$.

(b) relative regular sequence with respect to M, if $(x_1, \ldots, x_i)IM :_M x_{i+1} \cap IM = (x_1, \ldots, x_i)M$ for $i = 0, \ldots, n-1$, where $I = (\underline{x})$.

(b*) relative* regular sequence with respect to M, if $(x_1, \ldots, x_i)M :_M x_{i+1} \cap IM = (x_1, \ldots, x_i)M$ for $i = 0, \ldots, n-1$, where $I = (\underline{x})$.

(c) proper sequence with respect to M, if $x_{i+1}H_j(x_1, \ldots, x_i; M) = 0$ for $i = 0, \ldots, n-1$, $j > 0$, where $H_j(x_1, \ldots, x_i; M)$ denotes the Koszul homology associated to the initial subsequence $\{x_1, \ldots, x_i\}$.

The notion (a) was introduced by Huneke ([Hu_1]), who has constructed many interesting classes of examples and proved some of its fundamental properties. Both (b) and (b*) were introduced by Fiorentini ([Fi]). As for (c), it was first defined in [HSV_1] for its usefulness in examining exactness in the approximation complexes. (see §12). There are also unconditioned versions of each of these notions - that is, independent of reorderings - but they will not be discussed here. Finally, in the enlarged context here - that of R-modules - the minimality condition (a_1) above does not play a major role. For this reason we sometimes refer to the following condition - equivalent to (a_2) - as a d-sequence also (or, a d*-sequence when a distinction is needed):

(a_2^*) $(x_1,\ldots,x_i)M \underset{M}{:} x_{i+1} \cap IM = (x_1,\ldots,x_i)M$ for $i = 0,\ldots,n-1$, where $I = (\underline{x})$.

Several implications hold between these kinds of sequences. We indicate in the diagram below those we shall be interested in:

$$\begin{array}{c} \text{d-sequence} \\ \Downarrow \\ \text{d*-sequence} \\ \Updownarrow \\ \text{relative* regular sequence} \\ \Downarrow \\ \text{relative regular sequence} \\ \Downarrow \\ \text{proper sequence} \end{array}$$

(6.1.1). The implication d*-sequence $\Rightarrow$ relative* regular sequence follows from (3.6). For the converse (*cf.* [HSV_1]), let

$a \in (x_1,\ldots,x_i)M : x_{i+1}x_k$; then

$ax_k \in (x_1,\ldots,x_i)M : x_{i+1} \cap IM = (x_1,\ldots,x_i)M$ and thus

$a \in (x_1,\ldots,x_i)M : x_k$.

(6.1.2). Fiorentini (*cf.* [Fi]) showed that for a relative regular sequence $\underline{x} = \{x_1,\ldots,x_n\}$ one has

$Z_j(x_1,\ldots,x_i;M) \cap IK = B_j(x_1,\ldots,x_i;M)$ for $i = 0,\ldots,n-1$ and $j > 0$, where K denotes the Koszul complex on the full sequence $\underline{x}$. The last implication, relative regular sequence $\Rightarrow$ proper sequence, then follows.

(6.1.3). (i) For $M = R$, d-sequences show features common to regular sequences and generating sets of projective ideals.

Thus, for instance,

(i_a) If $\dim R = r$, then every d-sequence has at most $r+1$ elements.

(i_b) For a Dedekind domain every ideal is generated by a d-sequence.

(ii) For other modules however the bound (i_a) ceases to exist, as when (R,m) is a local ring and $M = R/m$.

(6.1.4). There is another way to express condition (c) above. Using the recursive form of the Koszul homology modules (*cf.* [Mat]), (c) may also be written:

(c*) $\underline{x}$ is a <u>proper sequence</u> with respect to M, if $I.H_j(x_1,\ldots,x_i;M) = 0$ for $i = 0,\ldots,n-1$, $j > 0$, and $H_j(x_1,\ldots,x_i;M)$ with the earlier meaning.

Contrary to d-sequences for $M = R$, there may not exist dimension dependent bounds for the length of (minimally generated) proper sequences. Thus, if (R,m) is an Artinian local ring and $m^2 = 0$, then m is generated by a proper sequence.

(6.1.5). One way to, possibly, get regular sequences from d-sequences or proper sequences emerges directly from the definition (c*): Passing to the localization R_{x_n} it follows that $H_1(x_1,\ldots,x_{n-1};M)_{x_n} = 0$. Thus, if $(x_1,\ldots,x_{n-1})M_{x_n} \neq M_{x_n}$, then one obtains a regular M_{x_n}-sequence of length $n-1$ in the ideal $(x_1,\ldots,x_n)R_{x_n}$.

Before we make use of these remarks, let us show that d-sequences - especially for certain straight-up types - are rather ubiquitous. For simplicity we consider the case of $M = R$ since the extension to arbitrary modules is immediate.

Proposition 6.2. Let I be an ideal of height n of the Noetherian ring R. Then I contains a d-sequence (with respect to R) of length (and height) n.

Proof. Let $\{P_1,\ldots,P_r\}$ be the minimal prime ideals of I. We shall use the following notation. For an R-module M, denote by z(M) the union of the associated primes of M which are properly contained in one of the P_i.

We may assume $n > 0$. Let $y_1 \in I \setminus z(R)$, and stabilize it, that is, pick m large enough so that $0:y_1^m = 0:y_1^{m+1}$. Put $x_1 = y_1^m$; if $n = 1$, $\{x_1\}$ will do.

If $n > 1$, consider the ideal $J_1 = 0:(0:x_1)$. We claim that J_1 is not contained in any prime ideal associated to R and properly contained in some P_i. Otherwise, for such a prime P we would have: if x_1 is regular in R_P, $0:x_1 = 0$, which is impossible; if x_1 is not regular in R_P, P contains some associated prime of R, which conflicts with the choice of y_1 - and hence of x_1. Thus we may pick $y_2 \in I \cap J_1 \setminus \{z(R) \cup z(R/(x_1))\}$ and stabilize it - meaning now take m large enough so that $0:y_2^m = 0:y_2^{m+1}$ and $(x_1):y_2^m = (x_1)\cdot y_2^{m+1}$. Put $x_2 = y_2^m$.

In general, if $x_1,\ldots,x_s$ have been chosen and $n > s$, one picks x_{s+1} in the following manner. Again it is easy to see that $J_s = (x_1,\ldots,x_{s-1}):((x_1,\ldots,x_{s-1}):x_s)$ is not contained in any prime properly inside one of the P_i. Now pick y_{s+1} in $I \cap J_1 \cap \ldots \cap J_s \setminus \{z(R) \cup z(R/(x_1)) \cup \ldots \cup z(R/(x_1,\ldots,x_s))\}$, and stabilize it. This means picking m large enough so that for each initial subsequence $\{x_1,\ldots, x_\ell\}$, $\ell \leq s$, $(x_1,\ldots,x_\ell):y_{s+1}^m = (x_1,\ldots,x_\ell):y_{s+1}^{m+1}$. Finally put $x_{s+1} = y_{s+1}^m$.

We claim that $\{x_1,\ldots,x_n\}$ is a d-sequence. Since the chosen elements will generate an ideal of height n, the minimality condition is ensured. Note also that any sequence $\{x_1,\ldots,x_s,x_n\}$, $s < n-1$, is, by induction, a d-sequence since it contains at most n-1 elements and the construction leading to it is the same that leads to the sequence $\{x_1,\ldots,x_s,x_{s+1}\}$. Therefore the only step to check is the equality

$$(x_1,\ldots,x_{n-2}):x_{n-1}x_n = (x_1,\ldots,x_{n-2}):x_n.$$

Let t be an element in the fist ideal; then $tx_n \in (x_1,\ldots,x_{n-1}):x_n$ and since $x_n \in J_{n-1}$, we have $tx_n^2 \in (x_1,\ldots,x_{n-2})$. The assertion now follows from the stabilization step.

Remark. A similar construction was used by Bass ([Ba]) to find bounds for various finitistic dimensions of Noetherian rings.

(b) Syzygetic ideals

Among the notions of syzygetic ideals used in the past, we single out:

(a) I is syzygetic$_1$ if $\delta_1(I)$ (= $\delta_1(I;R)$) = 0.

(b) I is syzygetic$_2$ if $\delta_t(I) = 0$ for $t \geq 1$ (cf. §4). This means that for some set $\underline{x}$ of generators of I,

$Z_t(\underline{x}) \cap IK = B_t(\underline{x})$, $t > 0$. (It will be the case that this definition does not depend on the generating set.)

(c) I is syzygetic$_3$ if the associated complex $M(I;R)$ is acyclic.

Proposition 6.3. (c) => (b) => (a). Furthermore, none of the inverse implications is valid.

Proof. To show (c) => (b) we go back to the definition of the approximation complex $M^*(I;R)$ (cf. §4). The exact sequence

$$0 \longrightarrow \Delta \longrightarrow M \longrightarrow M^* \longrightarrow 0$$

has, in degree t, the following tail

$$\begin{array}{ccccccccc} 0 & \longrightarrow & \delta_t(\underline{x}) & \longrightarrow & H_t & \longrightarrow & H_t^* & \longrightarrow & 0 \\ & & \downarrow & & \downarrow & & \downarrow & & \\ 0 & \longrightarrow & \delta_{t-1}(\underline{x}) \otimes S_1 & \longrightarrow & H_{t-1} \otimes S_1 & \longrightarrow & H_{t-1}^* \otimes S_1 & \longrightarrow & 0. \end{array}$$

From this diagram, we see that $\delta_1(\underline{x}) = H_1(M_1)$. Assume that we have proved $\delta_{t-1}(\underline{x}) = 0$ for $t \geq 2$. Since the diagram has exact rows and the middle vertical complex is acyclic, it follows that $\delta_t(\underline{x}) = 0$ also.

Note also that the same argument shows that if $\delta_{t-1}(\underline{x}) = 0$, then $\delta_t(\underline{x}) = \ker(H_t(M_t) \longrightarrow H_t(M_t^*))$. Since $\delta_1(\underline{x})$ is independent of $\underline{x}$, it follows that if the $\delta_j(\underline{x})$, $j < t$ vanish, then $\delta_t(\underline{x})$ will be independent of the generating set.

Remark. In §12 it will be proved that for proper sequences the complex $Z(I;R)$ is acyclic. Thus, by (4.4), $H_j(M) = 0$ for $j > 1$. From the argument above it then follows that for ideals generated by proper sequences we have syzygetic$_1$ => syzygetic$_2$.

Let us now consider the inverse implications.

Example 6.4. Let P be a prime ideal of $R = k[[x,y,z]]$ needing at least 5 generators; e.g. pick P to be one of the so-called Macaulay primes. According to §7, the homology modules of the complex K are Cohen-Macaulay, meaning here they are torsion-free R/P-modules. Since P is generically a complete intersection, $(H)_P = (H^*)_P$, and thus $\delta_t(P) = 0$ for $t > 0$. We claim that $M(P;R)$ is not acyclic. Otherwise it would lead to the equality of the Rees and symmetric algebras of P and this would contradict (2.4). Thus P is syzygetic$_2$ but not syzygetic$_3$.

Example 6.5. Let $R = k[u,v,x,y]$, with the relations $ux = vy$, $uy = vx = u^2 = uv = v^2 = 0$, k = field. R is the symmetric algebra of a module over $k[u,v]$. Let $I = (x,y)$. Since I is the augmentation ideal of R, it follows by (2.3) that I is of linear type - in particular $\delta_1(I) = 0$. Note however that $H_2 = Rux = Z_2 = \delta_2(I)$, and thus I is not syzygetic$_2$.

(c) Coarse Cohen-Macaulayfication.

We shall now make use of (6.2) to obtain a crude form of Cohen-Macaulayfication. For a ring R, by the Cohen-Macaulayfication of *Spec*(R) we mean a pair $\{X,f\}$ of a Cohen-Macaulay scheme X and a proper mapping $f\colon X \longrightarrow \mathit{Spec}(R)$ which is an isomorphism at the Cohen-Macaulay points of *Spec*(R). Some special - but still hard - cases have been dealt with by Brodmann and Faltings (see [Br],[Fa]), and we shall consider here another approach to one of its facets.

Theorem 6.6. Let (R,m) be a Noetherian local ring of dimension n, and let $\underline{x} = \{x_1,\ldots,x_{n-1}, x_n = z\}$ be a system of parameters which is a d-sequence (such systems always exist by (6.2)). Let S be the subring of R_z generated by R and the x_i/z, $i < n$, and let P be the prime ideal $(m, x_1/z, \ldots, x_{n-1}/z)$. Then S_P is a Cohen-Macaulay ring of dimension n.

Proof. Write $I = (\underline{x})$ and let C be the Rees algebra $C = R[It] = \sum_j I^j t^j$. Denote the elements $x_1t,\ldots, x_{n-1}t, x_nt = zt$ respectively by $\zeta_1,\ldots,\zeta_{n-1},\zeta$. Note that in C_ζ the element x_n is regular and $C_\zeta/(x_n) = gr_I(R)_\zeta$.

Since $\{x_1,\dots,x_n\}$ is a d-sequence, by [Hu_3] (see (10.10) for a converse and fuller discussion) the corresponding 1-forms $\{\xi_1,\dots,\xi_{n-1}, \xi_n = \xi\}$ in $gr_I(R)$ is also a d-sequence. Thus, by (6.1.5) and the choice of the x_i's, $\{\xi_1,\dots,\xi_{n-1}\}$ form a regular sequence in $gr_I(R)_\xi = gr_I(R)_\zeta$. Altogether we have that $\{x_n,\zeta_1,\dots,\zeta_{n-1}\}$ is a regular sequence in C_ζ.

Finally, as $C_\zeta = S[\zeta,\zeta^{-1}]$ - S is the 0-th homogeneous piece of C_ζ - the rest of the assertion follows from standard facts on the dimension of blowing-up rings.

§7. Depth of the Koszul homology modules.

The applications in §5 and, more interestingly, in §9, of the approximation complexes to derive arithmetical properties of the blowing-up rings require detailed knowledge of the depths of the homology modules $H_i(I;R)$. In this section we first show how much simpler it is to estimate those depths for ideals which are generated by proper sequences with high residual depths. We then prove a result of Huneke ([Hu_5]) on the behaviour of the Cohen-Macaulayness of the $H_i(I;R)$ under linkage. This allows us to organize around it several earlier known cases of ideals with Cohen-Macaulay Koszul homology, e.g. perfect ideals of height 2 and Gorenstein ideals of height 3.

(a) Ideals generated by proper sequences.

Let $\underline{x} = \{x_1,\dots,x_n\}$ be a proper sequence with respect to the ring R itself. One has the following exact sequences, derived from the definition (*cf.* §6):

$$0 \longrightarrow H_1(x_1) \longrightarrow R \xrightarrow{x_1} R \longrightarrow R/(x_1) \longrightarrow 0$$

$$0 \longrightarrow H_1(x_1,\ldots,x_i) \longrightarrow H_1(x_1,\ldots,x_{i+1}) \longrightarrow R_1 \xrightarrow{x_{i+1}} R_i \longrightarrow R_{i+1} \longrightarrow 0$$

where $R_i = R/(x_1,\ldots,x_i)$, and, for $j > 1$,

$$0 \longrightarrow H_j(x_1,\ldots,x_i) \longrightarrow H_j(x_1,\ldots,x_{i+1} \longrightarrow H_{j-1}(x_1,\ldots,x_i) \longrightarrow 0.$$

From these sequences we can obtain despth estimates for $H_j(I;R)$ in terms of $\operatorname{depth}(R/(x_1,\ldots,x_i))$. The following is an extreme case.

Proposition 7.1. Let I be an ideal generated by a proper sequence $\underline{x} = \{x_1,\ldots,x_n\}$. Assume:

(a) $\operatorname{depth}(R/(x_i,\ldots,x_i)) \geq \operatorname{depth} R - i, \quad 1 \leq i \leq n-1.$

(b) $\operatorname{depth} R/I \geq \operatorname{depth} R - n - 1.$

Then $\operatorname{depth} H_i(I;R) \geq \operatorname{depth} R - n + i, \quad 1 \leq i \leq n.$

Proof. It is immediate from depth-chasing along the exact sequences above.

Remarks. (i) One instance where the residual depth conditions above are satisfied is that of monomial ideals. In fact, if $I = (x_1,\ldots,x_n)$, where the x_i's are monomials in some set of indeterminates, it follows from [Ta] that the projective dimension of $R/(x_1,\ldots,x_i)$ is at most i. Thus, by the Auslander-Buchsbaum equality ([Mat]), one has that $\operatorname{depth} R/(x_1,\ldots,x_i)$ is, locally, at least $\operatorname{depth} R - i$. In [$HSV_1$] necessary and sufficient condiions were given for a sequence of monomials to be a d-sequence. For 'square-free' monomials, Kühl ([Kü]) has shown that proper sequences are the same as d-sequences.

(ii) Another case to which (7.1) would apply is that of an almost complete intersection, already treated in (5.4).

(b) Koszul homology and linkage.

We shall now discuss a theorem of Huneke ([Hu_5]), and some of its consequences.

The notion of 'linkage' that we shall deal with is the following ([PS_2]). Let R be a Cohen-Macaulay local ring and let I and J be two ideals of R. I and J are said to be linked if there exists a regular sequence $\underline{x} = \{x_1,\ldots,x_n\}$ in $I \cap J$ such that $I = (\underline{x}):_R J$ and $J = (\underline{x}):_R I$. (Notation: $I \underset{(\underline{x})}{\sim} J$, or simply $I \sim J$.) Note that the definition implies $\mathrm{Ass}(R/I) \subset \mathrm{Ass}(R/(\underline{x}))$, so that I (and also J) is height unmixed.

The linkage class of the ideal I, L(I), consists of all ideals J for which there exists a sequence of linkages $I \sim L_1 \sim L_2 \sim \ldots \sim L_n \sim J$. Of interest here will be the subset $L^e(I)$ of ideals linked to I through an even number of linkages.

Several important linkage classes have been studied:

(7.2.1) ([BE],[Kun]) Let R be a Gorenstein local ring and assume $I \sim J$. Then I is a Gorenstein ideal if and only if J is an almost complete intersection.

(7.2.2) ([PS_2]) Let R be a regular local ring and let I be a Cohen-Macaulay ideal of height two. Then I is in the linkage class of a complete intersection.

(7.2.3) ([BE],[Wat]) Let R be a regular local ring and let I be a Gorenstein ideal of height three. Then I is in the linkage class of a complete intersection.

Note that in these two cases, it follows from (7.2.1) that I is even linked to either a complete intersection or to an almost complete intersection.

A basic property on the behaviour of the Cohen-Macaulay character under linkage is the following:

(7.2.4) <u>Dubreil's theorem</u>. (cf. [PS_2]) Let R be a Gorenstein local ring and let I be a Cohen-Macaulay ideal. If J is linked to I, then J is a Cohen-Macaulay ideal.

Otherwise stated, if R is a Gorenstein local ring, then the Cohen-Macaulayness of $H_0(I;R)$ is preserved under linkage. Unfortunately the higher Koszul homology modules do not always show this behaviour. The following, however, is sufficient for several applications.

<u>Theorem 7.3</u>. ([Hu_5]) Let R be a Cohen-Macaulay local ring and suppose L is an ideal of grade n. Let $\underline{x} = \{x_1,\ldots,x_n\}$ and $\underline{y} = \{y_1,\ldots,y_n\}$ be two regular sequences in L and set $I = (\underline{x}):L$ and $J = (\underline{y}):L$. If $H_i(I;R)$ are Cohen-Macaulay for $0 \leq i \leq m$, then $H_i(J;R)$ are Cohen-Macaulay for $0 \leq i \leq m$.

The assertion is to the effect that the Cohen-Macaulayness of the Koszul homology of I behaves well under <u>even</u> linkage. There are examples ([Hu_5]) showing that this is not always so under linkage in general.

From (7.1) it follows that if R is Cohen-Macaulay and I is a Cohen-Macaulay ideal that is an almost complete intersection, then $H_i(I;R)$ are Cohen-Macaulay (only $i = 0,1$ need be considered). Taking into account the remark following (7.2.3), we have:

<u>Corollary 7.4</u>. Let R be a regular local ring and let I be either a Cohen-Macaulay ideal of height two or a Gorenstein ideal of height three. Then the Koszul homology modules $H_i(I;R)$ are Cohen-Macaulay.

Remark. Both cases in (7.4) had been treated, directly, in [AH] and [Hu_4], respectively.

We sketch in detail the proof of (7.3). Despite its elementary nature, it displays a number of useful techniques on dealing with the Koszul homology modules, which may be pertinent in other circumstances.

(7.5) The dimension of the Koszul homology modules. Let R be a Cohen-Macaulay local ring of dimension d and let I be an ideal of height n. If $H_i(I;R) \neq 0$, then its dimension is $d-n$.

Proof. Let $\underline{x} = \{x_1,\ldots,x_n\}$ be a maximal regular sequence in I and let $K(I;R)$ be the Koszul complex on a set of r generators of I. First, since the $H_i(I;R)$ are annihilated by I, $\dim H_i(I;R) \leq d-n$.

The first non-vanishing homology module is (cf. [Ma]): $H_{r-n}(I;R) \cong (\underline{x}):I/(\underline{x})$. Its associated prime ideals are minimal primes of $(\underline{x})$, and so $\dim H_{r-n}(I;R) = d-n$.

To show $\dim H_k(I;R) = d-n$, $0 < k < r-n$, assume otherwise and let P be an associated prime of $H_{r-n}(I;R)$. Since $ht(P) = n$, localizing at P we get $H_k(I;R)_P = 0$ and $H_{r-n}(I;R)_P \neq 0$, thus contradicting the rigidity of the Koszul complex.

(7.6) The dependence of the Koszul homology modules on the generating set of I is contained in (4.2). Thus, if $\underline{x}'$ is the generating set obtained by adding a superfluous element to $\underline{x}$, say, $\underline{x}' = \{x_1,\ldots,x_n,0\}$, then

$$H_i(\underline{x}';R) = H_i(\underline{x};R) \oplus H_{i-1}(\underline{x};R).$$

In particular, if the $H_i(\underline{x};R)$ are Cohen-Macaulay for $0 \leq i \leq m$, so will be the $H_i(\underline{x};R)$ in the same range. Since any two generating sets for I may be thus compared to the set $\underline{x} \cup \underline{y}$, the Cohen-Macaulay condition, in a range as above, is seen as independent of the generating set.

(7.7) Let I be an ideal generated by the sequence $\underline{x} = \{x_1,\ldots,x_n\}$ and suppose x_1 is a regular element of R. We have then the exact sequences of Koszul complexes

$$0 \longrightarrow K(\underline{x};R) \xrightarrow{x_1} K(\underline{x};R) \longrightarrow K(\underline{x};R/(x_1)) \longrightarrow 0.$$

The last complex is the same as $K(\underline{x}^*;R^*)$, $\underline{x}^* = \{0,x_2^*,\ldots,x_n^*\}$, where '*' denotes the canonical homomorphism $R \longrightarrow R/(x_1) = R^*$.

Taking the homology sequence, we obtain the exact sequence:

$$0 \longrightarrow H_i(\underline{x};R) \longrightarrow H_i(\underline{x}^*;R^*) \longrightarrow H_{i-1}(\underline{x};R) \longrightarrow 0.$$

From (7.6) it follows that the passage $R \longrightarrow R^*$ preserves the Cohen-Macaulay property of $H_i(I;R)$, $0 \leq i \leq m$.

(7.8) We begin to prove (7.3). First, note that the case $n = 0$ is vacuous. For $n = 1$, consider $I = (x):L$ and $J = (x):L$.

(a) The condition $I = (x):L$ implies that $I \cong \mathrm{Hom}_R(L,R)$. Indeed the module $\mathrm{Hom}_R(L,R)$ may be identified to the set $R \underset{Q}{:} L = \{r \in Q \mid rL \subset R\}$, Q = total ring of fractions of R. The mapping $\phi: I \longrightarrow R \underset{Q}{:} L$ given by $\phi(a) = a/x$ is easily checked to be an isomorphism.

The hypothesis of (7.3), for $n = 1$, thus implies that $I \cong J$. Denote this isomorphism by α - which is realized as multiplication by a fixed element of Q still denoted by α.

(b) First we need a criterion for the condition $H_i(I;R)$ is Cohen-Macaulay. For that, let $K(I;R)$ be the Koszul complex on a fixed set of generators $\underline{x} = \{x_1,\ldots,x_n\}$. We claim that if $ht(I) = 1$, then $H_i(I;R)$ are Cohen-Macaulay $0 \leq i \leq m$, if and only if the cycles Z_i and the boundaries B_i of K are maximal Cohen-Macaulay modules, for $0 \leq i \leq m$. For the proof, it suffices to use induction on the defining sequences:

$$0 \longrightarrow B_i \longrightarrow Z_i \longrightarrow H_i(I;R) \longrightarrow 0$$

and

$$0 \longrightarrow Z_i \longrightarrow K_i \longrightarrow B_{i-1} \longrightarrow 0.$$

Assume then $J = \alpha I$, α a regular element of Q, and let $K(I;R)$ be the Koszul complex on the generating set $\underline{x}$ above. Denote by $K'(J;R)$ the Koszul complex corresponding to the set $\underline{y} = \{\alpha x_1,\ldots,\alpha x_n\}$ of generators of J.

We claim that $Z_i = Z_i'$ and $B_i \cong B_i'$. First, if we identify K and K' as R-modules, we have that $Z_i = Z_i'$, since $\alpha.d = d'$ and α is a regular element of Q. We can then complete the diagram

$$\begin{array}{ccccccccc} 0 & \longrightarrow & Z_i & \longrightarrow & K_i & \longrightarrow & B_i & \longrightarrow & 0 \\ & & \| & & \| & & & & \\ 0 & \longrightarrow & Z_i' & \longrightarrow & K_i' & \longrightarrow & B_i' & \longrightarrow & 0 \end{array}$$

and conclude that $B_i \cong B_i'$.

(7.9) Assume $n > 1$. To use induction on n, we first establish the following claim. Let $\underline{x} = \{x_1,\ldots,x_n\}$ and $\underline{y} = \{y_1,\ldots,y_n\}$ be regular sequences in L. There exist elements $z_1,\ldots,z_{n-1}$ satisfying the following conditions:

(a) $z_i = y_i + a_i$, $a_i \in (y_{i+1},\ldots,y_n)$

(b) $\{x_o,x_1,\ldots,x_k,z_1,\ldots,z_i\}$ is a regular sequence for $0 \leq k \leq n-i$ (here $x_o = 0$).

Suppose $z_1,\ldots,z_{i-1}$ have been chosen. By assumption

$\{x_o, x_1, \ldots, x_k, z_1, \ldots, z_{i-1}\}$ is a regular sequence for $k \leq n-(i-1) = n-i+1$. Thus it suffices to chose z_i as in (a) which is not contained in any associated prime of $R/(x_o, \ldots, x_k, z_1, \ldots, z_{i-1})$, $0 \leq k \leq n-i$. If P is such a prime, then $(y_i, y_{i+1}, \ldots, y_n)$ is not contained in P else $(y_1, \ldots, y_n)$ and grade $P \geq n$. It now follows that z_i can be chosen satisfying (a) and (b) (cf. [Kap, (Th. 124)].

Set $D_i = (x_1, \ldots, x_{n-i}, z_1, \ldots, z_i):L$ for $1 \leq i \leq n-1$. Set $D_o = (x_1, \ldots, x_n):L$ and $D_n = (y_1, \ldots, y_n):L$. Then $D_o = I$, and since $(z_1, \ldots, z_n) = (y_1, \ldots, y_n)$, $D_n = J$. By induction on j, we will show that $H_i(D_j;R)$ are Cohen-Macaulay for $0 \leq i \leq m$. For $j = 0$ this is our assumption. Suppose we have shown $H_i(D_{j-1};R)$ are Cohen-Macaulay for $0 \leq i \leq m$. The elements $x_1, \ldots, x_{n-j}, z_1, \ldots, z_{j-1}$ form a regular sequence in $D_{j-1} \cap D_j$. Denote by '*' the homomorphism $R \longrightarrow R/(x_1, \ldots, x_{n-j}, z_1, \ldots, z_{j-1})$. By the choice of the z_i, both $\{x_1, \ldots, x_{n-j}, z_1, \ldots, z_{j-1}, x_{n-j+1}\}$ and $\{x_1, \ldots, x_{n-j}, z_1, \ldots, z_j\}$ are regular sequences (R = local, so that regular sequences can be re-arranged). Hence z_j^* and x_{n-j+1}^* are regular elements in R*. Note $D_{j-1}^* = (x_{n-j+1}^*):L^*$ while $D_j^* = (z_j^*):L^*$. Now we use (7.7) and (7.8) to complete the induction.

Remark. Information on the depths of the Koszul homology modules $H_i(I;R)$ also plays a role in an open question regarding the homology of the conormal module I/I^2. Namely, let R be a local ring and let I be an ideal of finite projective dimension. In [Vas_1] and [Vas_2] the following conjecture was considered:

C_1(homological rigidity of the conormal module): The only possible values for the projective dimension of I/I^2 as an R/I-module are 0 and ∞.

This means that if $pd_{R/I}(I/I^2) < \infty$, then I must be a complete intersection. Some cases, involving ideals of small height, have been considered. The Koszul homology intervenes directly in the following exact sequence of §2:

$$0 \longrightarrow \delta_1(I) \longrightarrow H_1(I;R) \longrightarrow (R/I)^n \longrightarrow I/I^2 \longrightarrow 0.$$

If the conjecture holds for lower dimensions and $H_1(I;R)$ is Cohen-Macaulay then $\delta_1(I) = 0$ and thus one has that $pd_{R/I}(I/I^2) = 0$, or 1. But a theorem of Gulliksen (cf. [GL]) rules out the second possibility. This settles C_1 for those cases where $H_1(I;R)$ is Cohen-Macaulay.

§8. Dimension of symmetric algebras.

In this section we consider ways of expressing the dimension of the symmetric algebra of a module. These considerations will be enhanced in the case of an ideal, for they will mean a first step towards checking whether the ideal is of linear type (cf. §2). Furhter applications will bo given in §§9 and 10.

We set-up the notation: Throughout R is a Noetherian (frequently Cohen-Macaulay) ring and E is a finitely generated R-module endowed with a (generic) rank in the sense that $E \otimes_R Q$ is a free Q-module, where Q is the total ring of fractions of R. We will denote rk(E) the cardinality of the Q-basis of $E \otimes_R Q$.

Further notation will appear whenever a finite presentation

$$R^m \xrightarrow{\phi} R^n \longrightarrow E \longrightarrow 0$$

is given:

$I_t(\phi)$=: ideal of R generated by the t x t minors of the matrix of ϕ (this is independent of base choice).

$rk(\phi)$ =: the rank of ϕ = $rk(im(\phi))$ = largest r such that $I_r(\phi) \neq 0$.

$t_o(\phi)$ =: $\inf \{t \geq 1 \mid \text{grade } I_s(\phi) \geq rk(\phi)-s+1, \text{ for } s \geq t\}$.

Note that these depend on the presentation. To get absolute invariants (i.e. of the module E), one could replace $I_t(\phi)$ by the corresponding Fitting ideal of E and $t_o(\phi)$ by $t_o = t_o(E)$ =: $\inf\{ t_o(\phi) \mid R^m \xrightarrow{\phi} R^n \longrightarrow E \longrightarrow 0\}$.

If needed, we will work from a fixed presentation $R^m \xrightarrow{\phi} R^n \longrightarrow E \longrightarrow 0$ for which $t_o(\phi) = t_o(E)$.

Our first result shows that $t_o(E)$ is closely related to the dimension of Sym(E).

Proposition 8.1. Let R be a Cohen-Macaulay local ring and let E be an R-module with a rank. Then the following are equivalent:

(a) dim Sym(E) = dim R + rk(E).

(b) $t_o(E) = 1$.

(c) $v(E_P) \leq ht(P) + rk(E)$ for every prime P of R.

Proof. The equivalence of (a) and (b) is proved in [SV_2, (2.1),(2.3)]. The equivalence of (b) and (c) is a consequence of the more general result.

Lemma 8.2. Let R be a Cohen-Macaulay ring and let E be an R-module with a rank and a given presentation $R^m \xrightarrow{\phi} R^n \longrightarrow E \longrightarrow 0$. Then, for any given integer $k \geq 0$, the following conditions are equivalent:

(a) grade $I_t(\phi) \geq rk(\phi) - t + k + 1$, $1 \leq t \leq rk(\phi)$.

(b) $v(E_P) \leq ht(P) + rk(E) - k$, for every prime ideal P such that E_P is not R_P-free.

Proof. First, for any prime P, we bring the original presentation to the form:

$$R_P^{\ell} \oplus R_P^{r} \xrightarrow{id\oplus\phi'} R_P^{\ell} \oplus R_P^{v} \longrightarrow E_P \longrightarrow 0, \quad v = v(E_P), \ \ell = n-v.$$

The matrix ϕ' has entries in PR_P (hence $r = 0 \iff E_P$ is R_P-free). From this one clearly has: $I_t(\phi) \subset P \iff t \geq \ell+1$.

(a) $\Rightarrow$ (b): Let P be such that E_P is not R_P-free. Then $v(E_P) > rk(E)$, so $\ell+1 = n-v(E_P)+1 \geq n-rk(E) = rk(\phi)$ and $I_{\ell+1}(\phi) \subset P$. We then have from the assumption:

$ht(P) \geq \text{grade } I_{\ell+1}(\phi) \geq rk(\phi)-(\ell+1)+(k+1) = n-rk(E)-\ell+ k =$

$(n-\ell)-rk(E)+k = v(E_P)-rk(E)+k.$

(b) $\Rightarrow$ (a): Given $1 \leq t \leq rk(\phi)$, pick $P \supset I_t(\phi)$ such that $ht(P) = ht(I_t(\phi))$. In particular $I_{rk(\phi)}(\phi) \subset I_t(\phi)$ and hence E_P is not R_P-free. Therefore, by assumption and the above remark:

$\text{grade } I_t(\phi) = ht(P) \geq v(E_P)-rk(E)+k = n-\ell-rk(E)+k \geq$

$n-(t-1)-rk(E)+k = (n-rk(E))-t+k+1.$

Remark. In the above lemma, condition (b) may be replaced by:

(b') $v(E_P) \leq ht(P)+rk(E)-k$, for every prime P such that $ht(P) \geq k$, provided E is given to be free in codimension $\leq k-1$. Since E is always assumed to be free in codimension 0, (b) $\iff$ (b') for $k = 1$. We will apply this in the following corollary.

Corollary 8.3. Let R be a Cohen-Macaulay ring and let I be an ideal of height > 0. Assume:

(i) I is generically a complete intersection.

(ii) R/I is Cohen-Macaulay.

Then the following are equivalent:

(a) grade $I_t(\phi) \geq (n-1)-t+3$, $1 \leq t \leq n-ht(I)$.

(Note $R^m \xrightarrow{\phi} R^n \longrightarrow I \longrightarrow 0$ is a presentation of I.)

(b) $v(I_P) \leq ht(P)-1$, for every prime $P \supset I$, $ht(P) \geq ht(I)+1$.

Proof. Note that by assumption the R/I-module I/I^2 has a rank (= ht(I)).

It suffices to apply (8.2) along with the above remark, provided we show that:

1) Condition (a) is equivalent to

grade $I_t(\phi^*) \geq rk(\phi^*)-t+2$, $1 \leq t \leq rk(\phi^*)$, where

$(R/I)^m \xrightarrow{\phi^*} (R/I)^n \longrightarrow I/I^2 \longrightarrow 0$.

For this note that $I \subset \sqrt{I_{n-1}(\phi)} \subset \sqrt{I_t(\phi)}$, $1 \leq t \leq n-1$ and that $rk(\phi^*) = n-rk(I/I^2) = n-ht(I)$. Since R/I is Cohen-Macaulay, we get

grade $I_t(\phi)$ = grade $I_t(\phi^*) + ht(I)$, $1 \leq t \leq$ n-grade I.

2) Condition (b) is equivalent to:

$v((I/I^2)_P) \leq ht(P/I)+rk(I/I^2)-1$, for every prime $P \supset I$ with $ht(P) \geq ht(I)+1$.

But the equalities $v(I_P) = v((I/I^2)_P)$ and $ht(P/I) = ht(P)-ht(I)$ translate (2) into (b).

The conditions of (8.1) are realized when Sym(E) is a Cohen-Macaulay ring. More generally:

Proposition 8.4. Let R be a Cohen-Macaulay ring of finite Krull dimension and let E be an R-module with a rank. Then:

(i) If Sym(E) is unmixed (meaning: (0) is height unmixed), dim Sym(E) = dim R + rk(E).

(ii) The following conditions are equivalent:

(a) Sym(E) is R-torsion-free.

(b) Sym(E) is unmixed and $v(E_P) \leq ht(P) + rk(E) - 1$ for every prime P such that $ht(P) \geq 1$.

The proof of this result is essentially given in [SV_2, (2.4) and (3.3)] once the results there are translated by means of (8.2).

The main application of the above proposition will be given in the next section. For the moment, we single out some simple consequences.

Corollary 8.5. Let R be a Cohen-Macaulay ring and let I be an ideal of height > 0.

(a) If Sym(I) is Cohen-Macaulay, then $v(I_P) \leq ht(P)+1$ for every prime P.

(b) If I is generically a complete intersection and R/I and $Sym(I/I^2)$ are Cohen-Macaulay rings, then $v(I_P) \leq ht(P)$ for every prime $P \supset I$.

§9. Cohen-Macaulay blowing-up rings.

One of the major applications of the approximation complexes is to certain arithmetical aspects of the blowing-up rings. More exactly, the acyclicity of these complexes, along with a stronger depth requirement on the Koszul homology modules, implies that all blowing-up rings are Cohen-Macaulay. In a sense, the Cohen-Macaulayness of the algebra Sym(I) alone is sufficient, along with a local bound on the number of generators, to trigger the same property for the other blowing-up rings. This phenomenon transfers the burden of checking arithmetical properties to one single algebra; it is however under Cohen-Macaulayness of the Koszul homology modules that one meets most 'block' examples (*cf.* §10).

We first give a version that uses the approximation complex $\mathcal{M}$. As before, v(E) denotes the minimal number of generators of the module E and ht(I) stands for the height of the ideal I.

H_i is the i-th Koszul homology module relative to a set of generators of I.

Theorem 9.1. Let R be a Cohen-Macaulay ring and let I be an ideal of height > 0. Assume:

(i) $v(I_P) \leq ht(P)$ for every prime $P \supset I$.

(ii) $depth(H_i)_P \geq ht(P)-v(I_P)+i$, for every prime $P \supset I$ and every $0 \leq i \leq v(I_P)-ht(I_P)$.

Then:

(a) $Sym(I) \cong \mathcal{R}(I)$ and $Sym(I/I^2) \cong gr_I(R)$.

(b) Sym(I) and $Sym(I/I^2)$ are Cohen-Macaulay rings.

(c) If it is further the case that R is Gorenstein and the H_i are Cohen-Macaulay, then $Sym(I/I^2)$ is a Gorenstein ring.

Note that (ii) is in particular satisfied if the H_i are Cohen-Macaulay.

Proof. (i) and (ii) imply that $depth(H_i)_P \geq i$ for $0 \leq i \leq v(I_P)-ht(I_P)$. Since outside of this range $(H_i)_P$ is certainly free, we see that condition (b) of (5.1) is satisfied. Therefore the complex $\mathcal{M}$ is acyclic and this warrants (a) (cf. §4).

To show (b), it suffices to handle the case where R is local. Say, dim R = d. Then dim Sym(I) = dim $\mathcal{R}(I)$ = d+1 and dim $Sym(I/I^2)$ = dim $gr_I(R)$ = d. We have therefore to show that depth Sym(I) = d+1 and depth $Sym(I/I^2)$ = d.

To see this we apply (5.2b) to the complexes $\mathcal{Z}$ and $\mathcal{M}$, respectively (N.B. Since $\mathcal{M}$ is acyclic, so is $\mathcal{Z}$ (cf. §4), but this could be checked directly using (5.3)). Namely the depth conditions on the H_i imply the following depth estimates

$$depth(H_i \otimes (R/I)[T_1,\dots,T_n]) \geq d+i, \quad 0 \leq i \leq s$$

$$\mathrm{depth}(Z_i \otimes R[T_1,\ldots,T_n]) \geq d+1+i, \quad 0 \leq i \leq n-1,$$

where $n = v(I)$, $s = v(I)-\mathrm{ht}(I)$ and depths are taken with respect to the irrelevant maximal ideal of the polynomial ring $R[T_1,\ldots,T_n]$. This is exactly what is needed to apply (5.2b).

Finally, (c) follows from the self-duality in this context of $\mathcal{M}$. Namely, we must show that $\mathrm{Ext}_S^s(\mathrm{Sym}(I/I^2), K_S) = \mathrm{Sym}(I/I^2)$, where $s = v(I)-\mathrm{ht}(I)$, $S = (R/I)[T_1,\ldots,T_n]$ and $K_S = H_s \otimes S$ is the canonical module of S. By [HK] this is exactly the condition for the Cohen-Macaulay ring $\mathrm{Sym}(I/I^2)$ to be Gorenstein. The self-duality of $\mathcal{M}$ in turn is a consequence of the fact that, with H_i Cohen-Macaulay, the homology algebra $H = H_\bullet(\mathcal{K})$ is a Poincaré algebra (cf. [He$_2$]).

Remark. A major source of applications of the preceding theorem is provided by the theory of linkage (cf. §7), but there are other instances as well (cf. §5).

We next give a version that does not involve homology at all.

Theorem 9.2. Let R be a Cohen-Macaulay ring and let I be an ideal of height > 0. Assume:

(i) Sym(I) and R/I are Cohen-Macaulay rings.

(ii) I is generically a complete intersection.

Then the following conditions are equivalent:

(a) $\mathrm{Sym}(I) \cong \mathcal{R}(I)$.

(b) $\mathrm{Sym}(I/I^2)$ is a Cohen-Macaulay ring.

(c) $v(I_P) \leq \mathrm{ht}(P)$ for every prime $P \supset I$.

Moreover, the following conditions are equivalent:

(α) $\mathrm{Sym}(I/I^2)$ is R/I-torsion free.

(β) $v(I_P) \leq \mathrm{ht}(P)-1$ for every prime $P \supset I$ such that $\mathrm{ht}(P) \geq \mathrm{ht}(I)+1$.

(γ) ht $F_s(I) \geq s+1$, $ht(I)+1 \leq s \leq v(I)$ ($F_{\bullet}(I)$ = Fitting ideals of I).

Remark. In the latter set of equivalent conditions, if in addition I is a prime ideal then the ordinary and symbolic powers of I coincide. Such applications were the 'leitmotiv' for investigating the blowing-up rings ([SV_1]).

Proof. (sketch) Most of the ground has been prepared in the preceding section. To get (b) => (c) apply the equality $\dim \mathrm{Sym}(I/I^2) = \dim(R/I) + \mathrm{rk}(I/I^2)$ obtained by (8.4). To get (c) => (a) apply the implication (b) => (a) of (8.4) with $E = I/I^2$ and observe that the kernel $\mathrm{Sym}(I) \longrightarrow \mathcal{R}(I)$ is annihilated by a power of I, an ideal containing a nonzero divisor by assumption. The equivalence (α) <=> (β) is also a straightforward application of (8.4), while the equivalence (β) <=> (γ) is a reformulation of (8.3). Finally, the only implication that needs a different argument is (a) => (b). For this we assume R local and observe that, because of the isomorphisms $\mathrm{Sym}(I) \cong \mathcal{R}(I)$ and $\mathrm{Sym}(I/I^2) \cong gr_I(R)$, we may compute the depth of $\mathrm{Sym}(I/I^2)$ from the exact sequences

$$0 \longrightarrow I\mathcal{R}(I) \longrightarrow \mathcal{R}(I) \longrightarrow gr_I(R) \longrightarrow 0$$

and

$$0 \longrightarrow \mathcal{R}(I)_+ \longrightarrow \mathcal{R}(I) \longrightarrow R \longrightarrow 0,$$

with respect to the irrelevant ideal of $\mathcal{R}(I)$. Clearly, $I\mathcal{R}(I) \cong \mathcal{R}(I)_+$, so we get depth $gr_I(R)$ = depth R, as required.

Remark. The above proof actually shows that, without the assumption that either R/I be Cohen-Macaulay or that I be generically a complete intersection, the conditions "Sym(I) is Cohen-Macaulay" and "$v(I_P) \leq ht(P)$ for every prime $P \supset I$" imply $\mathrm{Sym}(I) \cong \mathcal{R}(I)$ and $\mathrm{Sym}(I/I^2)$ Cohen-Macaulay. Thus, the preceding theorem may be viewed as a natural non-homological substitute for (9.1).

In general it is not true that Sym(I) is Cohen-Macaulay if $\mathrm{Sym}(I/I^2)$ is Cohen-Macaulay, as easy examples show. Even if we add that I be generically a complete intersection, the implication fails to hold - an example will be given in the next section. We don't know however if this is the case if, moreover, R/I is Cohen-Macaulay.

§10. Cohen-Macaulay symmetric algebras.

In the preceding section we pointed to the interest of knowing when Sym(I) is a Cohen-Macaulay ring (cf. (9.2)). In the present one, we will make a few variations on this theme. We start with a criterion of acyclicity of the complex $\mathcal{Z}$.

Theorem 10.1. Let R be a Cohen-Macaulay ring and let I be an ideal of height > 0. Assume:

(a) $v(I_P) \leq \mathrm{ht}(P)+1$, for every prime ideal P.

(b) depth $(H_i)_P \geq \mathrm{ht}(P)-v(I_P)+i$, for every prime ideal P and every i such that $0 \leq i \leq v(I_P)-\mathrm{ht}(I_P)$.

Then Sym(I) is Cohen-Macaulay.

Proof. Consider the complex $\mathcal{Z}$ associated to I. By (5.3) we know that $\mathcal{Z}$ is acyclic. Depth-chasing in a way entirely similar to that in the proof of (9.1), yields depth $\mathrm{Sym}(I) \geq \dim R + 1$. On the other hand, by (8.1), $\dim \mathrm{Sym}(I) \leq \dim R + 1$. Therefore, Sym(I) is Cohen-Macaulay.

Note that the above theorem is analogous to (9.1), except that we stay away from the equality $\mathrm{Sym}(I) = \mathcal{R}(I)$.

We next give an application which extends some results of Rossi ([Ro]).

Corollary 10.2. Let (R,m) be a Cohen-Macaulay local ring of dimension > 0 and let I be an m-primary ideal. The following are equivalent:

(a) I is an almost complete intersection.

(b) Sym(I) is Cohen-Macaulay.

Proof. (a) $\Rightarrow$ (b) follows from (10.1), while the reverse implication issues from (8.5).

Next is a variant of (10.1), in terms of sequential conditions.

Proposition 10.3. Let R be a Cohen-Macaulay local ring and let I be an ideal of height > 0. Assume:

(i) $v(I_P) \leq ht(P)+1$ for every prime P.

(ii) I is generated by a proper sequence $\{x_1,\ldots,x_n\}$ satisfying depth $(R/(x_1,\ldots,x_j)) \geq$ depth $R-j$, $1 \leq j < n$ and depth $R/I \geq \dim R/I - 1$.

Then Sym(I) is Cohen-Macaulay.

Proof. Since $I = (x_1,\ldots,x_n)$ and the x_i form a proper sequence, the Z-complex is acyclic (cf. §12). The present depth estimates imply, moreover, that

depth $H_j(\underline{x};R) \geq \dim R + j - n$, for $1 \leq j \leq n$, (cf. (7.1)).

Now, depth-chasing as in the proof of (9.1), we get depth $(Z_i \otimes R[T_1,\ldots,T_n]) \geq \dim R + i + 1$. By (5.2b), depth Sym(I) $\geq \dim R + 1$. Finally, as before, (i) means that $\dim \mathrm{Sym}(I) = \dim R + 1$.

As a consequence, we can derive a handy criterion in the case I is an almost complete intersection.

Corollary 10.4. Let R be a Cohen-Macaulay local ring and

let I be an ideal of height > 0. Assume:

(a) I is an almost complete intersection.

(b) depth $R/I \geq \dim R/I - 1$.

Then Sym(I) is Cohen-Macaulay.

Proof. By (a), I is generated by $\{x_1,\ldots,x_{n-1},x_n\}$ where the first n-1 elements form a regular sequence; this is certainly a proper sequence. Furthermore, (b) provides for the only depth requirement needed. Finally, an almost complete intersection always satisfies the requirement $v(I_P) \leq ht(P)+1$, for any prime P.

Remark 10.5. In (10.4) I is not assumed to be generically a complete intersection. If that is further granted and R is moreover Gorenstein, then $Sym(I/I^2)$ is Cohen-Macaulay as well. In fact, in the presence of the condition depth $R/I \geq \dim R/I - 1$, it must then be the case that depth $R/I = \dim R/I$, so R/I is Cohen-Macaulay (we thank C. Huneke for this observation). But, if this is the case we can apply (9.2) to get $Sym(I) \cong \mathcal{R}(I)$ and $Sym(I/I^2)$ Cohen-Macaulay. It may be worthwhile to add that if I is a complete intersection only at its primes of minimal height, then R/I is not necessarily Cohen-Macaulay yet $Sym(I/I^2)$ is Cohen-Macaulay and $Sym(I) \cong \mathcal{R}(I)$ (cf. [SV_2;§5] for both an example and the proof of this fact.).

Remark 10.6. Criterion (i) of (10.3) is not superfluous in the presence of a generating set that is a proper sequence and satisfies the depth requirements stated there. For example, take I to be the maximal ideal of a local ring, whose square is zero. However, it may be the case that (i) is superfluous under the assumption that I be generically a complete intersection.

In the sequel we show, in the vein of [EH, (1.1)], under what conditions the Cohen-Macaulay property of the symmetric algebra is preserved under specialization.

Precisely, let R be a Cohen-Macaulay ring and let I be an ideal of height > 0. We assume an ideal N is given such that: (i) N is perfect, and (ii) I is not contained in any associated prime of N. We then ask when Sym(I) Cohen-Macaulay implies Sym(I/NI) Cohen-Macaulay. Note that (ii) implies that I/NI is (as R/N-module) with generic rank one. In the applications N will actually be generated by an R/I-sequence so that $I/NI \cong I+N/N \subset R/N$ in this case.

Proposition 10.7. Let I and N be given as above. If Sym(I) is Cohen-Macaulay, then the following are equivalent:

(a) Sym(I/NI) is Cohen-Macaulay.

(b) $v(I_P) \leq ht(P/N)+1$, for every prime $P \supset I+N$.

Proof. (a) => (b): Since $v(I_P) = v((I/NI)_P)$ holds, this implication follows directly from (8.1) and (8.4).

(b) => (a): By the usual argument (cf. $[PS_1]$) it suffices to show that $ht(N.Sym(I)) \geq ht(N)$. To arrive at this. we first appeal to (8.4) as applied to Sym(I), to get dim Sym(I) = dim R + 1. Then we apply (8.1) to Sym(I/NI) under condition (b). Putting the pieces together we find ht(N.Sym(I)) = ht(N).

Remark. Note that the conditions on N imply $Tor_1^R(I,R/N) = 0$, which, however, may not suffice to make I/NI isomorphic to an ideal of R/I. For this we need the stronger requirement $v(I_P) \leq ht(P/N)$ as in (8.4ii).

The following may be viewed as an application of (10.7) - but follows also from (8.5), (10.1) and the results of §7. It is analogous to [EH, (3.4)], but note the present hypotheses do not necessarily imply the equality of the symmetric and Rees algebras.

Corollary 10.8. Let R be a Cohen-Macaulay ring and let I be an ideal of finite projective dimension. Assume either

(i) $ht(I) = 2$ and R/I is Cohen-Macaulay, or

(ii) $ht(I) = 3$ and R/I (and R) are Gorenstein.

Then the following conditions are equivalent:

(a) Sym(I) is Cohen-Macaulay.

(b) $v(I_P) \leq ht(P)+1$, for every prime P.

Example. Consider the matrix

$$\phi = \begin{bmatrix} x & y & z & 0 \\ 0 & x & y & z \\ 0 & 0 & x & y \end{bmatrix} \quad (\text{where } R = k[[x,y,z]])$$

Here $I = I_3(\phi) = (x^3, x^2y, x(y^2-xz), y^3)$

$ht(I_3(\phi)) = 2$, $ht(I_2(\phi)) = 3$, $ht(I_1(\phi)) = 3$. Therefore, by (8.1) and the preceding corollary, we get that Sym(I) is Cohen-Macaulay. (Note, however, that $\mathrm{Sym}(I) \neq \mathcal{R}(I)$ as $v(I) = 4 > \dim R$.)

A similar example for case (ii) can be gotten with a 5 x 5 skew-symmetric matrix in a polynomial ring with 4 variables. It is quite conceivable that for any given $n \geq 3$ there exist examples of $n \times (n+1)$ matrices (resp. $2n-1 \times 2n-1$ skew-symmetric matrices) whose ideal of $n \times n$ minors (resp. $2n-2 \times 2n-2$ pfaffians) has a Cohen-Macaulay symmetric algebra and, moreover, not coinciding with the corresponding Rees algebra.

We now wish to point out that, contrarily to what happens in the favourable habitat of Rees algebras, symmetric algebras tend to be non-Cohen-Macaulay in some very standard generic situations.

(1) Determinantal ideals. Let (x_{ij}), $1 \leq i \leq m$, $1 \leq j \leq n$, be a matrix of indeterminates over a field. Assume $m \geq n+2$. Let I be the ideal of $n \times n$ minors. Then, we know that $\mathrm{Sym}(I) \neq \mathcal{R}(I)$. Moreover, if $\binom{m}{n} \geq mn+2$ then we know that $\mathrm{Sym}(I)$ is not Cohen-Macaulay because this blasts the number of generators condition. Say, $n = 2$. In this case, it is easy to see that I is a complete intersection in the punctured spectrum of R $(= k[[x_{ij}]])$. It follows that depth $H_1 = 0$, so that we are far from being able to apply any of the so-far given criteria for the acyclicity of either $\mathcal{M}$ or $\mathcal{Z}$. In fact, $\mathrm{Sym}(I)$ is never Cohen-Macaulay in this case: If $m \geq 6$ then $v(I) = \dim R + 3$; if $m \leq 5$, then $v(I_P) \leq \mathrm{ht}(P)$ for every $P \supset I$, so that $\mathrm{Sym}(I)$ is again not Cohen-Macaulay as otherwise it would be torsion-free by (8.4).

(2) Divisors in scrolls. (cf. [EH]) These are given as follows: Pick a generic 2×2 matrix

$$X = \begin{bmatrix} x_{11} & x_{12} \\ x_{21} & x_{22} \end{bmatrix}$$

and a 'semi-generic' 2×2 matrix

$$Y = \begin{bmatrix} y_{11} & y_{12} \\ y_{21} & y_{11} \end{bmatrix}$$

and consider the 2×4 matrix (X, YX). Let I be the ideal generated by the 2×2 minors of this matrix. We claim that $\mathrm{Sym}(I)$ is not Cohen-Macaulay (not even unmixed).

To see this, we first observe that the estimate $v(I_P) \leq \mathrm{ht}(P)$ holds for every prime $P \supset I$. Indeed, this is obvious if $\mathrm{ht}(P) \geq 4$. If $\mathrm{ht}(P) = 2$, then $P = I$ so that I_P is a complete intersection in R_P. Finally, if $ht(P) = 3$ then $(x_{11}, x_{12}, x_{21}, x_{22}) \not\subset P$; say $x_{11} \notin P$. By effecting an elementary transformation in the matrix (X, YX), we set it in the form

$$\begin{bmatrix} 1 & 0 & 0 & 0 \\ 0 & \#_1 & \#_2 & \#_3 \end{bmatrix}$$

from which it is clear that $v(I_P) \leq 3$.

On the other hand, Sym(I) is not torsion-free (in fact, the Plücker relation among the generators of I shows that $S_2(I) \neq I^2$). Therefore, by (8.4ii), Sym(I) is not Cohen-Macaulay.

It follows from (10.7) (checking (b)) that if I is the homogeneous ideal of the smooth rational non-normal quartic in $\mathbb{P}^3$, then Sym(I) is not Cohen-Macaulay (of course, this could also be checked directly as above).

As for $\mathrm{Sym}(I/I^2)$, the situations are different in (1) as in (2). We have:

(1)' Let $I \subset R = k[[x_{ij}]]$, $1 \leq i \leq 2$, $1 \leq j \leq 4$, be the ideal of the 2 x 2 minors of the generic 2 x 4 matrix over the field . Then $\mathrm{Sym}(I/I^2)$ is not Cohen-Macaulay.

<u>Proof</u>. We have $v(I) = 6 < 7 - \dim R \quad 1$. Also, as remarked in (1), I is a complete intersection of height three in the punctured spectrum of R. Therefore, for a prime P such that $4 \leq ht(P) \leq 7$, we have $v(I_P) = ht(I_P) = ht(I) \leq ht(P)-1$. Since R/I is Cohen-Macaulay, if $\mathrm{Sym}(I/I^2)$ were Cohen-Macaulay we could have, by (8.4ii), that $\mathrm{Sym}(I/I^2)$ is a domain. Clearly, we have $\dim \mathrm{Sym}(I/I^2) = \dim R$. On the other hand, we know that $\mathrm{Sym}(I/I^2) \neq gr_I(R)$. It follows that $\dim \mathrm{Sym}(I/I^2) \geq \dim gr_I(R) + 1 = \dim R + 1$, a contradiction.

(2)' Let $I \subset R = k[[x,y,z,w]]$ be the ideal of the smooth rational non-normal quartic in $\mathbb{P}^3$. Then $\mathrm{Sym}(I/I^2)$ is Cohen-Macaulay.

Proof. We use the $\mathcal{M}^*$-complex on a generating set of 4 elements:

$$0 \longrightarrow H_2^* \otimes S \longrightarrow H_1^* \otimes S \longrightarrow H_o \otimes S \longrightarrow \mathrm{Sym}(I/I^2) \longrightarrow 0.$$

Note that $H_2^* = H_2$ = canonical module of R/I, and thus has depth 2 (cf. [HK]). H_1^*, on the other hand, is torsionfree, and thus has depth ≥ 1. As the complex satisfies the conditions of (5.2), as in (5.1) we conclude that it is exact. Counting depths, we have depth $\mathrm{Sym}(I/I^2) \geq 4 = \dim \mathrm{Sym}(I/I^2)$, as asserted.

Remark. (2) gives an example of a generically complete intersection ideal I for which $\mathrm{Sym}(I/I^2)$ is Cohen-Macaulay but Sym(I) is not. Note, however, that R/I is not Cohen-Macaulay. Does there exist a similar example with R/I Cohen-Macaulay?

To conclude, a word on the property of being Gorenstein for the symmetric algebra. It was shown in §9 that if $I \subset R$ is generically a complete intersection in a Gorenstein ring and if one is given appropriate conditions that make the complex $\mathcal{M}$ acyclic, then $\mathrm{Sym}(I/I^2)$ is a Gorenstein ring. The situation is dramatically diverse for Sym(I). Here it easy to see, by localization at a minimal prime of height = ht(I), that Sym(I) Gorenstein implies $\mathrm{ht}(I) \leq 2$.

It is conceivable that this bound on the height of I works in general, even if I is not generically a complete intersection. This is the case if m is the maximal ideal of a hypersurface (communicated by E. Rossi) and, conversely, we can use the acyclicity of $\mathcal{Z}$, together with the high depths of its components in this case, to conclude that if $\mathrm{ht}(m) = 2$ then $\mathrm{Sym}(m)$ is Gorenstein (this has also been shown by Rossi).

§11. The syzygetic reduction groups.

In sections 2 and 6 we studied the $\delta_i(I)$'s and some of their various interpretations. This section is devoted to extending such notions to obtain a whole array of measures ofcomparison.

Here and in later sections it will be necessary to consider the approximation complexes with coefficients in a module (cf. §4).

Let I be an ideal of the Noetherian ring R and let M be a finitely generated R-module. On the Koszul complex $\mathcal{K}(I;M)$ - built on some generating set of I - one can define a natural filtration $\mathcal{F}_*$:

$$\mathcal{F}_j K_i(I;M) = \begin{cases} I^{j-i}K_i(I;M), & \text{for } j \geq i \\ \\ K_i(I;M), & \text{for } j \leq i. \end{cases}$$

Definition 11.1. The modules $\delta_i^j(I;M) = H_i(\mathcal{F}_{i+j}\mathcal{K}(I;M))$ are called the syzygetic reduction groups of I with respect to M.

Note that $\delta_i^j(I;M) = H_i(I;M)$ for $j \leq i$, and $\delta_1^1(I;R) = \delta_1(I)$ in the terminology of §2.

Of course, the $\delta_i^j(I;M)$ may depend on the choice of the generators of I, since this is the case of $\mathcal{K}(I;M)$ itself. However, we will see below that a certain invariance property holds.

There are two other obvious ways to describe the $\delta_i^j(I;M)$. For simplicity, let $\mathcal{K} = \mathcal{K}(I;M)$, $Z = Z(I;M)$ and $B = B(I;M)$.

Lemma 11.2. For $j \geq 1$, we have

(a) $\delta_i^j(I;M) = Z_i \cap I^jK_i/I^{j-1}B_i$.

(b) $\delta_i^j(I;M) = \text{im}(H_i(I;I^jM) \longrightarrow H_i(I;I^{j-1}M))$.

Lemma 11.3. If $\delta_i^j(I;M) = 0$ for some i and j, then $\delta_{i+1}^j(I;M)$

does not depend on the chosen generators of I.

Proof. By standard arguments we may restrict ourselves to show the following: If $\underline{x} = \{x_1,\ldots,x_n\}$ generates I, then the canonical mapping

$$\phi: H_{i+1}(\mathcal{F}_{i+j+1}K(\underline{x};M)) \longrightarrow H_{i+1}(\mathcal{F}_{i+j+1}K(\underline{x},0;M))$$

is an isomorphism.

In $K_1(\underline{x};0;M)$ we have a basis $\{e_1,\ldots,e_n,e_{n+1}\}$ with $\partial e_i = x_i$ for $i = 1,\ldots,n$ and $\partial e_{n+1} = 0$. We first show that ϕ is injective: In fact, let $z \in \mathcal{F}_{i+j+1}K(\underline{x};M)$ be a cycle and suppose that $z = \partial(b_o + b_1 \wedge e_{n+1})$ with $b_o, b_1 \in \mathcal{F}_{i+j+1}K(\underline{x};M)$. Then necessarily $\partial b_1 = 0$ and $z = \partial b_o$.

To see that ϕ is surjective, let $z = z_o + z_1 \wedge e_{n+1}$ be an $(i+1)$-cycle of $\mathcal{F}_{i+j+1}K(\underline{x};0;M)$ with $z_o \in \mathcal{F}_{i+j+1}K_{i+1}(\underline{x};M)$, and $z_1 \in \mathcal{F}_{i+j}K_i(\underline{x};M)$. It then follows that z_1 is also a cycle and hence a boundary, since we assume that $\delta_i^j(I;M) = 0$. Pick $b \in \mathcal{F}_{i+j}K_{i+1}(\underline{x};M)$ such that $z_1 = \partial b$. Then we have $b \wedge e_{n+1} \in \mathcal{F}_{i+j+1}K_{i+2}(\underline{x};0;M)$ and $\partial(b \wedge e_{n+1}) = z_1 \wedge e_{n+1}$. Hence z is homologous to z_o which lies in the image of ϕ.

Definition 11.4. The ideal I is called j-syzygetic with respect to M, if $\delta_i^k(I;M) = 0$ for all i and all k with $1 \leq k \leq j$.

By the previous lemma it is clear that this notion depends only on the ideal and not on the chosen generators. It was first introduced by Huneke ($[Hu_4]$). The most interesting case will be when I is j-syzygetic for all j.

Remark. (a) For any ideal I and any finitely generated module M there exists an integer n_o such that I is j-syzygetic with respect to I^nM for all j and all $n \geq n_o$. This follows from a simple application of the Artin-Rees lemma to the complexes $F_k K(I;M)$ (cf. [AB]).

(b) It is clear from the definition that if I is j-syzygetic with respect to M for all j, then so is I with respect to IM.

Let us give another interpretation of the syzygetic reduction groups: Having chosen a set of generators $\{x_1,\dots,x_n\}$ of I we obtain a natural R-algebra presentation

$$S = R[e_1,\dots,e_n] \longrightarrow R(I), \quad e_i \longrightarrow x_i t = x_i^*$$

of the Rees algebra of I. We fix this presentation. The irrelevant ideal $S_+ = (e_1,\dots,e_n)$ of S acts in a natural way on the $R(I)$-module $R(I;M) = \bigoplus_j I^j M$ and the $gr_I(R)$-module $gr_I(M) = \bigoplus_j I^j M/I^{j+1}M$, which corresponds to the actions of the irrelevant ideal of $R(I)$, resp. $gr_I(R)$.

Lemma 11.5.

(a) $$H_i(S_+;R(I;M))_{i+j} = \begin{cases} \delta_i^j(I;M), & \text{for } j > 0 \\ Z_i(I;M), & \text{for } j = 0. \end{cases}$$

(b) There exists a long exact sequence

$$\dots \longrightarrow \delta_i^{j+1}(I;M) \longrightarrow \delta_i^j(I;M) \longrightarrow H_i(S_+;gr_I(M))_{i+j} \longrightarrow \delta_{i-1}^{j+1}(I;M) \longrightarrow \dots$$

Proof. (a) is immediate.

(b) We have

$$K(S_+;gr_I(M)) \cong grK(I;M) = \bigoplus_j F_j K(I;M)/F_{j+1}K(I;M).$$

One now considers the long exact homology sequence corresponding to the isomorphism.

Some of these syzygetic groups may also be interpreted in terms of the syzygy part of the Koszul homology (cf. [Si]; see also §6).

In our terminology $H_i^*(I;M) \cong \operatorname{coker}(\delta_i^1(I;M) \longrightarrow \delta_i^0(I;M))$. Since the mapping above is always injective, we obtain

Corollary 11.6. $H_i^*(I;M) \cong H_i(S_+;gr_I(M))_i$.

Next we show that the syzygetic reduction groups measure the homology of the approximation complex $\mathcal{M}(I;M)$. It is convenient to consider $\mathcal{M}(I;M)$ with its natural augmentation mapping

$$\partial_o': \mathcal{M}(I;M) \longrightarrow gr_I(M),$$

where ∂_o' is the composition

$$H_o(I;M) \otimes S \longrightarrow H_o(\mathcal{M}(I;M)) \xrightarrow{\beta} gr_I(M).$$

Thus we obtain an extended complex

$$\mathcal{N}: \ \ldots \longrightarrow \mathcal{M}_1(I;M) \longrightarrow \mathcal{M}_o(I;M) \longrightarrow gr_I(M) \longrightarrow 0$$

It is then clear that $H_{-1}(\mathcal{N}) = 0$ and that $H_o(\mathcal{N}) = 0$ if and only if β is an isomorphism - that is, if I is of linear type with respect to M (4.9).

Our aim is to express the first non-vanishing homology of $\mathcal{N}$ in terms of the δ_i^j.

Theorem 11.7. Suppose that for some $i \geq 0$ we have $H_\ell(\mathcal{N}) = 0$ for all $\ell < i$, then

(a) $H_\ell(S_+;gr_I(M)) \cong H_\ell(I;M)(\ell)$, for $\ell \leq i$.

(b) $$(H_i(\mathcal{N})/S_+H_i(\mathcal{N}))_j = \begin{cases} 0, & \text{for } j < i \text{ and } j = i+1 \\ H_{i+1}(S_+;gr_I(M))_j, & \text{for } j > i+1. \end{cases}$$

Note that, from the definition, one has

$$(H_i(\mathcal{N})/S_+H_i(\mathcal{N}))_i = H_i(\mathcal{N})_i \cong \begin{cases} 0, & \text{for } i = 0 \\ \ker(H_i(I;M) \xrightarrow{\partial'} H_{i-1}(I;M)\otimes S_1), & \text{for } i > 0. \end{cases}$$

Before proving the theorem we draw some consequences. First we reprove a result of M. Kühl ([Kü], (2.7)).

Corollary 11.8. I is an ideal of linear type with respect to M, if and only if $\delta_1^j(I;M) = 0$ for $j \geq 1$.

Proof. By the theorem $H_o(\mathcal{N}) = 0$ if and only if $\delta_1^1(I;M) = \ker(H_1(I;M) \longrightarrow H_o(I;M)\otimes S_1) = 0$ and $H_1(S_+;\mathcal{gr}_I(M))_j = 0$ for $j > 1$. By (11.5) we have the exact sequence

$$\delta_1^{j+1}(I;M) \longrightarrow \delta_1^j(I;M) \longrightarrow H_1(S_+;\mathcal{gr}_I(M))_{i+j} \longrightarrow 0.$$

But, as remarked earlier, $\delta_1^k(I;M) = 0$ for $k >> 0$, by the Artin-Rees lemma, and thus the assertion.

Corollary 11.9. The following conditions are equivalent:

(a) $\mathcal{M}(I;M)$ is acyclic.

(b) I is j-syzygetic with respect to M for all $j \geq 1$.

(c) $H_i(S_+;\mathcal{gr}_I(M)) \cong H_i(I;M)(i)$ for all $i \geq 0$.

Proof. The equivalence of (b) and (c) follows immediately from (11.5).

In §4 we have seen that I is of linear type with respect to M if $\mathcal{M}(I;M)$ is acyclic. Hence $\mathcal{M}(I;M)$ is acyclic if and only if $\mathcal{N}$ is exact. Now the equivalence of (a) and (c) follows easily from (11.7).

Proof of (11.7). By assumption we have the exact sequence

$$0 \longrightarrow U_i \longrightarrow N_i \longrightarrow \cdots \longrightarrow N_o \longrightarrow N_{-1} \longrightarrow 0.$$

Since $H_\ell(S_+;N_k) = 0$ for $k \geq 0$ and $\ell > 0$, this exact sequence gives rise to isomorphisms of graded modules

(1) $H_\ell(S_+;\mathcal{gr}_I(M)) \cong H_o(S_+;N_\ell)$ for $\ell \leq i$,

(2) $H_{i+1}(S_+;\mathcal{gr}_I(M)) \cong \ker(H_o(S_+;U_i) \longrightarrow H_o(S_+;N_i))$.

Since $H_o(S_+;N_\ell) \cong H_\ell(I;M)(\ell)$, the assertion (a) of the theorem follows from (1). (2), on the other hand, implies

(3) $H_{i+1}(S_+;\mathcal{gr}_I(M))_j \cong H_o(S_+;U_i)_j$ for $j > i$.

For all j we also have the exact sequence

(4) $H_o(S_+;N_{i+1})_j \longrightarrow H_o(S_+;U_i)_j \longrightarrow H_o(S_+;H_i(N))_j \longrightarrow 0.$

Again, since $H_o(S_+;N_{i+1})_j = 0$ for $j > i+1$, we obtain from (4) $H_o(S_+;U_i)_j \cong H_o(S_+;H_i(N))_j$ for $j > i+1$, which in conjunction with (3) proves part of (b). It is clear that $H_o(S_+;H_i(N))_j = 0$ for $j < i$, since $H_i(N)$ is a subquotient of N_i. It remains therefore to show that $H_o(S_+;H_i(N))_{i+1} = 0$. This follows from the commutative diagram with exact rows

$$\begin{array}{ccccccc} H_o(S_+;N_{i+1})_{i+1} & \longrightarrow & H_o(S_+;U_i)_{i+1} & \longrightarrow & H_o(S_+;H_i(N))_{i+1} & \longrightarrow & 0 \\ \| & & \| & & & & \\ \delta^0_{i+1}(I;M) & \longrightarrow & H_{i+1}(S_+;\mathcal{gr}_I(M))_{i+1} & \longrightarrow & \delta^1_i(I;M) & \longrightarrow & \delta^0_i(I;M) \end{array}$$

In fact, since $\delta^1_i(I;M) \longrightarrow \delta^0_i(I;M)$ is trivially injective, it follows that $H_o(S_+;N_{i+1})_{i+1} \longrightarrow H_o(S_+;U_i)_{i+1}$ is surjective, and $H_o(S_+;H_i(N))_{i+1} = 0$, as asserted.

§12. Sequences and the approximation complexes.

In §3 we have seen that ideals which are generated by d-sequences are ideals of linear type. In this section we want to show much more. If (R,m,k) is a Noetherian local ring whose residue field k is infinite, then for an ideal I and a finitely generated R-module M, we show that the following are equivalent:

(a) $\mathcal{M}(I;M)$ is acyclic.

(b) I is generated by a d-sequence with respect to M.

The implication (b) => (a) has been shown already in [HSV_1] - without the restriction on k. Also, under some additional Cohen-Macaulay conditions, Huneke ([Hu_4]) proved the equivalence of (a) and (b).

The notion of a d-sequence that we shall use is (6.1.a_2^*), as it is convenient to skip the minimality condition.

For several reasons, we need some other characterizations of d-sequences.

Lemma 12.1. Let $\underline{x} = \{x_1,\dots,x_n\}$ be a sequence of elements generating the ideal I. The following conditions are equivalent:

(a) $\underline{x}$ is a d-sequence with respect to M.

(b) $(x_1,\dots,x_i)M : x_{i+1} \cap IM = (x_1,\dots,x_i)M$, for $i = 1,\dots,n-1$.

(c) $(x_1,\dots,x_i)IM : x_{i+1} \cap IM = (x_1,\dots,x_i)M$

and $(x_1,\dots,x_i)IM : x_{i+1} \cap I^2M = (x_1,\dots,x_i)IM$, for $i = 1,\dots,n-1$.

We have picked only a few of the possible characterizations of d-sequences which are listed in [Kü, (1.2)]. The implication (a) => (b) is due to Huneke ([Hu_1]). The equivalence of (a) and (c) was first observed by Kühl ([Kü]). Note that a sequence satisfying the first equation of (c) is a relative regular sequence with respect to M (cf. (6.1)). The second equation of (c) says that $\underline{x}$ is a d-sequence with respect to IM. Thus we may rephrase (12.1) as follows.

Corollary 12.2. The following conditions are equivalent:

(a) $\underline{x}$ is a d-sequence with respect to M.

(b) $\underline{x}$ is a relative regular sequence with respect to M and a d-sequence with respect to IM.

(c) For all $k \geq 0$, $\underline{x}$ is a relative regular sequence with respect to I^kM and a d-sequence with respect to $I^{k+1}M$.

In fact, (c) follows by induction from (a) and (b).

Before proving (12.1), let us draw one interesting consequence of (12.2). We will use a result of Fiorentini ([Fi]).

Theorem 12.3. If I is generated by a relative regular sequence with respect to M, then I is 1-syzygetic with respect to M.

From (12.2) it now follows immediately

Corollary 12.4. If I is generated by a d-sequence with respect to M, then I is j-syzygetic with respect to M for all j, and hence $\mathcal{M}(I;M)$ is acyclic (by (11.9)).

This is a new proof of [HSV, (5.6)] and also a new proof of the fact (cf. §3) that if I is generated by a d-sequence with respect to M, then I is of linear type with respect to M.

Proof of (12.1). (a) => (b): If $a \in (x_1,\ldots,x_i)M : x_{i+1} \cap IM$ then $aIM \subset (x_1,\ldots,x_i)M$. In fact, if $k \geq i+1$, then $ax_{i+1}x_k \in (x_1,\ldots,x_i)M$ and hence $ax_k \in (x_1,\ldots,x_i)M$, i.e. $ax_k \in (x_1,\ldots,x_i)M$ for all k. Now suppose $a = \Sigma a_j x_j$, $1 \leq j \leq k$, $a_i \in M$. If $k \geq i+1$, then $ax_k \in (x_1,\ldots,x_i)M \subset (x_1,\ldots,x_{k-1})M$, and hence $a_k x_k^2 \in (x_1,\ldots,x_{k-1})M$. But this implies that $a_k x_k \in (x_1,\ldots,x_{k-1})M$. Thus we can rewrite $a = \Sigma a_j' x_j$, $1 \leq j \leq k-1$. By inverse induction on k we conclude that $a \in (x_1,\ldots,x_i)M$. The other inclusion is trivial.

(b) => (c): The first equality of (c) follows trivially from (b). To prove the second one we show that

(#) $(x_1,\ldots,x_i)M \cap I^2M = (x_1,\ldots,x_i)IM$ for $i = 1,\ldots,n$.

From (#) the second equality follows easily. In fact,

$$(x_1,\ldots,x_i)IM : x_{i+1} \cap I^2M = ((x_1,\ldots,x_i)IM : x_{i+1} \cap IM) \cap I^2M =$$
$$(x_1,\ldots,x_i)M \cap I^2M = (x_1,\ldots,x_i)IM.$$

(b) => (#): Let $a \in (x_1,\ldots,x_i)M \cap I^2M$, $a = \sum^{k} a_j x_j$, $a_j \in IM$. Suppose that $k \geq i+1$; then $a_k x_k \in (x_1,\ldots,x_{k-1})M$ and hence $a_k \in (x_1,\ldots,x_{k-1})M : x_k \cap IM = (x_1,\ldots,x_{k-1})M$. Thus $a = \sum^{k-1} a_j' x_j$. By induction it follows that $(x_1,\ldots,x_i)M \cap I^2M \subset (x_1,\ldots,x_i)IM$. Again, the other inclusion is trivial.

(c) ⟹ (a): We are going to show that for all $k \geq i+1$ we have $(x_1,\ldots,x_i)M : x_{i+1}x_k \subset (x_1,\ldots,x_{k-1})M : x_k$. This implies easily that $\underline{x}$ is a d-sequence.

Let $a \in (x_1,\ldots,x_i)M : x_{i+1}x_k$; then $ax_{i+1}x_k \in (x_1,\ldots,x_i)M$, and hence $ax_k^2 \in (x_1,\ldots,x_i)IM : x_{i+1} \cap I^2M = (x_1,\ldots,x_i)IM$ $(x_1,\ldots,x_{k-1})IM$. It follows that $ax_k \in (x_1,\ldots,x_{k-1})IM : x_k \cap IM = (x_1,\ldots,x_{k-1})M$.

As it was pointed out in §6, if $\underline{x}$ is a relative regular sequence with respect to M, it follows from Fiorentini's theorem that $\underline{x}$ forms a proper sequence with respect to M.

<u>Theorem 12.5</u>. Let I be an ideal generatod by a proper sequence with respect to M. Then $Z(I;M)$ is acyclic.

<u>Proof</u>. (cf. [HSV$_1$]) We let $\{e_1,\ldots,e_n\}$ denote the standard basis of R^n, and the basis of $S_1(R^n)$ as well (cf. §4). K and Z will stand for the Koszul complex $K(\underline{x};M)$ and its module of cycles $Z(\underline{x};M)$.

Let $z \in Z$; since we view the elements of Z as polynomials in $e_1,\ldots,e_n$ with coefficients in Z, we can write

$$z = \Sigma\, z_{h,i} \otimes he_j^i$$

where the h's are distinct monomials in $e_1,\ldots,e_{j-1}$, and $\partial(z_{h,i}) = 0$ for all h,i. We prove by induction on j that $z = \partial'(b)$, with $b \in Z$.

$j = 1$: We write simply

$$z = \Sigma\, z_i \otimes e_1^i.$$

We claim that $z_i \in Z(x_1)$ for all i. Indeed, let $k = \inf\{\ell \mid z_i \in Z(x_1,\ldots,x_\ell)\}$. Assume $k > 1$. Each z_i can then be written as

$$z_i = z_i' + w_i' \wedge e_k, \quad \text{with } z_i', \ w_i' \in K(x_1,\ldots,x_{k-1}).$$

Since

$$0 = \partial'(z) = \Sigma\, w_i' \otimes e_1^i e_k + (\text{terms in } e_1,\ldots,e_{k-1}),$$

it follows that $w_i' = 0$. Thus $z_i \in K(x_1,\ldots,x_{k-1})$, a contradiction.

We now have

$$z = \Sigma\, a_i e_1 \otimes e_1^i, \quad a_i x_1 = 0.$$

Since $0 = \partial'(z) = \Sigma\, a_i e_1^{i+1}$, it follows that $z = 0$.

$j > 1$: As in the preceding case, it can be shown that $z_{h,i} \in Z(x_1,\ldots,x_j)$. We now prove by induction on $\deg_{e_j}(z) = t$, that $z = \partial'(b)$.

If $t = 0$, then by our induction hypothesis on j, we are done. Suppose then $t > 0$. We write

$$z_{h,t} = z_{h,t}' + w_{h,t}' \wedge e_j$$

as before (if $z_{h,t} \in Z_1(K)$, then $z_{h,t} = z_{h,t}' + w_{h,t}' e_j$, $w_{h,t}' \in R$

Then

$$0 = \partial'(z) = \sum_h w_{h,t}' \otimes h e_j^{t+1} + (\text{terms of lower } e_j\text{-degree}).$$

It follows that all $w_{h,t}' = 0$ and hence $z_{h,t} \in Z(x_1,\ldots,x_{j-1})$.

On the other hand, we have

$$\partial(z_{h,t} \wedge e_j \otimes h e_j^{t-1}) = x_j z_{h,t} \otimes h e_j^{t-1}.$$

Since $\underline{x}$ is proper, there exists $b_{h,t} \in K(x_1,\ldots,x_{j-1})$ with $\partial(b_{h,t}) = x_j z_{h,t}$. Let

$$b = \sum_h (z_{h,t} \wedge e_j \otimes h e_j^{t-1} - b_{h,t} \otimes h e_j^{t-1})$$

Then $b \in Z$ and $\deg_{e_j}(z - \partial'(b)) < t$. By induction $z - \partial'(b)$ is a boundary and the proof is complete.

As this proof indicates, the notion of a proper sequence is closer to the definition of the approximation complexes than the other kinds of sequences. The next proposition shows related aspects. Let $H_+(x_1,\dots,x_i;M) = \bigoplus_{j>0} H_j(x_1,\dots,x_i;M)$.

Proposition 12.6. Let $\underline{x} = \{x_1,\dots,x_n\}$ be a sequence generating the ideal I. The following conditions are equivalent:

(a) $\underline{x}$ is a proper sequence with respect to M.

(b) $IH_+(x_1,\dots,x_i;M) = 0$ for $i = 1,\dots,n-1$.

(c) The canonical homomorphisms

$H_+(x_1,\dots,x_i;M) \longrightarrow H_+(x_1,\dots,x_{i+1};M)$ are injective for $i = 1,\dots,n-1$.

It is quite surprising that (12.4) and (12.5) have a converse - for R local with infinite residue field. In other words, the M-complex recognizes ideals which are generated by d-sequences, while the Z-complex recognizes ideals which are generated by proper sequences To prove this we first need two technical results.

Lemma 12.7. Let $S = S_o \oplus S_1 \oplus \dots$ be a positively graded S_o-algebra generated by the 1-forms $x_1,\dots,x_n$. Let M be a graded S-module with $(M/S_+M)_i = 0$ for $i \neq 0$. The following conditions are equivalent:

(a) $\{x_1,\dots,x_n\}$ is a d-sequence with respect to M.

(b) $\{x_1,\dots,x_n\}$ is a relative regular sequence with respect to M.

(c) $H_j(x_1,\dots,x_i;M)_k = 0$ for $j > 0$, $k \neq j$ and all $i = 1,\dots n$.

(d) $H_1(x_1,\dots,x_i;M)_k = 0$ for $k \neq 1$ and all $i = 1,\dots,n$.

Proof. (a) => (b) has been remarked before.

(b) => (c) (cf. [Kü, (2.3)]): By Fiorentini's theorem, $x_1,\dots,x_i$ is 1-syzygetic for all i. Hence if z is a homogeneous cycle of $\deg(z) > j$ representing an element of $H_j(x_1,\dots,x_i;M)$, then its coefficients lie in S_+M and thus are all boundaries. This shows that $H_j(x_1,\dots,x_i;M)_k = 0$ for $k > j$. That $H_j(x_1,\dots,x_i;M)_k = 0$ for $k < j$ is trivial.

(d) => (a): Let $a \in (x_1,\dots,x_{i-1})M : x_i \cap S_+M$ be a homogeneous element; then $ax_i = \Sigma\, a_jx_j$, $1 \leq j \leq i-1$, where the a_j are homogeneous elements of the same degree as a. It follows that $z = \sum^{i-1} a_je_j - ae_i$ is a homogeneous cycle of degree > 1. By our assumption z is a boundary of $K(x_1,\dots,x_i;M)$, which proves that $a \in (x_1,\dots,x_{i-1})M$.

<u>Lemma 12.8</u>. Let S and M be as in (12.7). Assume that S_o is local with infinite residue field and assume that

$H_i(S_+;M)_k = 0$ for $j > 0$ and $k \neq j$.

Then S_+ can be generated by a d-sequence with respect to M.

<u>Proof</u>. By (12.7) it suffices to construct a sequence $x_1,\dots,x_n$ of 1-forms generating S_+ such that

(#) $H_j(x_1,\dots,x_i;M)_k = 0$ for $j > 0$, $k \neq j$ and all $i = 1,\dots,n$.

Since the residue field of S_o is infinite, we can choose a generating sequence $x_1,\dots,x_n$ such that $H_j(x_1,\dots,x_i; M)_k = 0$ for $j > 0$, $k >> 0$ and all $i = 1,\dots,n$.

We now show by induction on $\ell = n-i$ that this sequence even satisfies (#). If $\ell = 0$, then this is the assumption of the lemma. Suppose now that $\ell > 0$ and consider the long exact sequence

$$\dots \longrightarrow H_{j+1}(x_1,\dots,x_{i+1};M)_{k+1} \longrightarrow H_j(x_1,\dots,x_i;M)_k \xrightarrow{x_{i+1}}$$
$$H_j(x_1,\dots,x_i;M)_{k+1} \longrightarrow H_j(x_1,\dots,x_{i+1};M)_{k+1} \longrightarrow \dots$$

(here $i = n-\ell$).

By the induction hypothesis we have $H_j(x_1,\dots,x_{i+1};M)_k = 0$ for $k \neq j$. It follows that $H_j(x_1,\dots,x_i;M)_k \cong H_j(x_1,\dots,x_i;M)_{k+1}$ for $k > j$. However, by the choice of the sequence, we have $H_j(x_1,\dots,x_i;M)_k = 0$ for $k >> 0$. Thus we conclude that

$H_j(x_1,\ldots,x_i;M)_k = 0$ for $k > j$. Since, trivially, $H_j(x_1,\ldots,x_i;M)_k = 0$ for $k < j$, the assertion follows.

Theorem 12.9. Let (R,m,k) be a local ring with infinite residue field. Let I be an ideal of R and M a finitely generated R-module.

(1) The following conditions are equivalent:

(a) I is generated by a d-sequence with respect to M.

(b) I is generated by a relative regular sequence with respect to M and is of linear type with respect to M.

(c) I is generated by a proper sequence with respect to M and is of linear type with respect to M.

(d) $\mathcal{M}(I;M)$ is acyclic.

(2) The following conditions are equivalent:

(a) I is generated by a proper sequence with respect to M.

(b) $\mathcal{Z}(I;M)$ is acyclic.

Proof. The implications (1) (a) => (b) => (c) => (d) and (2) (a) => (b) follow from our previous considerations. It remains to show (1) (d) => (a) and (2) (b) => (a).

The acyclicity of $\mathcal{M}(I;M)$, resp. $\mathcal{Z}(I;M)$, may be used to compute $H(S_+;gr_I(M))$ and $H(S_+;\mathrm{Sym}(I;M))$. (By $\mathrm{Sym}(I;M)$ we mean $H_o(\mathcal{Z}(I;M))$.) From (11.7) we know that

$$H_i(S_+;gr_I(M))_j = \begin{cases} 0 & \text{for } j \neq i \\ H_i(I;M) & \text{for } j = i. \end{cases}$$

Similarly we have, for the $\mathcal{Z}$-complex,

$$H_i(S_+;\mathrm{Sym}(I;M))_j = \begin{cases} 0 & \text{for } j \neq i \\ Z_i(I;M) & \text{for } j = i. \end{cases}$$

Using (12.8) we see that the irrelevant ideal of $gr_I(R)$ can be generated by a d-sequence with respect to $gr_I(M)$. The same holds true for the irrelevant ideal of Sym(I) with respect to Sym(I;M). The proof of the theorem is completed with the next result.

Theorem 12.10. Let $x_1,\ldots,x_n$ be a sequence generating the ideal I and let $\bar{x}_1,\ldots,\bar{x}_n$, resp. $x_1^*,\ldots,x_n^*$, be the corresponding sequence generating the irrelevant ideal of $gr_I(R)$, resp. Sym(I).

(1) The following conditions are equivalent:

(a) $\{x_1,\ldots,x_n\}$ is a d-sequence with respect to M.

(b) $\{\bar{x}_1,\ldots,\bar{x}_n\}$ is a d-sequence with respect to $gr_I(M)$.

(2) ([Kü,(2.15)]) The following conditions are equivalent:

(a) $\{x_1,\ldots,x_n\}$ is a proper sequence with respect to M.

(b) $\{x_1^*,\ldots,x_n^*\}$ is a d-sequence with respect to Sym(I;M).

Proof. (1) (a) => (b) (cf. [Hu$_3$]): By (12.7) it suffices to show that $H_1(\bar{x}_1,\ldots,\bar{x}_i;gr_I(M))_j = 0$ for $i = 1,\ldots,n$ and $j \neq 1$. Let $x = \Sigma\bar{a}_k e_k$ be a homogeneous cycle of degree $j > 1$, i.e. $\bar{a}_k \in I^{j-1}M/I^jM$. Let $a_k \in I^{j-1}M$ be representatives of the $\bar{a}_k$. We then have $a_i \in (x_1,\ldots,x_{i-1})I^{j-1}M : x_i \cap I^{j-1}M + I^jM$. Since $\underline{x}$ is a d-sequence we conclude from (12.2) that $a_i \in (x_1,\ldots,x_{i-1})I^{j-2}M + I^jM$. In other words, $\bar{a}_i \in (\bar{x}_1,\ldots,\bar{x}_{i-1})gr_I(M)_{j-2}$. Thus the cycle z is homologous to a cycle $\Sigma\, \bar{a}'_k e_k$, $1 \leq k \leq i-1$. By induction on i the assertion follows.

(b) => (a): We show that $\{x_1,\ldots,x_n\}$ satisfies

$$((x_1,\ldots,x_i)IM : x_{i+1}) \cap I^kM = (x_1,\ldots,x_i)I^{k-1}M \quad \text{for} \quad i = 1,\ldots,n-1$$

and all $k \geq 1$. By (12.1c), it will follow that $\underline{x}$ is a d-sequence.

Since $H_1(x_1,\ldots,x_i;gr_I(M))_k = 0$ for $k > 1$, the natural map $H_o(F_{k+1}K(x_1,\ldots,x_i;M)) \longrightarrow H_o(F_kK(x_1,\ldots,x_i;M))$ is injective. This says that

$$(x_1,\dots,x_i)I^{k-1}M \cap I^{k+1}M = (x_1,\dots,x_i)I^kM$$

for $i = 1,\dots,n-1$ and $k > 1$. By induction on k one derives from this equality

$$(\#)\quad (x_1,\dots,x_i)IM \cap I^{k+1}M = (x_1,\dots,x_i)I^kM \quad \text{for all } k \geq 0.$$

Now let $a \in (x_1,\dots,x_i)IM : x_{i+1} \cap I^kM$; then by (#)

$ax_{i+1} \in (x_1,\dots,x_i)IM \cap I^{k+1}M \subset (x_1,\dots,x_i)I^kM$, and thus

$ax_{i+1} = \sum^i b_jx_j$, $b_j \in I^kM$. Let $\bar{a} = a + I^{k+1}M$, $\bar{b}_j = b_j + I^{k+1}M$;

then $\overline{ax}_{i+1} = \sum^i \bar{b}_j\bar{x}_j$ and $\sum^i \bar{b}_je_j - \bar{a}e_{i+1}$ is a homogeneous cycle of degree > 0. By (12.7) it is therefore a boundary in $K(\bar{x}_1,\dots,\bar{x}_i;gr_I(M))$, which implies that $a \in (x_1,\dots,x_i)I^{k-1}M + I^{k+1}M$. Write $a = a' + a''$, $a' \in (x_1,\dots,x_i)I^{k-1}M$, $a'' \in I^{k+1}M$. We have $a''x_{i+1} \in (x_1,\dots,x_i)IM \cap I^{k+2}M$, hence as before $a'' \in (x_1,\dots,x_i)I^kM + I^{k+2}M$. It follows that $a \in (x_1,\dots,x_i)I^{k-1}M + I^{k+2}M$.

Repeating this argument we see that

$$a \in \bigcap_{r\geq 1} ((x_1,\dots,x_i)I^{k-1}M + I^{k+r}M) = (x_1,\dots,x_i)I^{k-1}M,$$

as required.

(2) (a) => (b): From the presentation

$$Z_1(I;M) \longrightarrow Z_o(I;M) \longrightarrow \mathrm{Sym}(I;M) \longrightarrow 0$$

it follows immediately that $H_1(\underline{x}^*;\mathrm{Sym}(I;M))_j = 0$ for $j > 1$. Suppose we knew that the natural mappings

$$\gamma_i\colon H_1(x_1^*,\dots,x_i^*;\mathrm{Sym}(I;M)) \longrightarrow H_1(\underline{x}^*;\mathrm{Sym}(I;M))$$

are all injective for $i = 1,\dots,n-1$. We would then have $H_1(x_1^*,\dots,x_i^*;\mathrm{Sym}(I;M))_j = 0$ for $i = 1,\dots,n-1$ and $j > 1$, which by (12.7) implies that $\underline{x}^* = \{x_1^*,\dots,x_n^*\}$ is a d-sequence with respect to $\mathrm{Sym}(I;M)$.

The injectivity of the maps γ_i follows once we have shown that $e_{i+1}H_1(e_1,\ldots,e_i;Sym(I;M)) = 0$ for $i = 1,\ldots,n-1$, or, equivalently, that

$$e_{i+1}(U \cap (e_1,\ldots,e_i)Z_o(I;M))_j \subset ((e_1,\ldots,e_i)U)_{j+1}$$

for $i = 1,\ldots,n-1$ and all $j \geq 1$. Here U denotes the image of $\partial':Z_1(I;M) \longrightarrow Z_o(I;M)$.

We prove the above inclusion by induction on j. For $j = 1$ it amounts to showing that

$$e_{i+1}Z_1(x_1,\ldots,x_i;M) \subset (e_1,\ldots,e_i)U,$$

since $(U \cap (e_1,\ldots,e_i)Z_o(I;M))_1 = Z_1(x_1,\ldots,x_i;M)$. (Here $Z_1(x_1,\ldots,x_i;M)$ is understood to be a submodule of $Z_o(I;M)$.)

To prove this inclusion, let $z \in Z_1(x_1,\ldots,x_i;M)$; then $x_{i+1}z = \partial u$, $u \in K_2(x_1,\ldots,x_i;M)$, since we assume that $\underline{x}$ is a proper sequence with respect to M. In the Z-complex we have $\partial(e_{i+1} \otimes z) = x_{i+1}z = \partial u = \partial(\partial'(u\otimes 1))$, and hence $\partial(e_{i+1}\otimes z - \partial'(u\otimes 1)) = 0$. This shows that

$$e_{i+1}\otimes z - \partial'(u\otimes 1) \in Z_1(I;M) \otimes R[e_1,\ldots,e_i]$$

and it follows that

$$e_{i+1}z = \partial'(e_{i+1}\otimes z - \partial'(u\otimes 1)) \in (e_1,\ldots,e_i)U,$$

as required.

Now let $j > 1$ and $z \in (U \cap (e_1,\ldots,e_i)Z_o(I;M))_j$. We can write $z = \sum^{n} e_k z'_k$, with $z'_k \in U_{j-1}$, since U is generated by elements of degree 1. It follows that $e_n z'_n \in (e_1,\ldots,e_{n-1})Z_o(I;M)$. Since $e_1,\ldots,e_n$ is a regular sequence on $Z_o(I;M) = M \otimes R[e_1,\ldots,e_n]$, we then see that $z'_n \in (e_1,\ldots,e_{n-1})Z_o(I;M)$.

We thus have $z'_n \in (U \cap (e_1,\ldots,e_{n-1})Z_o(I;M))_{j-1}$ and by the induction hypothesis we know that $e_n z'_n \in ((e_1,\ldots,e_{n-1})U)_j$. Therefore we can write $z = \sum^{n-1} e_k z''_k$, $z''_k \in U_{j-1}$. By induction we obtain a presentation

$z = \sum^{i} e_k z_k$ with $z_k \in U_{j-1}$, so that $e_{i+1}z \in ((e_1,\ldots,e_i)U)_{j+1}$.

(b) => (a): We have the following identifications:

$K_j(x_1,\ldots,x_i;M) = K_j(x_1^*,\ldots,x_i^*;\mathrm{Sym}(I;M))_j$ and

$Z_j(x_1,\ldots,x_i;M) = H_j(x_1^*,\ldots,x_i^*;\mathrm{Sym}(I;M))_j$.

Using them we have that for $z \in Z_j(x_1,\ldots,x_i;M)$, $x_{i+1}z = x_{i+1}^*z \in B_j(x_1^*,\ldots,x_i^*;\mathrm{Sym}(I;M))_{j+1}$, since $\underline{x}^*$ is assumed to be a d-sequence. Hence there exists $u \in K_{j+1}(x_1^*,\ldots,x_i^*;\mathrm{Sym}(I;M))_{j+1} = K_{j+1}(x_1,\ldots,x_i;M)$ with $x_{i+1}z = x_{i+1}^*z = \partial u$. This proves that $\underline{x}$ is a proper sequence.

In the proof of (12.10), (2) (a) => (b) we used only that portion of the definition of a proper sequence which refer to $j = 1$ (cf. §6). Thus we obtain

Corollary 12.11. ([Kü, (2.16)]) The following conditions are equivalent:

(a) $\{x_1,\ldots,x_n\}$ is a proper sequence with respect to M.

(b) $x_{i+1}H_1(x_1,\ldots,x_i;M) = 0$ for $i = 1,\ldots,n-1$.

§13. Linear resolutions.

In this section we are mainly concerned with the study of the M-complex associated with the maximal ideal of a local ring. Assume for now that (R,m,k) is a regular local ring and that M is a finitely generated R-module. We are going to study the complex

$$M(m;M)\colon 0 \longrightarrow H_n(m;M)\otimes S[-n] \longrightarrow \ldots \longrightarrow H_1(m;M)\otimes S[-1] \longrightarrow H_0(m;M)\otimes S \longrightarrow 0$$

contructed with respect to a minimal system of generators $\{x_1,\ldots,x_n\}$ of m.

Recall that $S = R[e_1,\ldots,e_n]$ and that $M(m;M)_i = H_i(m;M) \otimes_R R[e_1,\ldots,e_n] \cong k[e_1,\ldots,e_n]^{b_i}$, where $b_i = \dim_k H_i(m;M)$.

Thus $M(m;M)$ may be considered to be a complex of free

$gr_m(R)$-modules, since $gr_m(R) \cong k[e_1,\ldots,e_n]$, where the isomorphism is given by $x_i + m^2 \longrightarrow e_i$, for $i = 1,\ldots,n$.

To M we may assign a different complex of free $gr_m(R)$-modules: Let

$$(G,d): \quad \ldots \longrightarrow G_2 \xrightarrow{d_2} G_1 \xrightarrow{d_1} G_o \longrightarrow 0$$

be a minimal free R-resolution of M. We define a natural filtration on (G,d) by setting

$$F_jG_i = \begin{cases} G_i & \text{for } j < i \\ m^{j-i}G_i & \text{for } j \geq i. \end{cases}$$

The associated graded complex

$$gr(G): \ \ldots \longrightarrow gr(G_2) \xrightarrow{gr(d_2)} gr(G_1) \xrightarrow{gr(d_1)} gr(G_o) \longrightarrow 0$$

is then a complex of free $gr_m(R)$-modules.

Theorem 13.1. $M(m;M) \cong gr(G)$.

Proof. We proceed by induction on pd(M), the projective dimension of M. If M is free, the assertion is clear. Suppose now $pd(M) > 0$, and consider the exact sequence

$$0 \longrightarrow N \longrightarrow G_o \longrightarrow M \longrightarrow 0, \quad N = d_1(G_1).$$

Since G_o is free and $N \subset mG_o$, we get isomorphisms

$$H_{i+1}(m;M) \underset{\sigma}{\cong} H_i(m;N), \quad \text{for all } i \geq 0.$$

Here σ is the connecting homomorphism in the long exact homology sequence.

One checks readily that for all $i \geq 0$ the following diagram commutes:

$$\begin{array}{ccc} H_{i+2}(m;M) \otimes S & \xrightarrow{\partial'} & H_{i+1}(m;M) \otimes S \\ \sigma\otimes id \downarrow & & \downarrow \sigma\otimes id \\ H_{i+1}(m;N) \otimes S & \xrightarrow{\partial'} & H_i(m;N) \otimes S \end{array}$$

Using this fact and the induction hypothesis, all that remains is to prove that

$$gr(G_1) \xrightarrow{gr(d_1)} gr(G_o)$$

is isomorphic to

$$H_1(m;M) \otimes S \xrightarrow{\partial'} H_o(m;M) \otimes S.$$

To see this we chose bases (f_k), $k = 1,\ldots,b_1$ of G_1 and (g_ℓ), $\ell = 1,\ldots,b_o$ of G_o. We then have

$d_1(f_k) = \Sigma\alpha_{k\ell}g_\ell$, where each $\alpha_{k\ell}$ can be written as $\alpha_{k\ell} = \Sigma\,\alpha^i_{k\ell}x_i$. If we now put $a_{ki} = \Sigma\,\alpha^i_{k\ell}g_\ell$, then $d_1(f_k) = \Sigma\,a_{ki}x_i$.

Let f^*_k, resp. g^*, denote the initial forms of f_k, resp. g_ℓ, in $gr(G)$ and denote by $\tilde{r}$ the residue class modulo m of an element $r \in R$.

We then obtain

$$gr(d_1)(f^*_k) = \sum_\ell\Big(\sum_i \tilde{\alpha}^i_{k\ell}e_i\Big)g^* \quad (\text{recall that } e_i = x_i + m^2).$$

In order to describe ∂' we observe that the elements $[\sum_i \rho(a_{ki})e_i]$, $k = 1,\ldots,b_1$ form a basis of $H_1(m;M)$ and the elements $[\rho(g_\ell)]$, $\ell = 1,\ldots,b_o$ form a basis of $H_o(m;M)$. (Here the brackets [ ] denote the homology class of a cycle and $\rho\colon G_o \longrightarrow M$ is the augmentation map.)

We now compute ∂' explicitly:

$$\partial'[\sum_i \rho(a_{ki})e_i] = \sum_i[\rho(a_{ki})] \otimes e_i = \sum_i\sum_\ell \tilde{\alpha}^i_{k\ell}[\rho(g_\ell)] \otimes e_i =$$

$$\sum_\ell([\rho(g_\ell)] \otimes \sum_i \tilde{\alpha}^i_{k\ell}e_i.$$

The assertion follows.

Definition. The R-module M has a linear resolution if the associated complex $gr(G)$ is acyclic.

Corollary 13.2. (a) If the R-module M has a linear resolution, so does mM.

(b) For an R-module, there exists an integer n_o so that $m^n M$ has a linear resolution for $n \geq n_o$.

Proof. Both assertions follow from the remark following (11.4), (12.8) and the theorem above.

Combining (13.1) and (12.9) we get

Corollary 13.3. Let (R,m) be a regular local ring with infinite residue field, and let M be a finitely generated R-module. The following conditions are equivalent:

(a) The maximal ideal m is generated by a d-sequence with respect to M.

(b) M has a linear resolution.

If the equivalent conditions hold, then

$$M(m;M):\ 0 \longrightarrow H_n(m;M) \otimes S[-n] \longrightarrow \cdots \longrightarrow H_o(m;M) \otimes S \longrightarrow 0$$

is a minimal free homogeneous $gr_m(R)$-resolution of $gr_m(M)$, and the Hilbert function $H_M(z)$ of M is given by the formula

$$H_M(z) = P(z)/(1-z)^n,\ P(z) = \Sigma(-1)^{i+1} b_i z^i,\ b_i = \dim_k H_i(m;M).$$

Moreover, depth M = depth $gr_m(M)$.

Now let (R,m,k) be any local ring (not necessarily regular). We are going to give a description of $M(m;R)$. The homology of $M(m;R)$ does not change if we pass to the m-adic completion of R. Hence we may assume that R is complete and thus choose a presentation $R = A/I$, where (A,n,k) is a regular local ring and $I \subset n^2$.

Let G be a minimal free resolution of R as an A-module.

Corollary 13.4. $H_i(M(m;R)) \cong H_i(gr(G))$.

Examples 13.5. (a) Let $I \subset k[x_{ij}]$ be the ideal generated by the maximal minors of an m x n matrix of indeterminates. Let $A = k[[x_{ij}]]$, $R = A/IA$. Since the resolution of this ideal is known to be linear ([EN]), we see that

$H_i(M(m;R)) = 0$ for $i > 1$, and

$H_1(M(m;R)) = I$.

(b) Let $I \subset k[x_{ij}]$ be the ideal generated by the $2n \times 2n$ Pfaffians of a skew-symmetric $(2n+1) \times (2n+1)$ matrix in the indeterminates x_{ij}. Again let $A = k[[x_{ij}]]$, $R = A/IA$. Using the structure theorem of [BE], we see that

$H_1(M(m;R)) = I$,

$H_2(M(m;R)) = 0$, and

$H_3(M(m;R)) = gr_m(R)$.

This example shows that, the M-complex is not rigid, in contrast to the ordinary Koszul complex.

Quite generally we have that $M_1(m;R) \longrightarrow M_0(m;R)$ is the zero mapping. Hence the following conditions are equivalent:

(a) $H_1(M(m;R)) = 0$.

(b) $M(m;R)$ is acyclic.

(c) R is a regular local ring.

Thus for a non-regular local ring (R,m), we cannot have much better exactness of $M(m;R)$ than $H_i(M(m;R)) = 0$ for $i > 1$.

Bringing in (12.7), we have

Theorem 13.6. Let (R,m) be a local ring with infinite residue field. The following conditions are equivalent:

(a) m is generated by a proper sequence.

(b) $H_i(M(m;R)) = 0$ for $i > 1$, and the canonical homomorphism $H_1(M(m;R)) \longrightarrow \mathrm{Sym}(m)$ is injective.

Suppose again that R has a presentation $R = A/I$, with (A, n, k) regular, $I \subset n^2$, and k infinite. In general $gr_n(I) \neq \operatorname{Ker}(gr_n(A) \longrightarrow gr_m(R))$. Making an extra assumption on $gr_n(I)$ we get

Theorem 13.7. Suppose that there exists an integer $d \geq 1$ such that $gr_n(I)[-d] = \operatorname{Ker}(gr_n(A) \longrightarrow gr_m(R))$. Then the following conditions are equivalent:

(a) The maximal ideal m of R is generated by a proper sequence.

(b) I has a linear resolution.

If the equivalent conditions hold, then

$$(\#) \quad 0 \longrightarrow H_n(m;R) \otimes S[-n-d+1] \xrightarrow{\partial'} \cdots \xrightarrow{\partial'} H_1(m;R) \otimes S[-d] \longrightarrow H_o(m;R) \otimes S \longrightarrow 0.$$

is a minimal free homogeneous $gr_n(A)$-resolution of $gr_m(R)$, and depth R = depth $gr_m(R)$.

Remark. What the hypothesis on $gr_n(I)$ means is that

$$I \cap n^\nu = \begin{cases} I & \text{for } \nu < d \\ n^{\nu-d} I & \text{for } \nu \geq d. \end{cases}$$

In other words, I is generated minimally by a standard basis $f_1, \ldots, f_m$ of degree $\nu(f_i) = d$ for all i.

Proof of (13.7). (a) => (b): Let G be a minimal free A-resolution of R. By (13.4) we have $H_i(M(m;R)) \cong H_i(gr(G))$, and by (13.6) we know that $H_i(M(m;R)) = 0$ for $i > 1$. This shows that

$$\cdots \longrightarrow gr(G_2) \longrightarrow gr(G_1) \longrightarrow 0$$

is acyclic, that is, I has a linear resolution.

(b) => (a): If I has a linear resolution then arguing as above we get that $H_i(M(m;R)) = 0$ for $i > 1$ and $H_1(M(m;R)) = gr_n(I)$. Using the isomorphism

$gr_n(I)[-d] = \ker(gr_n(A) \longrightarrow gr_m(R))$ we get that the complex (#) is a homogeneous $gr_n(A)$-resolution of $gr_m(R)$.

As an application of these methods we give a generalization of a result of Steurich ([St]), which itself represented an extension of results of [Sa] and [Sc].

Theorem 13.8. Let (A,n) be a regular local ring with infinite residue field, and let $R = A/I$ and denote by m its maximal ideal. Assume that for some $d \geq 2$ we have:

(i) $I \subset n^d$.

(ii) There exists a system of parameters $\underline{x}$ of R which forms a relative regular sequence, such that $m^d = (\underline{x})m^{d-1}$.

Then I has a linear resolution.

Proof. $\underline{x}$ forms part of a minimal system of generators of m and can be completed to one such: $\{x_1,\ldots,x_n,y_1,\ldots,y_m\}$.

We claim that this sequence is proper. According to the result of Kühl (12.11), it suffices to show that

(a) $mH_1(x_1,\ldots,x_i;R) = 0$ for $i = 1,\ldots,n$;

(b) $mH_1(x_1,\ldots,x_n,y_1,\ldots,y_j;R) = 0$ for $j = 1,\ldots,m$.

(a) follows from (b), since we assume that $\underline{x}$ is a relative regular sequence and, by (b), $mH_1(x_1,\ldots,x_n;R) = 0$.

To prove (b), choose $z \in Z_1(\underline{x},\underline{y};R)$, $z = \sum_i a_ie_i + \sum_j b_jf_j$, $\partial e_i = x_i$, $\partial f_j = y_j$. Since $I \subset n^d$ we conclude that all $a_i, b_j \in m^{d-1}$. For $x \in m$, $xb_j \in m^d = (\underline{x})m^{d-1}$ and thus $xb_j = \sum b_{ji}x_i$ with $b_{ji} \in m^{d-1}$. It follows that $z - \partial(\sum_{i,j} b_{ji}e_i\wedge f_j) = \sum a_i'e_i$ where all $a_i' \in m^d = (\underline{x})m^{d-1}$. Since $\underline{x}$ is a relative regular sequence it follows that xz is a boundary.

Next we show that I is generated minimally by a standard basis $g_1,\ldots,g_r$, with $\upsilon(g_i) = \text{degree}(g_i) = d$.

Let $K = K(m;R)$, F the m-adic filtration on K. Our proof thus far shows that $H_1(F_iK) = H_1(K)$ for $i \leq d$ and $H_1(F_{d+1}K) = 0$. This shows that $H_1(K(m^*;gr_m(R))_i = 0$ for $i \neq d$, and $H_1(m;R) \cong H_1(F_dK) \cong H_1(m^*;gr_m(R))_i \cong H_1(m^*;gr_m(R))$, where m^* is the irrelevant maximal ideal of $gr_m(R)$.

This shows that $I^* = \ker(gr_n(A) \longrightarrow gr_m(R))$ is generated by forms of degree d. In other words, I has a standard basis $g_1,\ldots,g_r$, with degree $(g_i) = d$, generating I minimally since $v(I) = \dim H_1(m;R) = \dim H_1(m^*;gr_m(R)) = v(I^*)$.

References

[AB] M. Auslander and D. Buchsbaum, Codimension and Multiplicity, Annals of Math. 68 (1958), 625-657; Corrections, 70 (1959), 395-397.

[AH] L. Avramov and J. Herzog, The Koszul algebra of a codimension 2 embedding, Math. Zeit. 175 (1980), 249-260.

[Ao] Y. Aoyama, A remark on almost complete intersections, Manuscripta Math. 22 (1977), 225-228.

[Ba] H. Bass, Injective dimension in Noetherian rings, Trans. Amer. Math. Soc. 102 (1961), 18-29.

[BE_1] D. Buchsbaum and D. Eisenbud, What makes a complex exact?, J. Algebra 25 (1973), 259-268.

[BE_2] D. Buchsbaum and D. Eisenbud, Algebra structures for finite free resolutions, and some structure theorems for ideals of codimension 3, Amer. J. Math. 99 (1977), 447-485.

[Br] M. Brodmann, A Macaulayfication of unmixed domains, J. Algebra 44 (1977), 221-234.

[EH] D. Eisenbud and C. Huneke, Cohen-Macaulay Rees algebras and their specializations, preprint, 1980.

[EN] J. Eagon and D.G. Northcott, Ideals defined by matrices and a certain complex associated with them, Proc. Royal Soc. 269 (1962), 188-204.

[Fa] G. Faltings, Über Macaulayfizierung, Math. Ann. 238 (1978), 175-192.

[Fi] M. Fiorentini, On relative regular sequences, J. Algebra 18 (1971), 384-389.

[GL] T. Gulliksen and G. Levin, "Homology of Local Rings", Queen's Paper in Pure and Applied Math. 20, Queen's University, Kingston, 1969.

[He_1] J. Herzog, Ein Cohen-Macaulay Kriterium mit Anwendungen auf den Konormalenmodul und den Differentialmodul, Math. Zeit. 163 (1978), 149-162.

[He_2] J. Herzog, Komplexe, Auflösungen und Dualität in der lokalen Algebra, Habilitationsschrift, Universität Regensburg, 1974.

[HK] J. Herzog and E. Kunz, "Der kanonische Modul eines Cohen-Macaulay Rings", Lecture Notes in Math. 238, Springer, Heidelberg, 1971.

[Ho] M. Hochster, Criteria for equality of ordinary and symbolic powers of an ideal, Math. Zeit. 133 (1973), 53-65.

[HR] M. Hochster and L.J. Ratliff, Jr., Five theorems on Macaulay rings, Pacific J. Math. 44 (1973), 147-172.

[HSV_1] J. Herzog, A. Simis and W.V. Vasconcelos, Approximation complexes and blowing-up rings.I, J. Algebra, to appear.

[HSV_2] J. Herzog, A. Simis and W.V. Vasconcelos, Approximation complexes and blowing-up rings.II, preprint, 1981.

[Hu_1] C. Huneke, The theory of d-sequences and powers of ideals, Adv. in Math., to appear.

[Hu_2] C. Huneke, On the symmetric and Rees algebras of an ideal generated by a d-sequence, J. Algebra 62 (1980), 268-275.

[Hu_3] C. Huneke, Symbolic powers of prime ideals and special graded algebras, Comm. in Algebra 9 (1981), 339-366.

[Hu_4] C. Huneke, The Koszul homology of an ideal, preprint, 1981.

[Hu_5] C. Huneke, Linkage and the Koszul homology of ideals, preprint, 1981.

[Kap] I. Kaplansky, "Commutative Rings", The University of Chicago Press, Chicago, 1974.

[Kun] E. Kunz, The conormal module of an almost complete intersection, Proc. Amer. Math. Soc. 73 (1979), 15-21.

[Kü] M. Kühl, Über die Symmetrische Algebra eines Ideals, Dissertation, Universität Essen, 1981.

[Le] K. Lebelt, Freie Auflösungen äußerer Potenzen, Manuscripta Math. 21 (1977), 341-355.

[Mat] H. Matsumura, "Commutative Algebra", Benjamin-Cummings, New York, 1980.

[Mats] T. Matsuoka, On almost complete intersections, Manuscripta Math. 21 (1977), 329-340.

[Mi] A. Micali, Sur les algèbres universelles, Ann. Inst. Fourier 14 (1964), 33-88.

[MR] A. Micali and N. Roby, Algèbres symétriques et syzygies, J. Algebra 17 (1971), 460-469.

[MSS] A. Micali, P. Salmon and P. Samuel, Integrité et factorialité des algèbres symétriques, Atas IV Colóquio Brasileiro de Matemática, 61-67, São Paulo, 1965.

[PS_1] C. Peskine and L. Szpiro, Dimension projective finie et cohomologie locale, IHES Publ. Math. 42 (1973), 323-395.

[PS_2] C. Peskine and L. Szpiro, Liaison des varietés algébriques, Inventiones Math. 26 (1973), 271-302.

[Ra] L.J. Ratliff, Jr., On the prime divisors of zero in form rings, Pacific J. Math. 70 (1977), 488-517.

[Ro] M.E. Rossi, On symmetric algebras which are Cohen-Macaulay, Manuscripta Math. 34 (1981), 199-210.

[Sa] J. Sally, Cohen-Macaulay rings of maximal embedding dimension, J. Algebra 56 (1979), 168-183.

[Sal] P. Salmon, Sulle algebre graduate relative ad un ideale, Symposia Mathematica 8 (1972), 269-293.

[Sch] P. Schenzel, Uber die freien Auflösungen extremaler Cohen-Macaulay-Ringe, J. Algebra 64 (1980), 93-101.

[Ser] J.P. Serre, "Algèbre Locale - Multiplicités", Lecture Notes in Math. 11, Springer, Heidelberg, 1965.

[Si] A. Simis, Koszul homology and its syzygy-theoretic part, J. Algebra 55 (1978), 28-42.

[SV_1] A. Simis and W.V. Vasconcelos, The syzygies of the conormal module, Amer. J. Math. 103 (1981), 203-224.

[SV_2] A. Simis and W.V. Vasconcelos, On the dimension and integrality of symmetric algebras, Math. Zeit. 177 (1981), 341-358.

[St] M. Steurich, On rings with linear resolutions, J. Algebra, to appear.

[Va] G. Valla, On the symmetric and Rees algebras of an ideal, Manuscripta Math. 30 (1980), 239-255.

[Vas_1] W.V. Vasconcelos, On the homology of I/I^2, Comm. in Algebra 6 (1978), 1801-1809.

[Vas_2] W.V. Vasconcelos, The conormal bundle of an ideal, Atas V Escola de Algebra, 111-166, IMPA, Rio de Janeiro, 1979.

[Wat] J. Watanabe, A note on Gorenstein rings of embedding codimension 3, Nagoya Math. J. 50 (1973), 227-232.

[Wey] J. Weyman, Resolutions of the exterior and symmetric powers of a module, J. Algebra 58 (1979), 333-341.

SOME REMARKS ON NORMAL FLATNESS AND MULTIPLICITY IN COMPLEX SPACES

Monica Idà - Mirella Manaresi

University of Bologna
Bologna (Italy)

The purpose of this note is to show how, using the properties of the ring of holomorphic functions on a compact, semianalytic, Stein subset D of a complex space (studied by Frisch [4], Kiehl [14], Greco - Traverso [7], Idà - Manaresi [13]) one can transpose results for excellent schemes to results for complex spaces, and, particularly, one can give elementary proofs of some facts on normal flatness and multiplicity of complex spaces, which are stated many times in literature, but not satisfactorily proved.

In particular, in section 1 one studies the morphism of ringed spaces $(U, \mathcal{O}_{X|U}) \to (\mathrm{Spec}\ \mathcal{O}_{X,D}, \mathcal{O}_{X,D})$, where U is an open subset of a complex space X and is contained in the compact, semianalytic, Stein subset D of X (see prop. (1.1)). Its properties are used to show that:

(A) for a large class of properties P of local rings, if the P-locus of an excellent ring is open, then the P-locus of a complex space is open too (see cor. (1.2));

(B) the locus of normal flatness of a complex space X along a closed reduced analytic subset Y is a Zariski open subset of Y (see th. (1.5)).

This research was done when the authors were members of G.N.S.A.G.A. of the C.N.R.

Part (B) is stated in [10], partially proved in [12] and completely proved in [15] and [16] by methods different from the ones suggested in [10]. We give an elementary proof of (B) by using the method outlined by Hironaka in [10], where he refers to a work which was never published.

In section 2 we use (A) to give a simple and purely algebraic proof (see prop. (2.3)) of the fact that the function "multiplicity" is constructible on a complex space; note that, by multiplicity of a complex space X in one of its points x we mean the multiplicity of its local ring $\mathcal{O}_{X,x}$. Using this fact it is possible to define the multiplicity e(Y) of a reduced irreducible analytic subset Y of a complex space X, and, using the results of section 1, one proves that e(Y) is equal to $e((\mathcal{O}_{X,x})_{\mathfrak{p}})$, where x is an arbitrary point of Y and $\mathfrak{p}$ is an arbitrary minimal prime of the ideal of Y_x in $\mathcal{O}_{X,x}$ (see th. (2.6)).
The analogue local result (see prop. (2.5)) is stated in [17], prop. (4.1), but the proof outlined there by Lipman uses different methods.

In the appendix we give an elementary proof of the fact that the algebraic definition of multiplicity used here coincides with the geometric one (see Whitney [22], 7J, 7P) which is more often used in the literature about complex spaces. We use techniques similar to the ones used by Draper in [2], where he proves that a further geometric definition of multiplicity and the algebraic one coincide.

NOTATIONS AND PRELIMINARIES.

1) By complex space we always mean a reduced complex analytic space.
2) If R is a local ring, e(R) will denote the multiplicity of R.
   If X is a complex space, and $x \in X$, we set $e(x) := e(\mathcal{O}_{X,x})$.
3) If X is a complex space, by "Zariski open subset of X" we mean the complementary of a closed analytic subset of X.
4) Let D be a compact semianalytic Stein subset of a complex space X; let $\mathcal{O}_{X,D}$ denote the ring of D-germs of holomorphic functions. For a summary of the properties of analytic D-germs and of the ring $\mathcal{O}_{X,D}$ we send to [13]. We just recall that:
   i) $\mathcal{O}_{X,D}$ is excellent (see [7], (6.7));
   ii) the canonical map: $D \to \operatorname{Specm}(\mathcal{O}_{X,D})$, $x \mapsto \mathfrak{m}_x := \{f \in \mathcal{O}_{X,D} \mid f(x)=0\}$ is a bijection (see [14], lemma 2, and [7], (6.2)).

   If Z is a D-germ, by $\mathcal{I}(Z)$ we mean the radical ideal of $\mathcal{O}_{X,D}$

associated to Z.

5) If D is a compact, semianalytic, Stein subset of a complex space X, we shall denote either an analytic D-germ or one of its representatives in a Stein neighbourhood of D by the same symbol; the same for a D-germ of holomorphic function and its representatives.

By 1), the analytic D-germs that we consider are always reduced.

1. MORPHISM OF RINGED SPACES

In this section we determine some properties of the ring of the holomorphic functions on a compact semianalytic Stein subset of a complex space, and we use them to prove that, if P is a property of local rings such that the P-locus of an excellent ring is open, then also the P-locus of a complex space is open. We then use the same techniques to give an elementary algebraic proof of a result about normal flatness stated by Hironaka in [10] and [12].

We recall that, if P is a property of local rings, and A is a ring [resp. X is a complex space], the set

$$P(A) := \{q \in \operatorname{Spec}(A) \mid A_q \text{ has } P\}$$

$$[\text{resp. } P(X) := \{x \in X \mid \mathcal{O}_{X,x} \text{ has } P\}]$$

is called *P-locus* of A [resp. *P-locus* of X].

We also recall that (see [18], p. 249), if A and B are two rings, a morphism $\phi: A \to B$ is said to be *regular* if it is flat and its fibers are geometrically regular.

Remark. We shall only concider properties P of local rings, verifying the following condition:

(*) If (A,m) and (B,n) are two excellent rings, $f: A \to B$ is a regular morphism, and $mB = n$, then A has P if and only if B has P.

PROPOSITION 1.1. Let X be a complex space, D a compact semianalytic Stein subset of X and U an open subset of X contained in D. Set $A := \mathcal{O}_{X,D}$, and let $\phi: U \to \operatorname{Spec}(A)$ be the natural map. Then:

i) ϕ is continuous with respect to the Zariski topology of U (in particular, then, with respect to the usual one);

ii) ϕ induces a morphism of ringed spaces;

iii) if $x \in U$ and m_x is the corresponding maximal ideal of A, the local homomorphism $A_{m_x} \to \mathcal{O}_{X,x}$ induced by ϕ is regular, and $m_x\mathcal{O}_{X,x}$ is the maximal ideal of $\mathcal{O}_{X,x}$;

iv) if P is a property of local rings verifying (*), one has: $P(U) = \phi^{-1}(P(A))$.

Proof. Let us consider the natural map (see [14], lemma 2)

$$\phi: U \hookrightarrow \mathrm{Spec}(A)$$
$$x \mapsto m_x = \{f \in A \mid f(x)=0\}.$$

A closed subset of Spec(A) is an analytic D-germ, and the generators of the associated ideal have some holomorphic functions in a neighbourhood of D as representatives; restricting these functions to U, we get a closed analytic subset of U. Then i) follows.

For each $f \in A$, the map:

$$\tilde{\phi}: \Gamma(D(f),A) \simeq A_f \to \Gamma(D(f),\phi_*(\mathcal{O}_X|U)) \simeq \Gamma(U-\{f=0\},\mathcal{O}_X|U)$$

$$g = \frac{h}{f^n} \mapsto g|_{U-\{f=0\}}$$

is a ring homomorphism, so ii) holds.

Moreover, [13] th. 2 (a) and (b) gives iii).

The rings A and $\mathcal{O}_{X,x}$ are excellent, hence as iii) holds, by assumption A_{m_x} has P if and only if $\mathcal{O}_{X,x}$ has P. So iv) follows.

Remark. We recall that in [1] (1.1) Bingener already proved that the morphism $A_{m_x} \to \mathcal{O}_{X,x}$ is flat and induces an isomorphism between the completions.

COROLLARY 1.2. If P is a property of local rings verifying (*), such that the P-locus of an excellent ring is open, then the P-locus of a complex space is a Zariski open subset.

Proof. For each $x \in X$, let D_x be a compact, semianalytic, Stein neighbourhood of x; set $U_x := \mathrm{Int}\, D_x$, and $A_x := \mathcal{O}_{X,D_x}$.

As A_x is excellent, by assumption $P(A_x)$ is open in $\mathrm{Spec}(A_x)$. By prop. (1.1) iv) one has $P(U_x) = \phi^{-1}(P(A_x))$, and by (1.1) i), $\phi^{-1}(P(A_x))$ is a Zariski open subset of U_x.

Moreover, one has $P(X) = \bigcup_{x \in X} P(U_x)$, then $X - P(X)$ is closed in X and locally analytic.

REMARK 1.3. Corollary (1.2) applies, for example, to the following properties:

1) P = regular (see [18], th. 51 p. 155 and [8], (7.8.3));
2) P = Gorenstein (see [21], th. 2.8 and [6] cor. (1.5));
3) P = Cohen-Macaulay (see [8] (6.3.3) and (7.8.3));
4) P = S_n Serre condition, see [8] (5.7.2) (see [18], cor. 2 p. 154 and [8], (7.8.3));
5) P = R_n Serre condition, see [8] (5.8.2) (see [18], (21.E) p. 156 and [8], (7.8.3));
6) P = normal (see 4) and 5), and [18], th. 39 p. 125).

We recall that (see [10], def. 1 p. 135) if X is a noetherian scheme [resp. a complex space], $Y \subset X$ a closed subscheme [resp. a closed analytic subset] which is defined by an ideal sheaf $J \subset \mathcal{O}_X$, X is said to be *normally flat* along Y at a point $x \in Y$ if $\mathrm{Gr}_{J_x}(\mathcal{O}_{X,x})$ is a free $\mathcal{O}_{Y,x}$-module (that is if each J_x^i/J_x^{i+1} is free). The point $x \in Y$ is said to be *admissible* for Y if x is regular for Y and X is normally flat along Y at x.

LEMMA 1.4. Let X be a complex space, let U be an open subset of X contained in a compact semianalytic Stein subset D of X, and let Z be an analytic D-germ, defined by an ideal $J \subset \mathcal{O}_{X,D} =: A$.

Let $x \in U \cap Z$, let $\mathfrak{m}_x$ be the maximal ideal of A corresponding to x, and let $I \subset \mathcal{O}_{X,x}$ be the ideal associated to the germ of Z at x. Then one has:

i) $\mathrm{Gr}_I(\mathcal{O}_{X,x})$ is $\mathcal{O}_{Z,x}$-flat if and only if $\mathrm{Gr}_{JA_{\mathfrak{m}_x}}(A_{\mathfrak{m}_x})$ is $(A_{\mathfrak{m}_x}/JA_{\mathfrak{m}_x})$-flat;

ii) x is admissible for Z if and only if $\mathfrak{m}_x$ is admissible for Spec(A/J).

Proof. The canonical morphism $A_{\mathfrak{m}_x} \to \mathcal{O}_{X,x}$ is faithfully flat (see [7], (6.2) ii)), and, moreover, by Cartan-Serre theorem A (with an argument similar to [13], th. 2 (a)), one has $(JA_{\mathfrak{m}_x})\mathcal{O}_{X,x} = I$; this and [10] (see ch. II, lemma 9 p. 198) give i).

As condition (∗) holds for P = regular (see remark (1.3) 1)), ii) follows by i).

THEOREM 1.5 (Hironaka [10] (II) p. 136, [12] cor. (2.8), (3.16)). Let X be a complex space, and Y a closed reduced analytic subset of X. Then the set of points of Y where X is normally flat along Y [resp. admissible for Y] is a Zariski open dense subset of Y.

Proof. For each $x \in Y$, let D_x be a compact, semianalytic, Stein neighbourhood of x, let $U_x := \text{Int}\, D_x$, $A_x := \mathcal{O}_{X,D_x}$. Let J_x be the ideal of A_x associated to the D_x-germ induced by Y. By [10] cor. p. 189 the locus N_x of normal flatness of $\text{Spec}(A_x)$ along $\text{Spec}(A_x/J_x)$ is open and dense. By prop. (1.1) i) applied to $\phi\colon Y \cap U_x \to \text{Spec}(A_x/J_x)$, $\tilde{N}_x := \phi^{-1}(N_x)$ is a Zariski open subset of $U_x \cap Y$, and by (1.4) i) $\tilde{N}_x$ is exactly the locus of normal flatness of X along Y in U_x. Setting $N := \bigcup_{x \in Y} \tilde{N}_x$, N is the locus of normal flatness of X along Y, and $Y - N$ is closed in Y, and locally analytic.

To show that N is dense in Y, it is enought to show that each irreducible component Z of Y is such that $Z \cap N \neq \emptyset$.
Let $x \in Z$, $x \notin \text{Sing}(Y)$. In the previous notations, we have $J_x = I(Z_{D_x})$.
Let $N_x = \text{Spec}(A_x/J_x) - V(\mathfrak{a})$; let us show that at least one closed point $m_y \in \text{Spec}(A_x/J_x)$ is in N_x, by which it follows that $N \cap Z \supset \phi^{-1}(m_y)$.
If for each $m_y \in \text{Specm}(A_x/J_x)$, m_y is not in N_x, then

$$\mathfrak{a} \subset \bigcap_{m_y \in \text{Specm}(A_x/J_x)} m_y = \{f \in A_x/J_x \mid f(y) = 0 \ \forall\, y \in Y \cap D_x\} = \{0\}, \text{ and}$$

then we have $N_x = \emptyset$, which is absurd.

Recalling that the set of admissible points for $\text{Spec}(A_x/J_x)$ is open and dense, as $\text{Spec}(A_x)$ is an excellent scheme (see [8] (7.8.3) and [10] th. 1 and cor. p. 188-189), and using (1.4) ii) instead of (1.4) i), we get the statement for admissible points.

Remark. A complete proof of this theorem has been given by Lejeune and Teissier in [15] (see (4.1.2) and (5.1.1)) and [16] (see (10.3) and (10.9)) using different methods.

2. MULTIPLICITY

In this section we show that, if Y is an analytic subset of a complex space, and $x \in Y$, the multiplicity e(x) is constant on a Zariski open subset of Y, and it is equal to $e((\mathcal{O}_{X,x})_p)$, where p is an arbitrary minimal prime of the ideal of Y in $\mathcal{O}_{X,x}$.

LEMMA 2.1. Let (A,m), (B,n) be two local rings, $f: A \to B$ a flat homomorphism, such that $mB = n$. Then $e(A) = e(B)$.

Proof. As $Gr(A) \otimes_{A/m} B/n \simeq Gr(B)$, A and B have the same Hilbert functions.

LEMMA 2.2. Let R be an excellent ring. Then the sets

$$E_n := \{p \in Spec(R) \mid e(R_p) \leq n\}$$

are open in Spec(R).

Proof. The result follows by Nagata criterion (see [18], (22.B) lemma 2 p. 158), once proved that, if $p \in E_n$, then:

i) $\forall\ p' \in Spec(R)$ such that $p' \subseteq p$, $p' \in E_n$;

ii) E_n contains a non empty open subset of $V(p)$.

i) Let $p \in E_n$. As R is excellent, R (hence R_p) is in particular universally catenarian and a G-ring, therefore by [19], th. (40.1) p. 153, one gets $e(R_{p'}) \leq e(R_p)$.

ii) Let U' be the set of points where Spec(R) is normally flat along $V(p)$; U' is an open dense subset of $V(p)$ (see [10], cor. p. 189). Let $U := U' \cap Reg(Spec(R/p))$.
As R/p is excellent, $Reg(Spec(R/p))$ is open, hence U is a non empty open subset of $V(p)$; moreover, for each $q \in U$, by [10] (see cor. 2, p. 186) one has $e(R_p) = e(R_q)$.

PROPOSITION 2.3. Let X be a complex space. Then the sets

$$\tilde{X}_n := \{x \in X \mid e(x) \leq n\}$$

are Zariski open subsets of X, hence the sets

$$X_n := \{x \in X \mid e(x) = n\}$$

are locally closed (in particular constructible(†)) subsets of X.

Proof. Let us consider the following property P of local rings:

A has P if and only if $e(A) \leq n$.

This property satisfies (✱) (see lemma (2.1)), hence the conclusion fol-fows by (1.2) and (2.2).

(†) For the definition of constructible subset, see [13], def. 3 and remark 4.

COROLLARY 2.4. Let X be a complex space, and let Y be an irreducible closed analytic subset of X. Then there exists a Zariski open dense subset $\Omega \subset Y$ such that the function $x \mapsto e(x)$ is constant on Ω. Denoting this constant value by $e(Y)$, one has $e(x) \geq e(Y) \quad \forall\, x \in Y$.

Proof. The conclusion follows immediatly by (2.3), setting $m := \min_{y \in Y} e(y)$ and $\Omega := \tilde{X}_m \cap Y$.

PROPOSITION 2.5. Let X be a complex space, let D be a compact, semianalytic, Stein subset of X and let $U := \text{Int } D$. Let Z be an irriducible D-germ, and let $x \in Z \cap U$ such that the germ Z_x induced by Z at x is irreducible. Denoting $A := \mathcal{O}_{X,D}$, $q := I(Z) \subset A$, $R := \mathcal{O}_{X,x}$, and $p_x := I(Z_x) \subset R$, we have:

i) $e(A_q) = e(R_{p_x})$;

ii) $e(z) = e(R_{p_x})$ for each $z \in U$, z admissible for Z.

Proof. i) This follows by lemma (2.1) and by [13], th. 2 (b).

ii) Let $T := \text{Spec}(A)$, $T' := \text{Spec}(A/q)$. The locus W' of the points admissible for T' is open in T' (see proof of (1.5)). By [10], cor. 2 p. 186 one gets

(1) $$e(A_{m_z}) = e(A_q) \qquad \forall\; z \in W'.$$

Moreover, by prop. (1.1) iii) and lemma (2.1), we have

(2) $$e(A_{m_z}) = e(\mathcal{O}_{X,z}) = e(z) \qquad \forall\; z \in U.$$

Finally, by prop. (1.1) i) applied to $\phi: U \cap Z \to T'$, $W := \phi^{-1}(W')$ is a Zariski open subset of $Z \cap U$, and by (1.4) ii) W is exactly the locus of the points admissible for $Z \cap U$, therefore by (1.5) W is dense in $Z \cap U$.

By i), (1) and (2) one gets $e(z) = e(R_{p_x}) \quad \forall\, z \in W$.

THEOREM 2.6. Let X be a complex space, let Y be a (closed) irreducible analytic subset of X. Then, for each $x \in Y$, and each minimal prime p_x of $\mathcal{O}_{X,x}$ associated to the germ Y_x (i.e. $p_x = I(Y_x)$ if Y_x is irreducible, or $p_x = I(Y_x^i)$, where Y_x^i is an irreducible component of Y_x), we have $e((\mathcal{O}_{X,x})_{p_x}) = e(Y)$.

Proof. The proof follows by prop. (2.4) and (2.5), with an argument similar to [13], th. 7.

REMARKS 2.7. i) By (2.4), (2.5) and (2.6), we have $\{x \in Y \mid e(x) = e(Y)\} \supset \{x \in Y \mid x \text{ is admissible for } Y\}$, but easy examples show that in general the inclusion is strict.

ii) By (2.4) and (2.6) we have, in particular, [17], prop. (4.1).

iii) Though (A.1), prop. (2.3) gives an elementary proof of [22] 8E p. 237.

APPENDIX.

Let $X \subset \mathbb{C}^n$ be a reduced analytic subset, let $x \in X$ and $d = \dim_x X$.

We recall that if P_N is a sufficiently small coordinate polycylinder near x, the generic projection $\pi_\alpha : P_N \to P_d^\alpha$ induced by $\tilde{\pi}_\alpha : \mathbb{C}^N \to H_\alpha^d$ with linear fibers (where $H_\alpha^d \simeq \mathbb{C}^d$ is a d-dimensional hyperplane and $P_d^\alpha \subset H_\alpha^d$ is a d-dimensional polycylinder) gives a branched covering $p_\alpha := \pi_\alpha|_X : X \to \pi_\alpha(X)$. Denoting by $\text{ord } p_\alpha$ the number of sheets of p_α, it is well known that for almost all α $\text{ord } p_\alpha$ is constant and reaches its minimum. Precisely (see [22], ch. VII, 7J and 7P), setting $L_\alpha := \tilde{\pi}_\alpha^{-1}(\tilde{\pi}_\alpha(x))$, there exists an open subset W of the Grassmann space G(N,N-d) such that $\text{ord } p_\alpha = \min_\beta(\text{ord } p_\beta)$ for each $L_\alpha \in W$.

One defines the *multiplicity* of X at x, denoted by $\mu_x(X)$, the integer $\min_\alpha(\text{ord } p_\alpha)$.

PROPOSITION A.1. (see also [2] (6.5) and [17] p. 5) Let X be a complex space, let $x \in X$. Then one has $\mu_x(X) = e(\mathcal{O}_{X,x})$.

Proof. Let $\dim_x X = d$. It is not restrictive to suppose $X \subset \mathbb{C}^N$ and $x = (0,\dots,0)$.

1° Step. Let us assume that the germ of X at 0 is irreducible.

We recall that, if $p_\alpha : (X,0) \to (\mathbb{C}^d,0)$ is a generic projection, the canonical morphism $\tilde{p}_\alpha : \mathcal{O}_{\mathbb{C}^d,0} \to \mathcal{O}_{X,0}$ induced by p_α is a finite morphism (see [3], p. 57) and, denoting respectively by $\mathcal{M}_{\mathbb{C}^d,0}$ and $\mathcal{M}_{X,0}$ the fields of the germs of meromorphic functions of $\mathbb{C}^d$ at 0 and X at 0, and denoting by $\tilde{p}_\alpha$ again the induced morphism $\mathcal{M}_{\mathbb{C}^d,0} \to \mathcal{M}_{X,0}$, one gets:

$\text{ord } p_\alpha = [\mathcal{M}_{X,0} : \tilde{p}_\alpha \mathcal{M}_{\mathbb{C}^d,0}]$ (see [9], th 10, p. 35).

By the above observations, one gets: $\mu_0(X) = \min_\alpha [\mathcal{M}_{X,0} : \tilde{p}_\alpha \mathcal{M}_{\mathbb{C}^d,0}]$.

Let $R = \mathcal{O}_{X,0}$ and m be its maximal ideal. Note that each parameter system $x_1,\ldots,x_d$ of R defines a generic projection p of a neighbourhood of 0 in X on a $\mathbb{C}^d$ (see [9] cor. 1 p. 27) and viceversa the local equations of any such projection form a parameter system for R (see [9] cor. 2 p. 37). Note that, if $x_1,\ldots,x_d$ is any parameter system of R, denoting by q the ideal of R generated by them, one has:

i) the canonical homomorphism $\phi: \mathbb{C}\{X_1,\ldots,X_d\} \to R$, $X_i \mapsto x_i$

is finite and injective (see [5] p. 120, satz 4);

ii) $e(q) = [L : \phi(K)]$, where L and K are respectively the fields of quotients of R and of $\mathbb{C}\{x_1,\ldots,x_d\} \simeq \mathbb{C}\{X_1,\ldots,X_d\}$, and the morphism induced between K and L is again denoted by ϕ.

The proof follows by $[23]_{II}$, cor 1 p.299, remembering that R is by assumption an integral domain, that i) holds, and that, setting $n = (x_1,\ldots,x_d)$, by i) one has $e_{\mathbb{C}\{x_1,\ldots,x_d\}}(n) = 1$.

Now, if $p_\alpha: (X,0) \to (\mathbb{C}^d,0)$ is any projection, and $x_1,\ldots,x_d$ is the parameter system of R which defines it, by ii) one has

$$\text{ord}\, p_\alpha = e(x_1,\ldots,x_d).$$

As for each parameter system $x_1,\ldots x_d$ of R one has $(x_1,\ldots,x_d) \subset m$, then one gets

$$\text{ord}\, p_\alpha = e(x_1,\ldots,x_d) \geq e(m) = e(R),$$

by which $\mu_0(X) = \min_\alpha \text{ord}\, p_\alpha \geq e(R)$.

On the other hand, by $[23]_{II}$, th. 22 p. 294 there exists an ideal q' of R generated by a parameter system such that

$$e(q') = e(R).$$

Let $y_1,\ldots,y_d$ be this parameter system, and let p be the projection defined by it. Then we have

$$e(R) = e(q') = \text{ord}\, p \geq \min_\alpha \text{ord}\, p_\alpha = \mu_0(X),$$

by which the conclusion follows.

2° *Step.* Let us now suppose that the germ of X at 0 is reducible.

Let $X_1,\ldots,X_r$ be the irreducible components of the germ of X at 0(†) and let $p_1,\ldots,p_r$ be the corresponding minimal primes of $\mathcal{O}_{X,0}$. By [19], (23.5), if $\dim_0 X_i = \dim_0 X = d$ for $i = 1,\ldots,k$ and

(†) In the following we shall denote by X, $\tilde{X}$, $X_1,\ldots, X_r$ either the analytic germs at 0 or their representatives in a suitable open neighbourhood of 0.

$\dim_0 X_i < \dim_0 X = d$ for $i = k+1,\ldots,r$ one has

(1) $$e(R) = \sum_{i=1}^{k} e(R/p_i).$$

Setting $\tilde{X} := X_1 \cup \ldots \cup X_k$, let us show that

(2) $$\mu_0(X) = \mu_0(\tilde{X}).$$

Let $p\colon X \to V \subset \mathbb{C}^d$ (where V is an open neighbourhood of 0 in $\mathbb{C}^d$) be a generic projection, such that $p(0) = 0$. Then $\dim_0 p(X_{k+1} \cup \ldots \cup X_r) < d$ (see [3] th. p. 133), so that one has $\operatorname{card} p^{-1}(y) = \operatorname{card} (p|_{\tilde{X}})^{-1}(y)$ in an open dense subset of V; so we get $\operatorname{ord} p = \operatorname{ord} p|_{\tilde{X}}$, by which (2) follows.

Now let us show that

(3) $$\mu_0(\tilde{X}) = \sum_{i=1}^{k} \mu_0(X_i).$$

If $\tilde{\pi}_\alpha\colon (\mathbb{C}^N,0) \to (\mathbb{C}^d,0)$ is any projection with linear fibers, set $p_\alpha := \tilde{\pi}_\alpha|_{\tilde{X}}$ and $L_\alpha := (\tilde{\pi}_\alpha)^{-1}(0)$. Let $W, W_1, \ldots, W_k$ be respectively the open dense subsets of the Grassmann space $G(N,N-d)$ such that if $L_\alpha \in W$ [resp. $L_\alpha \in W_i$ for $i = 1,\ldots,k$], then $\operatorname{ord} p_\alpha = \mu_0(\tilde{X})$ [resp. $\operatorname{ord} p_\alpha|_{X_i} = \mu_0(X_i)$ for $i = 1,\ldots,k$].
The open subset $\tilde{W} := W \cap W_1 \cap \ldots \cap W_k$ is dense in $G(N,N-d)$; if $L \in \tilde{W}$, and $p\colon (X,0) \to (\mathbb{C}^d,0)$ and $p_i\colon (X_i,0) \to (\mathbb{C}^d,0)$ for $i = 1,\ldots,k$ are the projections corresponding to L, we obviously have $p_i = p|_{X_i}$ for $i = 1,\ldots,k$, hence $\operatorname{ord} p = \sum_{i=1}^{k} \operatorname{ord} p_i$ for each $L \in \tilde{W}$, so that (3) is true.
By (1), (2), (3) and step 1, we have $e(\mathcal{O}_{X,0}) = \mu_0(X)$.

REFERENCES.

[1] J.BINGENER *Schemata über Steinschen Algebren.* Schriftenreihe Math. Inst. Univ. Münster, 2 Serie, Heft 10, 1976.

[2] R.DRAPER *Intersection Theory in Analytic Geometry.* Math. Ann. 180 (1969), 175-204.

[3] G.FISCHER *Complex Analytic Geometry.* Lect. Notes Math. 538, Springer-Verlag, Berlin Heidelberg 1976.

[4] J.FRISCH *Points de platitude d'un morphisme d'espaces analytiques complexes.* Invent. Math. 4 (1967), 118-138.

[5] H.GRAUERT - R.REMMERT *Analytische Stellenalgebren.* Grund. der Math. 176, Springer-Verlag, Berlin - Heidelberg - New York 1971.

|6| S.GRECO - M.G.MARINARI *Nagata's criterion and openess of loci for Gorenstein and complete intersection.* Math. Z. 160 (1978), 207-216.

|7| S.GRECO - C.TRAVERSO *On seminormal schemes.* Comp. Math. 40 (1980), 325-365.

|8| A.GROTHENDIECK - J.DIEUDONNE' *Eléments de géométrie algébrique; ch. IV, part II.* Publ. Math. IHES 24 (1975).

|9| R.C.GUNNING *Lectures on Complex Analytic Varieties. Finite Analytic Mappings.* Princeton Univ. Press, Princeton N.J., 1974.

|10| H.HIRONAKA *Resolution of singularities of an algebraic variety over a field of characteristic zero: I.* Ann. Math. 79 (1964), 109-203.

|11| H.HIRONAKA *Bimeromorphic smoothing of a complex analytic space.* University of Warwick, 1971.

|12| H.HIRONAKA *Introduction to the theory of infinitely near singular points.* Memorias de Matematica del Instituto "Jorge Juan" 28, Madrid, 1974.

|13| M.IDA' - M.MANARESI *On the branches of a complex space.* J. Reine Angew. Math. (Crelles Journal) 321 (1981), 173-178.

|14| R.KIEHL *Note zu der Arbeit von J.Frisch "Points de platitude d'un morphisme d'espaces analytiques complexes.* Invent. Math. 4 (1967), 139-141.

|15| M.LEJEUNE-JALABERT - B.TEISSIER *Quelques calculs utiles pour la résolution des singularités.* Seminaire de Centre de Mathématiques de l'Ecole Polytechnique, 1971.

|16| B.TEISSIER *Contributions a l'étude des singularités: a) du point de vue du Polygone de Newton, b) du point de vue dés deformations.* First part of the thesis submitted to the University Paris VII, in order to obtain the degree of Docteur és Sciences Mathématiques; in collaboration with M.LEJEUNE, 1973.

|17| J.LIPMAN *Multiplicity, reduction and blowing up.* Preprint.

|18| H.MATSUMURA *Commutative Algebra.* Benjamin 1971.

|19| M.NAGATA *Local Rings.* Interscience, New York 1962.

|20| R.NARASIMHAN *Introduction to the Theory of Analytic Spaces.* Lect. Notes Math. 25, Springer-Verlag, Berlin - Heidelberg 1966.

|21| R.Y.SHARP *The Euler characteristic of a finitely generated module of finite injective dimension.* Math. Z. 130 (1973), 79-93.

|22| H.WHITNEY *Complex Analytic Varieties.* Addison-Wesley, Reading, Mass. 1972.

|23| O.ZARISKI - P.SAMUEL *Commutative Algebra.* Springer-Verlag, Berlin - Heidelberg - New York, 1960.

INTEGRAL MORPHISMS AND HOMEOMORPHISMS OF AFFINE k-VARIETIES

Maria Grazia Marinari - Mario Raimondo

Istituto Matematico - University of Genoa
Genoa, Italy

INTRODUCTION. In the context of algebraic geometry over an algebraically closed field (and also in scheme theory) the study of such geometric notions as morphisms with finite fibers and birational homeomorphisms has been pursued owing to purely algebraic tools. In this paper we look into the geometric results that the algebraic notions mentioned above yield in the sphere of algebraic geometry over a field whatever, showing how these results are weakened if compared with the classical case.

Following A. Tognoli, by affine variety defined over a field k we mean a couple (X, $\mathcal{O}_X$) where X is homeomorphic to a Zariski closed subset of some k^n and $\mathcal{O}_X$ is the sheaf of germs of regular functions on X. By $\mathcal{V}_k$ (resp. $\mathcal{A}_k$) we denote the category of affine varieties defined over k (resp. the category of k-algebras which are rings of global sections for the sheaf of regular functions on an affine k-variety).

If $k \neq \bar{k}$ and X is $\mathcal{V}_k$, to any embedding $X \hookrightarrow k^n$ is associated an affine $\bar{k}$-variety $\tilde{X}$, moreover any morphism $f : Y \longrightarrow X$ in $\mathcal{V}_k$ can be lifted in $\mathcal{V}_{\bar{k}}$.

Whichever is k, there is a natural duality between $\mathcal{V}_k$ and $\mathcal{A}_k$.
We recall the following well-known facts that hold if the base field is algebraically closed.

i) For every X in $\mathcal{V}_k$ there exist (in $\mathcal{A}_k$) a normal variety $\bar{X}$ and a morphism $p : \bar{X} \longrightarrow X$ with the following universal property. Whenever Z is a normal variety and $f : Z \longrightarrow X$ is a morphism with f(Z) dense in X, then there exists a unique morphism $q : Z \longrightarrow \bar{X}$ such that $f = pq$. Such $\bar{X}$ is called the normalization of X and p is a finite morphism. A perfectly analogous result holds for algebraic schemes. Moreover any morphism with finite fibers factors through an open immersion and a finite morphism (Zariski Main Theorem).

ii) For every X in $\mathcal{V}_k$ there exist a weakly-normal variety X* and a morphism $w : X^* \longrightarrow X$ with the property that for every birational morphism $f : Y \longrightarrow X$ which is also a homeomorphism of topological spaces, there exists $h : X^* \longrightarrow Y$ such that w dominates f (w = fh).

In this note we start with the well-known fact that the scheme-theoretic normalization $\bar{\mathbf{X}}$ of the affine scheme $\mathbf{X} = (\text{Spec } \Gamma(X, \mathcal{O}_X), \mathcal{O}_X)$ does not correspond to an object in $\mathcal{V}_k$ if k is not algebraically closed. To this extent, in the first section we show that to $\bar{\mathbf{X}}$ it is associated in a canonical way an object $\bar{X}$ in $\mathcal{V}_k$ that has the the universal property expressed in i), yet in general the morphism p is not finite nor surjective. Its image X° is a sort of algebraic counterpart of the "quasi-regular real part of a complex space with real structure", introduced by Andreotti and Holm. Moreover we can give a version of the Z.M.T..

In the second section we show that it is always possible to seminormalize varieties in $\mathcal{V}_k$ (i.e. the seminormalization ${}^+A$ of a k-algebra A in $\mathcal{A}_k$ is in $\mathcal{A}_k$ too). We also point out the following two facts:

j) a homeomorphic birational morphism in $\mathcal{V}_k$ in general is not integral (even if it can be lifted to an integral morphism in $\mathcal{V}_{\bar{k}}$)

jj) a birational integral morphism $Y \longrightarrow X$ such that the induced $\text{Spec } \Gamma(Y, \mathcal{O}_Y) \longrightarrow \text{Spec } \Gamma(X, \mathcal{O}_X)$ is a homeomorphism needs not to be a universal homeomorphism.

Therefore, even if char(k) = 0 (in which case "semi-normalization" and "weak-normalization" coincide), in order to obtain an analogue of the property expressed in ii) one has to consider the scheme- theoretic hypotheses of Andreotti and Bombieri.

Finally, we wish to thank the referee who corrected some blunders (Propositio 1.9. and Example 2.5.iii) are essentially due to him).

0. DEFINITION

Let k be any field. If R is any k-algebra we denote by $\Omega(R)$ (resp. $\Omega_k(R)$) the set of all maximal ideals of R (resp. its, possibly empty, subset consisting of all the elements having residue field k). We will consider both these sets endowed with the induced topology of Spec R.

By $\mathcal{V}_k$ (resp. $\mathcal{A}_k$) we denote the category of affine k-varieties (resp. its antidual) (see [T] for a general reference). In [M-R_2] $\mathcal{A}_k$ has been characterized as full subcategory of the category of k-algebras with k-algebra homomorphisms. Precisely, an object of $\mathcal{A}_k$ is a k-algebra A for which the following hold:

" there exist finitely generated (f.g.) k-algebras R such that $(0_R) = \cap \underline{m}$ (the intersection running through the set $\Omega_k(R)$) and $A \simeq S^{-1}R$ with $S = \{ s \in R \mid s \notin \underline{m},\ \forall\, \underline{m} \in \Omega_k(R)\}$ " (★).

Given an A in $\mathcal{A}_k$, any $A \simeq S^{-1}R$ (as in (★)) is called affine representation of the affine k-variety $(X, \mathcal{O}_X) = (\Omega_k(A), \tilde{A})$ and, by abuse of notation, affine representation of A .

If A is in $\mathcal{A}_k$ for any affine representation $A \simeq S^{-1}R$ it is $S = R - \cup\, \underline{m}$ ($\underline{m} \in \Omega_k(R)$) and so $\Omega(A) = \Omega_k(A)$. If k is algebraically closed ($k = \bar{k}$) then $\mathcal{A}_k$ is the category of f.g. reduced k-algebras, so for an A in $\mathcal{A}_k$ any affine representation $A \simeq S^{-1}R$ is with $A \simeq R$. In general, if $k \neq \bar{k}$, it is possible to find "non-isomorphic" affine representation, i.e. $A \simeq S_1^{-1}R_1 \simeq S_2^{-1}R_2$ with $R_1 \neq R_2$; nevertheless it is $\Omega_k(R_i) \simeq \Omega(A)$.

For an ideal $\underline{a}$ contained in any k-algebra R we will use the following notation: $Z(\underline{a}) = \{ \underline{m} \in \Omega_k(R) \mid \underline{m} \supset \underline{a} \}$, $V(\underline{a}) = \{ p \in \text{Spec } R \mid \underline{p} \supset \underline{a}\}$.

By integral closure of any reduced ring A, unless contrarily specified, we mean its integral closure $\overline{A}$ in Fract(A).

By U(A) we mean the group of units of A.

We recall that a ring A is Nagata if it is noetherian and for every $\underline{p} \in \operatorname{Spec} A$, $A/\underline{p}$ has the property that for every finite extension L of $\operatorname{Fract}(A/\underline{p})$ the integral closure of $A/\underline{p}$ in L is finite over $A/\underline{p}$.

We stress the fact that any A in $\mathcal{A}_k$ is Nagata as it is the ring of fractions of f.g. k-algebras.

Finally, for any X in $\mathcal{V}_k$, we will write Γ_X for $\Gamma(X, \mathcal{O}_X)$.

1. DERIVATION

Let A be in $\mathcal{A}_k$, for every affine representation $A \simeq S^{-1}R$ then $\overline{A} \simeq S^{-1}\overline{R}$ where $\overline{R}$ is a f.g. k-algebra; unfortunately $\overline{A}$ is not always in $\mathcal{A}_k$ as the following well-known examples show.

EXAMPLES 1.1. i) Let $X = \{ (x,y) \in \mathbb{R}^2 \mid y^2 = x^3 - x^2 \}$. Then $A = \Gamma_X \simeq S^{-1}\mathbb{R}[t^2+1, t(t^2+1)]$. Since $(t^2+1) \in \Omega(S^{-1}\mathbb{R}[t])$ we obtain that $\overline{A}$ is not in $\mathcal{A}_{\mathbb{R}}$. We remark that in the fine topology of $\mathbb{R}^2$, X has an isolated (hence singular (cf. [W])) point which "gets lost" in $\Omega_{\mathbb{R}}(\overline{A})$.

ii) Let $X = \{(x,y,z) \in \mathbb{R}^3 \mid x^2 z = y^2\}$. Then $A = \Gamma_X \simeq S^{-1}\mathbb{R}[u, uv, v^2]$. Now, for each $\alpha \in \mathbb{R}$, $(u, uv, v^2 + \alpha^2) \in \Omega_{\mathbb{R}}(\mathbb{R}[u,uv,v^2])$ and to this ideal corresponds $(u, v^2 + \alpha^2) \in \Omega(\overline{A}) - \Omega_{\mathbb{R}}(\overline{A})$. Therefore $\overline{A}$ is not in $\mathcal{A}_{\mathbb{R}}$. The surface X is the real part of the "Cayley umbrella" and it consists of the z-axis for $z \leq 0$ and of the surface $x^2 = y^2/z$ for $z > 0$. We point out that the singular locus of X consists of the whole z-axis whose negative part (umbrella stick) "disappears" passing to $\Omega_{\mathbb{R}}(\overline{A})$.

iii) Let $X = \{ (x,y) \in \mathbb{R}^2 \mid y^2 = x^3 + x^2 \}$. It is $A = \Gamma_X \simeq S^{-1}\mathbb{R}[t^2 - 1, t(t^2 - 1)]$. If S' denotes the saturated multiplicative part of S in $\mathbb{R}[t]$, then $S' = \{ f \in \mathbb{R}[t] \mid f(t) \neq 0 \ \forall t \in \mathbb{R}\}$. Therefore $\overline{A}$ is in $\mathcal{A}_{\mathbb{R}}$ as it is exactly the ring of regular functions of the real line.

To give an analogue of the classical theory we need to introduce, for every X in $\mathcal{V}_k$ the following subset:

$$X^\circ = \{ x \in X \mid \Omega_k(\overline{\mathcal{O}_{X,x}}) \neq \emptyset \} .$$

LEMMA 1.2. For every X in $\mathcal{V}_k$, X° is a dense subset.

Proof. Let $\Gamma_X \simeq S^{-1}R$ be an affine representation and let $X = \operatorname{Spec} R$. By definition R is a f.g. reduced k-algebra with $\underline{I}(\Omega_k(R)) = (0_R)$. Therefore $X = V(\underline{I}(\Omega_k(R)))$ (i.e. $\Omega_k(R)$ is dense in X) and Reg X is a non-empty subset of X. On the whole: $\Omega_k(R) \cap \operatorname{Reg} X$ is a non-empty open subset of $\Omega_k(R)$, which, modulo identifying $\Omega_k(R)$ with X, is contained in X°. So our contention follows. □

PROPOSITION 1.3. Let A be in $\mathcal{A}_k$ and let $A \simeq S^{-1}R$ be an affine representation. Then:

i) $\overline{R}$ has the property that $\cap\, \overline{\underline{m}}$ $(\overline{\underline{m}} \in \Omega_k(\overline{R})) = (0_{\overline{R}})$

ii) if $\overline{S} = \{ s \in \overline{R} \mid s \notin \overline{\underline{m}}, \forall\, \overline{\underline{m}} \in \Omega_k(\overline{R})\}$, then $\overline{S}^{-1}\overline{R}$ is an object in $\mathcal{A}_k$ which depends only on A (i.e. it is independent of the chosen affine representation).

Proof. i) Let $\nu : \operatorname{Spec} \overline{R} \longrightarrow \operatorname{Spec} R$ be the canonical map and let $X = \Omega(A)$. Consider $\underline{b} = \cap\, \overline{\underline{m}}_y$ with $\overline{\underline{m}}_y \in \Omega_k(\overline{R})$ and $\nu(y) \in X^\circ$. We have $\underline{b} \cap R = \cap\, \underline{m}_x$ $(x \in X^\circ) = (0_R)$ as X° is dense in X. If A is an integral domain this implies that $\underline{b} = (0_{\overline{R}})$. In the general case, by definition of affine representation, we have that ${}_S(0_R)$ is a radical ideal (namely $(0_R) = \cap\, \underline{m}$ $(\underline{m} \in \Omega_k(R))$. If $(0_R) = \bigcap_{i=1} \underline{p}_i$ is the reduced primary decomposition, then for each i, $i \in \{1,\ldots,s\}$ there exists an $f_i \in \bigcap_{i\neq j} \underline{p}_j$, $f_i \notin \underline{p}_i$ so that $R_{f_i} \simeq R_{f_i}/\underline{p}_i R_{f_i}$ thus $p_i R_{f_i}$ is the intersection of all the maximal ideal in $\Omega_k(R_{f_i})$ that contain it (cf. [T] p.32). Therefore our contention follows from the integral case.

ii) Let $A \simeq S_1^{-1}R_1 \simeq S_2^{-1}R_2$ be two affine representations. An isomorphism

$\phi: S_1^{-1}R_1 \simeq S_2^{-1}R_2$ extends in a natural way to $\overline{\Phi} : \overline{S}_1^{-1}\overline{R}_1 \simeq \overline{S}_2^{-1}\overline{R}_2$. □

DEFINITION 1.4. If A is in $\mathcal{A}_k$ and $A \simeq S^{-1}R$ is an affine representation, then the k-integral closure of A (in symbols ${}^k\overline{A}$) is the object in $\mathcal{A}_k$ that has affine representation ${}^k\overline{A} \simeq \overline{S}^{-1}\overline{R}$.

We remark that the above definition makes sense by (1.3). We also point out that, given an A in $\mathcal{A}_k$, for every k-algebra homomorphism $\phi:\overline{A}\longrightarrow B$ with B in $\mathcal{A}_k$, then there exists a unique $\tilde{\phi}: {}^k\overline{A} \longrightarrow B$ which extends ϕ (cf. [M-R_2] p.68). More precisely, if B is an overring of A then ϕ extends to an isomorphism Φ : Fract($\overline{A}$) $\longrightarrow$ Fract(B) which induces $\tilde{\phi}$. Therefore we can say that ${}^k\overline{A}$ is the minimal overring of $\overline{A}$ which is in $\mathcal{A}_k$. So, if $\overline{A}$ is already in $\mathcal{A}_k$, then ${}^k\overline{A} = \overline{A}$. Finally we observe that for every X in $\mathcal{V}_k$ the affine k-variety $\Omega({}^k\overline{\Gamma}_X)$ will simply be denoted by $\overline{X}$, as no confusion is reasonable in $\mathcal{V}_k$. We say that an X in $\mathcal{V}_k$ is normal if $X = \overline{X}$.

We now restate the above results in terms of varieties.

THEOREM 1.5. Given an X in $\mathcal{V}_k$ then there exist in $\mathcal{V}_k$ a variety $\overline{X}$ and a morphism $p : \overline{X} \longrightarrow X$, such that:

i) $\overline{X}$ is a normal variety,

ii) the morphism p is birational, has finite fibers and $\mathrm{Im}(p) = X^\circ$,

iii) any morphism $f : Y \longrightarrow X$ in $\mathcal{V}_k$ with Y normal and f(Y) dense in X factors uniquely through $\overline{X}$.

Proof. The proofs of i) and ii) are given in (1.3.). As for iii) we have $\phi: \Gamma_X \longrightarrow \Gamma_Y$ that is injective since f is dominant, and therefore the integral closure A of Γ_X into Fract(Γ_Y) contains $\overline{\Gamma}_X$, and also $A \hookrightarrow \Gamma_Y$. So that we get a homomorphism $\overline{\Gamma}_X \hookrightarrow \Gamma_Y$ and by (1.3.i)) and the preceeding remark we obtain a uniquely determined map $\Gamma_{\overline{X}} \longrightarrow \Gamma_Y$. □

REMARKS 1.6. i) The universal property given in (1.5.iii)) says that $\overline{X}$ is the normalization of X in $\mathcal{V}_k$.

ii) The normalizations in (1.1.i)) and (1.1.ii)) are respectively the real line and the real plane. The subsets $X°$ consist of the parametric parts i.e. respectively $\{(x,y) \in \mathbb{R}^2 \mid x=t^2+1,\ y=t(t^2+1),\ t \in \mathbb{R}\}$ and $\{(x,y,z) \in \mathbb{R}^3 \mid x=u,\ y=uv,\ z=v^2,\ (u,v) \in \mathbb{R}^2\}$.

iii) In general the morphism p is not surjective and the ring $\Gamma_{\overline{X}}$ is not integral over Γ_X. Moreover it can happen (cf.2. .) that p is surjective but $\Gamma_{\overline{X}}$ is not integral over Γ_X.

In order to make clearer the differences between the case $k \neq \overline{k}$ and $k = \overline{k}$, we introduce the following definitions (which coincide if $k = \overline{k}$).

DEFINITION 1.7. A morphism $f : Y \longrightarrow X$ in $\mathcal{V}_k$ is called:
i) integral if the induced ring homomorphism $\Gamma_X \hookrightarrow \Gamma_Y$ is integral,
ii) almost integral if there exist affine representations $\Gamma_X \simeq S_X^{-1}R_X$, $\Gamma_Y \simeq S_Y^{-1}R_Y$ such that the induced $R_X \hookrightarrow R_Y$ is integral (so that R_Y is a finite R_X-module).

REMARKS 1.8. i) If $f : Y \longrightarrow X$ is integral then it is surjective and closed. The induced ring homomorphism $\phi : \Gamma_X \hookrightarrow \Gamma_Y$ is integral by definition. Then the morphism ϕ^*: Spec $\Gamma_Y \longrightarrow$ Spec Γ_X is closed and surjective. Its restriction to $Y = \Omega(\Gamma_Y)$ is just f as the Cohen-Seidenberg theorem holds for ϕ.

ii) For every X in $\mathcal{V}_k$ the morphism $p : \overline{X} \longrightarrow X$ is almost integral but in general it is not integral. For instance in (1.1.i)) the element $1/(t^2+1) \in \Gamma_{\overline{X}}$ is not integral over Γ_X.

We end this section by showing that integral morphisms have some of the "classical" properties.

PROPOSITION 1.9. A morphism $\phi : A \hookrightarrow B$ in $\mathcal{A}_k$ is integral if and only if it is finite (i.e. B is a finite A-module).

Proof. It has only to be checked that integral implies finite. We can choose affine representations $A \simeq S'^{-1}R'$, $B \simeq \tilde{S}^{-1}\tilde{R}$ such that $\tilde{R}$ is a f.g. R'-algebra (cf.[M-R_1] p.130). By assumption every element in B is integral over A, since only a finite number of denominators are involved, possibly inverting a finite number of elements in R' we can find R such that the induced $\psi: R \hookrightarrow \tilde{R}$ is integral, $\psi(S) \subset \tilde{S}$ and $S^{-1}R \simeq A$. As ψ is finite also $S^{-1}\tilde{R}$ is finite over $S^{-1}R$. On the other hand the map $S^{-1}\tilde{R} \hookrightarrow \tilde{S}^{-1}\tilde{R}$ is integral with $S^{-1}\tilde{R}$ a Nagata ring. Therefore $\tilde{S}^{-1}\tilde{R}$ is finite over $S^{-1}\tilde{R}$. Thus $B \simeq \tilde{S}^{-1}\tilde{R}$ is finite over $S^{-1}R \simeq A$. □

THEOREM 1.10. Let A and B be integral domains in $\mathcal{A}_k$. If $\phi : A \hookrightarrow B$ is a k-homomorphism with finite fibers then there exist in $\mathcal{A}_k$ an object C and maps $A \xrightarrow{\psi} C \xrightarrow{\chi} B$ such that: $\phi=\chi\psi$, ψ is almost integral and χ is an open immersion.

Proof. Let A' be the integral closure of A in B. Fract(A), Fract(A') and Fract(B) are f.g. field extensions of k. As A' is integral over A, Fract(A') is algebraic over Fract(A) and so it is a finite extension. Therefore A' is finite over A since A is a Nagata ring, say $A' = A[\xi_1,\dots,\xi_n]$ We can write $B = N^{-1}A[\xi_1,\dots,\xi_n,\xi_{n+1},\dots,\xi_s]$ with suitable $\xi_{n+i} \in B$ and $N \subset B$. Observe that A' turns out to be the integral closure of A in $A[\xi_1,\dots,\xi_s]$. Apply to $A \hookrightarrow A[\xi_1,\dots,\xi_s]$ the Zariski Main Theorem to get an open immersion $A' \hookrightarrow B$. Now Spec B can be identified with an open subspace of Spec A', $\Omega_k(B)$ is by assumption dense in Spec B, $\Omega_k(A')$ "contains" $\Omega_k(B)$ and so it is dense in Spec A'. A reasoning like that of (1.4.) gives us that $(S')^{-1}A'$ is in $\mathcal{A}_k$ (with $S' = \{s' \in A' \mid s' \notin \underline{m}',\ \forall \underline{m}' \in \Omega_k(A')\}$). □

REMARK 1.11. We finally observe that, if k is a completely valued field, then any integral morphism in $\mathcal{V}_k$ is proper in the norm topology. Namely, let $f : Y \longrightarrow X$ be an integral morphism and let $\Gamma_X \simeq S_X^{-1}k[X_1,\dots,X_m]/\underline{I}_X$ and $\Gamma_Y \simeq S_Y^{-1}k[Y_1,\dots,Y_n]/\underline{I}_Y$. By assumption we obtain that every Y_i satisfies a relation of the type:

$$Y_i^{d_i} + \sum_{h=1}^{n} A_{ih} Y_i^{d_i-h} = 0$$

Then direct computations show that the "coordinates" of the points in the inverse image of any compact contained in X satisfy a finite number of inequalities.

2. SEMINORMALIZATION

We briefly recall the definitions and principal properties of the seminormalization. If A is a ring and B is an overring of A integral over it; the seminormalization of A in B is:

$$_B^+A = \{ b \in B \mid \forall \underline{p} \in \text{Spec } A \ , \ b(\underline{p}) \in A_{\underline{p}} + R(B_{\underline{p}}) \} \quad .$$

If $B = \overline{A}$ then we write ^+A instead of $_{\overline{A}}^+A$ and A is called seminormal if $A = {}^+A$. Moreover, the ring $_B^+A$ is characterized (cf. [Tr] §1) as the largest subring A' of B containing A and such that:

i) for every $\underline{p} \in$ Spec A there exists exactly one $\underline{p}' \in$ Spec A' over $\underline{p}$ (i.e. such that $\underline{p}' \cap A = \underline{p}$) ,

ii) the canonical homomorphism $k(\underline{p}) \longrightarrow k(\underline{p}')$ is an isomorphism.

If A is a Nagata ring and $S \subset A$ is a multiplicative system then $^+(S^{-1}A) = S^{-1}(^+A)$ (cf. [G-Tr] p.329).

THEOREM 2.1. If A is in $\mathcal{A}_k$ then also ^+A is in $\mathcal{A}_k$.

Proof. Let $A \simeq S^{-1}R$ be an affine representation so that $^+A \simeq S^{-1}(^+R)$. Consider the radical ideal $\underline{a}' = \cap\, \underline{m}'$ $(\underline{m}' \in \Omega_k(^+R))$ and let $\underline{a}' = \bigcap_{i=1}^{n} \underline{p}_i'$ be its primary reduced decomposition. It is $\underline{a}' \cap R = (0_R)$ and, if $\underline{p}_i = \underline{p}_i' \cap R$ then $\bigcap_{i=1}^{n} \underline{p}_i = (0_R)$ is the primary reduced decomposition. If not, there exists $\underline{p}_i \supset \underline{p}_{j(i)}$ for some $i, j(i) \in \{1,\dots,n\}$ liftable to ^+R via going-up, this is impossible because of the property i) above. Now let $^+S = \{ s' \in {}^+R \mid s' \notin \underline{m}', \forall \underline{m}' \in \Omega_k(^+R)\}$, clearly $S \subset {}^+S$. We have to show that $^+A \simeq (^+S)^{-1}\,{}^+R$, for this we only need to check that for every $\underline{p}' \in \text{Spec}\,{}^+R$ with $\underline{p}' \cap {}^+S \neq \emptyset$ then $\underline{p}' \cap S \neq \emptyset$. As

${}^+S = {}^+R - \cup \underline{m}'$ $(\underline{m}' \in \Omega_k({}^+R))$ (resp. $S = R - \cup \underline{m}$ $(\underline{m} \in \Omega_k(R))$) and $\Omega_k(R)$ is in a one-to-one correspondence with $\Omega_k({}^+R)$, our contention follows. □

By the above theorem, without any assumption on the base field k, for every X in $\mathcal{V}_k$ there exists an affine k-variety ${}^+X$ (associated to ${}^+\Gamma_X$) which can be called seminormalization of X.

FACT 2.2. If the base field is algebraically closed and of characteristic zero, then, for every X in $\mathcal{V}_k$ the seminormalization morphism $n: {}^+X \longrightarrow X$ is characterized by the property that for every birational morphism in $\mathcal{V}_k$ $f : Y \longrightarrow X$ which is also homeomorphism of topological spaces there exists in $\mathcal{V}_k$ a morphism $g : {}^+X \longrightarrow Y$ such that $n = fg$.

We want to investigate what happens in the case $k \neq \bar{k}$. We list some observations.

REMARKS 2.3. i) If X is an irreducible k-variety the ring Γ_X satisfies the "maximal chain condition on primes" (M.C.C.).
Namely, if $\Gamma_X \simeq S^{-1}R$ is an affine representation, for every $\underline{p} \in \mathrm{Spec}\Gamma_X$ there exists at least an $\underline{m} \in \Omega(\Gamma_X)$ such that $\underline{p} \subset \underline{m}$ and correspondingly unique $\underline{q} \in \mathrm{Spec}\ R$ and $\underline{n} \in \Omega_k(R)$ such that $\underline{p} = \underline{q}^e$ and $\underline{m} = \underline{n}^e$. So as R satisfies the M.C.C. the same does Γ_X.

ii) If $f : Y \longrightarrow X$ is an integral morphism of irreducible k-varieties, then for every $\underline{P} \in \mathrm{Spec}\ \Gamma_Y$ and $\underline{p} = \underline{P} \cap \Gamma_X$ we have: $ht(\underline{P}) = ht(\underline{p})$.

We recall here that if Z is any affine variety, an ideal $\underline{a} \subset \Gamma_Z$ is called a defining ideal if $\underline{a} = \underline{I}(V(\underline{a}))$.

PROPOSITION 2.4. Let $f : Y \longrightarrow X$ be a morphism of irreducible k-varieties which is integral, birational and homeomorphism of topological spaces and let $\phi^*: \mathrm{Spec}\Gamma_Y \longrightarrow \mathrm{Spec}\Gamma_X$ be the induced morphism. Then:

i) for each defining ideal $\underline{P} \in \operatorname{Spec} \Gamma_Y$ if $\underline{p} = \underline{P} \cap \Gamma_X$ then $k(\underline{P})$ is purely inseparable over $k(\underline{p})$.

ii) for each defining ideal $\underline{p} \in \operatorname{Spec} \Gamma_X$ there exists exactly one defining ideal $\underline{P} \in \operatorname{Spec} \Gamma_Y$ such that $\phi^*(\underline{P}) = \underline{p}$.

Proof. i) The "dimension formula" together with $\operatorname{ht}(\underline{P}) = \operatorname{ht}(\underline{p})$ gives that $k(\underline{P})$ is an algebraic extension of $k(\underline{p})$ (cf. [M] th.23) and so $[k(\underline{P}):k(\underline{p})] = n < \infty$. The induced morphism $\tilde{f} : V(\underline{P}) \longrightarrow V(\underline{p})$ is generally m to 1, where $m = [k(\underline{P}):k(\underline{p})]_s$. Let $U \subset V(\underline{P})$, $W \subset V(\underline{p})$ be the involved open subsets. It is $U' = U \cap Z(\underline{P}) \neq \emptyset$ and $W' = W \cap Z(\underline{p}) \neq \emptyset$, also the following diagram

$$\begin{array}{ccc} U' & \xrightarrow{\tilde{f}} & W' \\ \downarrow & & \downarrow \\ Y & \xrightarrow[f]{} & X \end{array}$$

commutes, therefore as f is injective it is $m = 1$.

ii) For every $\underline{m}_X \in \Omega(\Gamma_X)$, $\dim (\Gamma_Y \otimes_{\Gamma_X} k(\underline{m}_X)) = 0$. Therefore $\dim (\Gamma_Y \otimes_{\Gamma_X} k(\underline{p})) = 0$ for every $\underline{p} \in \operatorname{Spec} \Gamma_X$ (cf. [M] th.21). That is, the preimage of $\underline{p}$ via ϕ^* consists of a finite set of prime ideals of height equal to the height of $\underline{p}$. Now $\phi^*|_{\Omega(\Gamma_Y)} = f$ that is a homeomorphism. As any defining ideal is the intersection of the maximal ideals containing it we have that the image of $\operatorname{Spec}_d \Gamma_Y = \{ \underline{P} \in \operatorname{Spec} \Gamma_Y | \ \underline{P}$ is a defining ideal $\}$ via ϕ is $\operatorname{Spec}_d \Gamma_X$. Also, if $\phi^*(\underline{q}_1) = \phi^*(\underline{q}_2) = \underline{p}$, then $Z(\underline{q}_1) = \{ x \in X \mid f(x) \in Z(\underline{p}) \} = Z(\underline{q}_2)$ so that if $\underline{q}_1$ and $\underline{q}_2$ are both defining ideals then $\underline{q}_1 = \underline{q}_2$. □

EXAMPLES 2.5. i) Let $\phi : S_X^{-1} \mathbb{R} [X_1,X_2,X_3] \longrightarrow S_Y^{-1} \mathbb{R} [Y_1,Y_2,Y_3]$ be the $\mathbb{R}$-algebra homomorphism defined by $\phi(X_1) = Y_1$, $\phi(X_2) = Y_2(Y_1^2+Y_2^2+Y_3^2)$, $\phi(X_3) = Y_3(Y_1^2+Y_2^2+Y_3^2)$ and let $f : \mathbb{R}^3 \longrightarrow \mathbb{R}^3$ be the corresponding morphism of real varieties. On one hand f is a homeomorphism of topological spaces, on the other hand f is not almost integral. For this we show that ϕ does not verify the "lying over". Let $\underline{p}_1 = (Y_1^2+Y_2^2+Y_3^2)$ and $\underline{p}_2 = (Y_1^2+Y_2^2, Y_3)$. We have $\underline{p}_1 \subsetneq \underline{p}_2$ but $\phi^{-1}(\underline{p}_1) = \phi^{-1}(\underline{p}_2) = (X_2,X_3)$

and there is no prime lying over $\underline{q} = (X_2)$.

ii) Let $B = \mathbb{C}[t]$, and let A' be the pull back of the following diagram:

$$\begin{array}{ccc} & & \mathbb{C}[t] \\ & & \downarrow f \\ \mathbb{C} & \xrightarrow[\Delta]{} & \mathbb{C}\times\mathbb{C}\times\mathbb{C} \end{array}$$

where $f(t) = (f(1), f(\exp((2/3)\pi i)), f(\exp((4/3)\pi i)))$. Then $A' \simeq \mathbb{C}[t^3-1, t(t^3-1), t^2(t^3-1)]$ is seminormal in B which is integral over it (cf. [G] p.188 and p.196). The $\mathbb{R}$-algebra $A = A' \cap \mathbb{R}[t]$ is therefore seminormal in $\mathbb{R}[t]$ (cf. [I] p. 102). It is easy to check that $\cap \underline{m}$ $(\underline{m} \in \Omega_{\mathbb{R}}(A)) = (0)$, then $X = \Omega_{\mathbb{R}}(A)$ is in $\mathcal{V}_{\mathbb{R}}$. The inclusion $A \hookrightarrow \mathbb{R}[t]$ induces a morphism which is clearly birational, homeomorphism of topological spaces and by construction almost integral, but it is not integral as $1/(t^2+t+1)$ is not so.

iii) Let $B = \mathbb{R}[X,Y]$, $\phi: B \to \mathbb{C}[T]$ the $\mathbb{R}$-algebras homomorphism defined by $\phi(X) = T$, $\phi(Y) = iT$ and let $A = \phi^{-1}(\mathbb{C}[T^3])$. Then A is a seminormal f.g. $\mathbb{R}$-algebra over which B is integral (cf. [G] p.188 and p.196) and $\cap \underline{m}' = (0)$ $(\underline{m}' \in \Omega_{\mathbb{R}}(A))$. The conductor of B in A is $\underline{c} = \ker\phi = (X^2+Y^2)B$. If $\underline{m} \in \Omega(B)$, $\underline{m} \not\supset \underline{c}$ then $B_{\underline{m}} \simeq A_{\underline{m}\cap A}$ so $B/\underline{m} \simeq A/\underline{m}\cap A$ that is $\underline{m} \in \Omega_{\mathbb{R}}(B)$ if and only if $\underline{m} \cap A \in \Omega_{\mathbb{R}}(A)$. On the other hand there is only one $\underline{m}' \in \Omega_{\mathbb{R}}(A)$ containing $\underline{c}$ and for every $\underline{m} \in \Omega(B)$ such that $\underline{m} \cap A = \underline{m}'$ then $\underline{m} = (X,Y) \in \Omega_{\mathbb{R}}(B)$. Let $S_A \subset A$ (resp. $S_B \subset B$) be the usual multiplicative system. By what we have seen above S_B is the saturated multiplicative part of $S_A \subset B$. The induced $S^{-1}A \hookrightarrow S^{-1}B$ (where $S = S_A$) is therefore integral birational and homeomorphism of affine schemes, but it is not radical since

$$[\mathrm{Fract}(S^{-1}B/S^{-1}\underline{c}) : \mathrm{Fract}(S^{-1}A/S^{-1}\underline{c})] = [\mathbb{C}(T) : \mathbb{C}(T^3)] = 3.$$

REMARK 2.6. We recall that the "geometric" Fact 2.2. is a particular case of a scheme theoretic result proved by Andreotti and Bombieri (cf [A-B]). In the general formulation instead of "homeomorphic birational morphism of varieties" appears "universal homeomorphism of schemes" i.e. surjective, integral and radical. Examples 2.5. show that if $k \neq \bar{k}$ ($k = \mathbb{R}$ too) the general hypotheses of [A-B] are needed in order to get the analogue in $\mathcal{V}_k$.

More precisely we have seen that there exist homeomorphic birational morphisms in $\mathcal{V}_k$ for which the correspondings homomorphisms in $\mathcal{A}_k$:

i) do not satisfy the "lying over" condition on prime ideals,

ii) are almost integral but not integral,

iii) are not radical though integral.

REFERENCES

[A-B] - A.Andreotti-E.Bombieri - Sugli omeomorfismi delle varietà algebriche, Ann.Sc.Norm.Sup.Pisa 23(1969).

[A-H] - A.Andreotti-P.Holm - Real and complex singularities, Proc.Nordic School NAVF Symposium in Mathematics Oslo (1976).

[EGA] - A.Grothendieck-J.Dieudonné - Eléments de Géométrie Algébrique, I.H.E.S. ch. IV 24(1965), 28(1966).

[G] - P.Gianni - Morfismi interi delle varietà algebriche, Boll.U.M.I. (5) 15-A (1978).

[G-Tr] - S.Greco-C.Traverso - On seminormal schemes, Comp.Math. 40(1980).

[I] - F.Ischebeck - Zwei Bemerkungen über seminormale Ringe, Math.Z. 152(1977).

[M-R_1] - M.G.Marinari-M.Raimondo - Fibrati vettoriali su varietà algebriche definite su corpi non algebricamente chiusi, Boll. U.M.I. (5) 16-A (1979).

[M-R_2] - M.G. Marinari-M.Raimondo - Properties of the regular functions ring of affine varieties defined over any field, Rend. Sem.Mat.Univ.Pol.Torino 37 (1979).

[M] - H.Matsumura - Commutative algebra, Benjamin N.Y. (1970).

[T] - A.Tognoli - Algebraic geometry and Nash functions, INDAM Ist. Math. III (1978).

[Tr] - C.Traverso - Seminormality and Picard group, Ann.Sc.Norm.Sup.

Pisa 24 (1970).

[W] - H.Whitney - Elementary structure of real algebraic varieties, Ann.of Math. 66 (1957).

FINITE MORPHISMS OF GORENSTEIN CURVES

Anna Oneto - Elsa Zatini
Università di Genova
Istituto di Matematica
GENOVA

INTRODUCTION. In this paper we consider a commutative diagram

$$\begin{array}{ccc} \overline{X} & \xrightarrow{\phi} & \overline{Y} \\ \pi \downarrow & & \downarrow \nu \\ X & \xrightarrow{f} & Y \end{array}$$

where f is a separable and finite morphism of projective curves, π , ν are normalizations and ϕ is the canonically induced morphism.

We consider the following objects which describe this situation :

(i) the conductor of Y in X $\gamma_{X/Y} = \mathcal{H}om_{\mathcal{O}_Y}(f_*\mathcal{O}_X, \mathcal{O}_Y)$

the conductor γ_X of X in $\overline{X}$

the conductor γ_Y of Y in $\overline{Y}$

(ii) the dualizing sheaves ω_X, ω_Y of X and Y respectively (see [S_1] and [K] for a definition in terms of regular differential forms)

(iii) the ramification of the morphism ϕ defined by the ideal

$$\mathcal{R}_{\overline{X}} = Ann_{\mathcal{O}_{\overline{X}}}(\Omega_{\overline{X}/\overline{Y}})$$

Relations between conductors,dualizing sheaves and ramification Ideal have been studied in [P] for particular morphisms of desingularization .

The first part of this work (§1) is devoted to the study of trace and cotrace morphisms (following the notation of [C]) induced by f on the sheaves of rational differentials. It is well known that the trace $\Omega(k(X)) \longrightarrow \Omega(k(Y))$ restricts to a 'trace' morphism between the dualizing sheaves $Tr : f_* \omega_X \longrightarrow \omega_Y$. Since the cotrace injection $\Omega(k(Y)) \longrightarrow \Omega(k(X))$ is easily proved to restrict to a morphism $f^* \omega_Y \longrightarrow \omega_X$ when f is flat (note 1.10), we ask whether this is true in a more general case. The answer is in general negative (example 1.7): theorem 1.8 gives a necessary and sufficient condition for the existence of such a morphism in the case Y Gorenstein .

In § 2 and § 3, where we consider only Gorenstein curves, the trace morphism turns out to be helpful to find relations between conductors and dualizing sheaves. More precisely : in 2.6 we prove that $\gamma_{X/Y}$ and the relative dualizing sheaf $\omega_{X/Y} = \omega_X \otimes_{\mathcal{O}_X} f^*\omega_Y^{-1}$ are isomorphic, in 2.7 we state the following isomorphism :

$$(1) \qquad \mathcal{R}_{\overline{X}} \otimes_{\mathcal{O}_{\overline{X}}} \tilde{\gamma}_X \cong \phi^* \tilde{\gamma}_Y \otimes_{\mathcal{O}_{\overline{X}}} \pi^* \omega_{X/Y}^{-1}$$

which extends to singular curves the classical Hurwitz's theorem ([H] IV,2) and generalizes to finite morphisms a relation involving conductors, which is well known in the birational case.

The isomorphism (1) gives on the stalks a numerical formula (3. 8) which shows how to find in the classical relation (3.9) between the arithmetical genera of X and Y the single contribution of each point x in X .

In the last part, following Kunz, who in [K] defines the dualizing sheaf ω_X to be the "generalized" complementary module of X respect to a noetherian normalization, we show (corollary 4.7) that the relative dualizing sheaf $\omega_{X/Y}$ is indeed the complementary module of X respect to Y (well defined also when Y is singular), provided, as above, that X and Y are Gorenstein curves.

Finally, using (1) we establish a link between the complementary modules of X and $\overline{X}$ (relative to Y and $\overline{Y}$ respectively) and the conductors γ_X and γ_Y, which is already known only when Y is non singular.

§ 0. *TERMINOLOGY*

In this paper the word "curve" will mean "integral one dimensional scheme, proper over an algebrically closed field k of characteristic 0" . The word "point" will always mean a closed point .

0.1 . If $\pi: \overline{X} \longrightarrow X$ is the normalization of a curve X, whose fractions field is k(X), we denote by:

Ω_X the sheaf of <u>relative differentials</u> of X over k

$\Omega(k(X))$ the constant sheaf of <u>rational differentials</u> .

0.2 . If X is a nonsingular curve on an algebrically closed field k the <u>residue</u> at $P\varepsilon X$ of any rational differential $w\varepsilon\Omega(k(X))$ is well defined. Let t be an uniformizing parameter in P, then dt is a basis of $\Omega(k(X))$ as a k(X)-vector space and we can write $w = fdt$, $f\varepsilon k(X)$, and $f = \sum_{n\gg-\infty} a_n T^n$ in the completion $\widehat{k(X)}_P$ of the field k(X) for the valuation v_P ($a_i \varepsilon k$). Hence we define $Res_P(w) = a_{-1}\varepsilon k$ (see $[S_1]$ chII, 7) .

0.3 . We denote by ω_X the <u>dualizing sheaf</u>, whose definition is given by Serre (see $[S_1]$) in the following form :

if $x\varepsilon X$ is a simple point, then $\omega_{X,x} =: \Omega(\mathcal{O}_x)$

if $x\varepsilon X$ is singular , then $\omega_{X,x} =: \{w\varepsilon\Omega(k(X))$ s.t. w is "regular in x"$\}$

where "w regular in x" means that $\sum_{P_i \xrightarrow{\pi} x} Res_{P_i}(hw) = 0$ for each $h\varepsilon\mathcal{O}_x$.

For another interesting construction of the dualizing sheaf see Kunz [K] . It is well known that $\omega_{X,x}$ is isomorphic to a fractional ideal of $\mathcal{O}_x$, which is principal if and only if $\mathcal{O}_x$ is a Gorenstein ring, i.e., if this is the case, for any basis τ of $\Omega(k(X))$ over k(X) there is a rational function $\beta\varepsilon k(X)$ such that for every $w\varepsilon\,\omega_{X,x}$ we have $w = \beta\cdot b\cdot\tau$ $b\varepsilon\mathcal{O}_x$.

0.4 . In the sequel we shall consider "canonical" the isomorphism related to a fixed basis τ; in other words, we shall use the following

Notation . Given a morphism $f: X \longrightarrow Y$ of Gorenstein curves, we fix once for all a basis τ of $\Omega(k(Y))$ over k(Y) (then τ is also a basis of $\Omega(k(X))$ over k(X)) :

then, if $x \in X$ and $y = f(x)$, we write

$$\omega_{X,x} = \beta \mathcal{O}_x \tau \qquad \beta \in k(X)$$

$$\omega_{Y,y} = \alpha \mathcal{O}_y \tau \qquad \alpha \in k(Y)$$

§ 1. *TRACE AND COTRACE*

In this section we consider the following situation.
Let $f: X \longrightarrow Y$ be a finite separable morphism of curves and let $n = \deg f = [k(X) : k(Y)]$. We have a commutative diagram

$$\text{(1.1)} \qquad \begin{array}{ccc} \overline{X} & \xrightarrow{\phi} & \overline{Y} \\ \pi \downarrow & & \downarrow \nu \\ X & \xrightarrow{f} & Y \end{array}$$

where π and ν are the normalizations and ϕ is the morphism induced (obviously $\deg \phi = n$).

Our purpose is now to study the behaviour of the morphisms "trace" and "cotrace" induced on the dualizing sheaves ω_X and ω_Y. We quote Zariski and Samuel [Z.S.] , Serre [S_1] and Chevalley [C] as references for this argument. We recall here some fundamental definitions and properties.

1.2. There is a trace map $Tr: k(X) \longrightarrow k(Y)$ satisfying the following conditions: (i) $Tr(h+g) = Tr(h)+Tr(g)$

(ii) $Tr(ah) = a\,Tr(h)$ if $a \in k(Y)$

(iii) If $a \in k(Y)$ then $Tr(a) = na$ $(n = [k(X) : k(Y)])$

DEFINITION 1.3. We call <u>trace</u> $Tr: \Omega(k(X)) \longrightarrow \Omega(k(Y))$ the homomorphism defined as follows : for every $w \in \Omega(k(X))$ $Tr(w) = Tr(h)\tau$ if $w = h\tau$ with $h \in k(X)$ (τ is the basis fixed in 0.4).

THEOREM 1.4. If X and Y are regular curves and if $f: X \longrightarrow Y$ is a separable covering, then for any $w \in \Omega(k(X))$ and for any $y \in Y$ we have:

$$\sum_{x_i \to y} \operatorname{Res}_{x_i}(w) = \operatorname{Res}_y(Tr(w)).$$

Proof. See [S_1] ch. II ,12. □

This last formula allows us to prove the following

THEOREM 1.5 (See [P]). If $f: X \longrightarrow Y$ is a proper morphism of curves then the trace map $\Omega(k(X)) \longrightarrow \Omega(k(Y))$ induces a trace morphism

$$Tr: \; f_*\omega_X \longrightarrow \omega_Y .$$

THEOREM 1.6 (See, e.g., [A.K.]). If $\pi : \bar{X} \longrightarrow X$ is the normalization of a curve X, then the trace morphism $Tr: \pi_* \omega_{\bar{X}} \longrightarrow \omega_X$ is injective. Moreover the $\mathcal{O}_X$-module $Ann_{\mathcal{O}_X}$ (coker $Tr: \pi_* \omega_{\bar{X}} \longrightarrow \omega_X$) is the classical conductor sheaf.

On the other hand, the natural injection $k(Y) \longrightarrow k(X)$ induces an injection $\Omega(k(Y)) \longrightarrow \Omega k(X))$ which we call "cotrace map", as in [C]. But in general there is not an analogue of (1.5) for it, as the following example shows:

EXAMPLE 1.7. In this example and in the following ones we shall denote by $k[\ ]_o$ the local rings of our curves at the points considered. Set $k = \mathbb{C}$. In $\mathbb{P}^2_k$ let X and Y be the curves of equations

$$xz - y^2 = 0 \quad \text{and} \quad x^3 - y^2 z = 0 \quad \text{respectively}$$

and let $f: X \longrightarrow Y$ such that $f(u^2, ut, t^2) = (u^4 t^2, u^6, t^6)$. Then deg f=2. In the origin O=(0,0,1) we get:

$$f^* \omega_{Y,o} = 1/u^4 \ k[u]_o du^2$$

$$\omega_{X,o} = 1/u \ k[u]_o du^2$$

Then $f^* \omega_{Y,o} \not\subset \omega_{X,o}$.

THEOREM 1.8. Let $f: X \longrightarrow Y$ be a finite separable morphism of curves. Assume Y Gorenstein. Then the trace $k(X) \longrightarrow k(Y)$ restricts to a morphism $Tr: f_* \mathcal{O}_X \longrightarrow \mathcal{O}_Y$ if and only if the cotrace map $\Omega(k(Y)) \longrightarrow \Omega(k(X))$ induces an injection $Ctr: f^* \omega_Y \longrightarrow \omega_X$.

Proof . ($\Rightarrow$) Let $x \in X$, $y = f(x)$. We define $Ctr: (f^* \omega_Y)_x \longrightarrow \omega_{X,x}$ via $w \otimes f \longrightarrow wf$. Since $\omega_{Y,y}$ is a free $\mathcal{O}_y$-module, it suffices to show that $w \in \omega_{Y,y}$ implies $Ctr(w) \in \omega_{X,x}$. With the same notations of 1.1, we set:
$\{p_1, \dots, p_r\} = \pi^{-1}(x)$, $\{p_1, \dots, p_m\} = (f \circ \pi)^{-1}(y)$ $(m \geq r)$, $\{q_1, \dots, q_j\} = \nu^{-1}(y)$.
We must prove that $\Sigma_1^r \operatorname{Res}_{p_i}(h\, Ctr(w)) = 0$ for each $h \in \mathcal{O}_x$.
Let $-k$ be the minimum order of poles of $hCtr(w)$ in $p_1, \dots, p_m$: it is possible to find a function $r \in (f_* \mathcal{O}_X)_y$ such that

$$r \equiv 1 \mod \mathfrak{m}^k_{p_i} \qquad \text{for } i = 1, \dots, r$$

$$r \equiv 0 \mod \mathfrak{m}^k_{p_i} \qquad \text{for } i = r+1, \dots, m$$

(where $\mathfrak{m}_{p_i}$ denotes the maximal ideal of the local ring O_{p_i}) and $rh\epsilon(f_*O_X)_y$.
Then $\mathrm{Res}_{p_i}(rhCtr(w)) = 0$ for $i= r+1,\dots,m$ and

$\mathrm{Res}_{p_i}(rhCtr(w)) = \mathrm{Res}_{p_i}(hCtr(w))$ for $i= 1,\dots,r$.

Since w is regular in y by hypothesis, we obtain:

$\Sigma_1^r \mathrm{Res}_{p_i}(hCtr(w)) = \Sigma_1^m \mathrm{Res}_{p_i}(rhCtr(w)) = \Sigma_1^j \mathrm{Res}_{q_i}(Tr(rh)w) = 0.$

($\Leftarrow$) From the hypothesis it follows that $(f_* f^*\omega_Y)_y \subset (f_*\omega_X)_y$ for each $y\in Y$. Since $(f_*f^*\omega_Y)_y = (f_*O_X)_y \otimes_{O_y} \omega_{Y,y}$ by the projection formula, using the convention 0.4, we write

$$\alpha\,(f_*O_X)_y\tau \subset (f_*\omega_X)_y.$$

Hence, $Tr(\alpha(f_*O_X)_y\tau) \subset \omega_{Y,y} = \alpha O_y\tau$ by 1.5, and finally the k(Y)-linearity of the trace map implies our claim. □

EXAMPLE 1.9. In $\mathbb{P}_k^2$ let X,Y be the curves of equations $x^5-y^2z^3= 0$ and $x^3-y^2z = 0$ respectively ($k= \mathbb{C}$) and let $f: X \longrightarrow Y$ such that $f(t^2u^3,t^5,u^5) = (t^6u^3,t^9,u^9)$. It is easy to see that f verifies the equivalent conditions of theorem 1.8 and degf=3. The only critical points are the origin $0 =(0,0,1)$ and $P_\infty =(0,1,0)$.

At the point 0 we have:

$$\omega_{X,o} = 1/t^6\, k[t^2,t^5]_o\, dt^3$$
$$\omega_{Y,o} = 1/t^6\, k[t^6,t^9]_o\, dt^3$$

hence $f^*\omega_{Y,o} = \omega_{X,o}$.

At the point P_∞ we have:

$$\omega_{X,P_\infty} = 1/u^{10}\, k[u^3,u^5]_o\, du^3$$
$$\omega_{Y,P_\infty} = k[u^3]_o\, du^3$$

hence $f^*\omega_{Y,P_\infty} \hookrightarrow \omega_{X,P_\infty}$.

Remark 1.10. (i) If $f: X \longrightarrow Y$ is birational, since the trace $k(X) \longrightarrow k(Y)$ is the identical map in this case, then the cotrace morphism is defined if and only if f is an isomorphism.

(ii) The equivalent conditions of theorem 1.8 are verified if Y is a non singular curve (this follows immediately from theorem 4 of [Z.S.] ch.V, §3) and more in general if $f: X \longrightarrow Y$ is a flat morphism. In fact, in this case, for any $y\in Y$ $B = (f_*O_X)_y$ is a free module over $A = O_y$ and it is possible to find a basis of k(X) over k(Y) such that it is also a basis of B as A-module. Therefore the assertion follows from the definition of the trace.

§ 2. *CONDUCTOR AND DUALIZING SHEAVES.*

In this section we introduce the notion of conductor sheaf for a finite, not necessarily birational, morphism $f: X \longrightarrow Y$ of curves (see definition 2.2). Our main purpose is to investigate the relation between the conductor of Y in X and the classical conductors of X and Y in their normalizations (see theorem 2.7). We begin by recalling some well known properties of the conductor which will be used later. For more details see for instance [H] ,[P] ,[A.K.] .

THEOREM 2.1. Let $\pi: Z \longrightarrow X$ be a birational morphism of Gorenstein curves and let $\gamma_{Z/X}$ ($\tilde{\gamma}_{Z/X}$) be the conductor sheaf considered as a module over $\mathcal{O}_X$ ($\mathcal{O}_Z$). Then:

(i) $\gamma_{Z/X} \cong Ann_{\mathcal{O}_X}(\text{coker } Tr(\pi_* \omega_Z \longrightarrow \omega_X))$
where $Tr: \pi_* \omega_Z \longrightarrow \omega_X$ is the trace morphism.

(ii) There is an isomorphism of $\mathcal{O}_X$-modules

$$\gamma_{Z/X} \otimes_{\mathcal{O}_X} \omega_X \cong \pi_* \omega_Z$$

(iii) There is an isomorphism of $\mathcal{O}_Z$-modules

$$\tilde{\gamma}_{Z/X} \otimes_{\mathcal{O}_Z} \pi^* \omega_X \cong \omega_Z$$

DEFINITION 2.2. Let $f: X \longrightarrow Y$ be a finite morphism of curves. The conductor of Y in X is the $f_* \mathcal{O}_X$-module $\gamma_{X/Y} = \mathcal{H}om_{\mathcal{O}_Y}(f_* \mathcal{O}_X , \mathcal{O}_Y)$. (If the morphism of 1.2 is birational, then the isomorphism

$$\mathcal{H}om_{\mathcal{O}_Y}(f_* \mathcal{O}_X, \mathcal{O}_Y) \cong Ann_{\mathcal{O}_Y}(f_* \mathcal{O}_X / \mathcal{O}_Y)$$

implies that $\gamma_{X/Y}$ is the classical conductor).

We fix now our attention to the situation of diagram 1.1 and we assume further that X, Y are Gorenstein curves.

DEFINITION 2.3. With the preceding assumptions, we call relative dualizing sheaf of X respect to Y the invertible $\mathcal{O}_X$-module

$$\omega_{X/Y} = \omega_X \otimes_{\mathcal{O}_X} f^* \omega_Y^{-1}$$

PROPOSITION 2.4. If X and Y are regular curves, then

$$\omega_{X/Y}^{-1} = Ann_{\mathcal{O}_X}(\Omega_{X/Y})$$

where $\Omega_{X/Y}$ is the usual sheaf of relative differentials of X over Y. (This ideal is usually called <u>ramification ideal of</u> $f: X \longrightarrow Y$).

Proof. By definition of the 0th-Fitting ideal, we have

$$F^{o}(\Omega_{X/Y}) = f^{*}\Omega_{Y} \otimes_{O_X} \Omega_{X}^{-1}$$

and, since $\Omega_{X/Y}$ is locally monogenous,

$$F^{o}(\Omega_{X/Y}) = Ann_{O_X}(\Omega_{X/Y}) \quad \square$$

2.5. Note that, if $f: X \longrightarrow Y$ is a finite morphism of Gorenstein curves, then $\omega_{X/Y,x} = O_{X,x}$ for almost all $x \in X$. This is clear by Proposition 2.4.

The relation between the conductor and the dualizing sheaf is stated by the following

<u>*THEOREM*</u> 2.6. (i) There is an isomorphism of $f_{*}O_X$-modules

$$\gamma_{X/Y} \cong f_{*}(\omega_{X/Y})$$

(ii) If f is birational, then

$$\tilde{\gamma}_{X/Y} \cong \omega_{X/Y}$$

Proof. (i) Since ω_Y is a free O_Y-module and duality holds (see [H] ch.III, ex.7.2), we have: $\mathcal{H}om_{O_Y}(f_{*}O_X, O_Y) \otimes_{O_Y} \omega_Y \cong \mathcal{H}om_{O_Y}(f_{*}O_X, \omega_Y) \cong f_{*}\omega_X$, so $\gamma_{X/Y} \cong f_{*}\omega_X \otimes_{O_Y} \omega_Y^{-1} \cong f_{*}(\omega_{X/Y})$.

For the assertion (ii) see (iii) of theorem 2.1.$\square$

Now we want to show how the ramification ideal of the morphism $\phi: \overline{X} \longrightarrow \overline{Y}$ is related to the conductor sheaves $\gamma_{X/Y}, \gamma_{\overline{X}/X}, \gamma_{\overline{Y}/Y}$.

<u>*THEOREM*</u> 2.7 (Hurwitz's theorem for Gorenstein curves).

Set $\gamma_X = \gamma_{\overline{X}/X}$, $\tilde{\gamma}_X = \tilde{\gamma}_{\overline{X}/X}$, $\gamma_Y = \gamma_{\overline{Y}/Y}$, $\tilde{\gamma}_Y = \tilde{\gamma}_{\overline{Y}/Y}$ and $\mathcal{R}_{\overline{X}}$ the ramification ideal of $\phi: \overline{X} \longrightarrow \overline{Y}$.

(i) There is an isomorphism of $O_{\overline{X}}$-modules

$$\mathcal{R}_{\overline{X}} \otimes_{O_{\overline{X}}} \tilde{\gamma}_X \cong \phi^{*}\tilde{\gamma}_Y \otimes_{O_{\overline{X}}} \pi^{*}\omega_{X/Y}^{-1}$$

(ii) There is an isomorphism of $\pi_{*}O_{\overline{X}}$-modules

$$\pi_{*}(\mathcal{R}_{\overline{X}} \otimes_{O_{\overline{X}}} \tilde{\gamma}_X) \cong \pi_{*}\phi^{*}\tilde{\gamma}_Y \otimes_{O_X} \omega_{X/Y}^{-1}$$

Proof . (i) From 2.4 and 2.1 (iii) we obtain

$$\mathcal{R}_{\overline{X}} \otimes_{O_{\overline{X}}} \tilde{\gamma}_X \cong \phi^{*}\Omega_{\overline{Y}} \otimes_{O_{\overline{X}}} \Omega_{\overline{X}}^{-1} \otimes_{O_{\overline{X}}} \Omega_{\overline{X}} \otimes_{O_{\overline{X}}} \pi^{*}\omega_X^{-1}$$

Applying 2.1 (ii) to the curve Y we get:

$$\mathcal{R}_{\overline{X}} \otimes_{\mathcal{O}_{\overline{X}}} \widetilde{\gamma}_X \cong \phi^*\widetilde{\gamma}_Y \otimes_{\mathcal{O}_{\overline{X}}} \pi^* f^*\omega_Y \otimes_{\mathcal{O}_{\overline{X}}} \pi^*\omega_X^{-1} \cong \phi^*\widetilde{\gamma}_Y \otimes_{\mathcal{O}_{\overline{X}}} \pi^*\omega_{X/Y}^{-1}$$

(ii) follows from (i) by applying projection formula. □

If the morphism $f: X \longrightarrow Y$ in 2.7 is birational, we obtain the following result of [G.V.] (Proposition 3.9):

COROLLARY 2.8. Let $f: X \longrightarrow Y$ be a birational morphism of Gorenstein curves, then

$$\widetilde{\gamma}_X \otimes_{\mathcal{O}_{\overline{X}}} \pi^*\widetilde{\gamma}_{X/Y} \cong \widetilde{\gamma}_Y$$

Proof. Since ϕ is an isomorphism, we obtain $\widetilde{\gamma}_X \cong \widetilde{\gamma}_Y \otimes_{\mathcal{O}_{\overline{X}}} \pi^*\omega_{X/Y}$ by 2.7 and the thesis follows from theorem 2.6 (ii). □

2.9. When the "cotrace" between the dualizing sheaves of Y and X is defined, we can deduce the isomorphism of 2.7 from the exact sequence of 3.2, like in the non singular case.

§ 3. *TWO NUMERICAL FORMULAS.*

Now we study the situation locally, so we fix a point $x \in X$. Set:

$$y = f(x)$$
$$\{ p_1, \ldots, p_r \} = \pi^{-1}(x)$$
$$\{ q_1, \ldots, q_j \} = \nu^{-1}(y)$$
$$\{ p_1, \ldots, p_r, p_{r+1}, \ldots, p_m \} = (\nu \circ \phi)^{-1}(y)$$

We fix the following notations.

3.1. The stalk $\omega_{X/Y,x}$ is isomorphic to a principal fractional ideal in $\mathcal{O}_x$, more precisely we can write:

$$\omega_{X/Y,x} = (\beta / \alpha) \, \mathcal{O}_x$$

where $\alpha, \beta \in k(X)$ are such that $\omega_x = \beta \mathcal{O}_x \tau$, $\omega_y = \alpha \mathcal{O}_y \tau$ and $\tau \in \Omega(k(Y))$ is the basis of $\Omega(k(Y))$ over $k(Y)$ as in 0.4.

3.2. We consider the following number

$$r_x = \sum_{p_i \to x} v_{p_i}(\alpha / \beta)$$

where v_{p_i} is the valuation relative to the D.V.R. $\mathcal{O}_{p_i}$. Obviously r_x does not depend on the choice of the basis τ. We observe that this number can be negative (see example 1.7 in which $r_x = -3$), but when the cotrace morphism is defined (see theorem 1.8 and remark 1.9), then r_x is always a positive integer. In fact in this case we have the following exact sequence:

$$0 \longrightarrow f^*\omega_Y \longrightarrow \omega_X \longrightarrow \mathrm{coker}(Ctr) \longrightarrow 0$$

Hence $\omega^{-1}_{X/Y} = \mathit{Ann}_{\mathcal{O}_X}\, \mathrm{coker}(Ctr)$ is a proper ideal of $\mathcal{O}_X$. We want to show how the number r_x is related to the singularities of X at x and of Y at y and to ramification at the points belonging to $\pi^{-1}(x)$.

3.3. If $p_i \in \pi^{-1}(x)$, let e_i be the ramification index of the morphism ϕ in p_i, defined in [H] from the relation $\mathfrak{m}_{\phi(p_i)}\, \mathcal{O}_{\overline{X},p_i} = \mathfrak{m}^{e_i}_{p_i}$.

Then we know that

$$\mathcal{R}_{\overline{X},p_i} = \mathit{Ann}_{\mathcal{O}_{\overline{X},p_i}} (\Omega_{\overline{X},p_i} / \phi^*\Omega_{\overline{Y},p_i}) = \mathfrak{m}^{e_i-1}_{p_i}$$

3.4. Let λ_x be the degree of the conductor $\gamma_{X,x}$ defined by

$$\lambda_x = l_{\mathcal{O}_x}(\overline{\mathcal{O}}_x / \gamma_x)$$

3.5. Let μ_i be the degree of the conductor $\tilde{\gamma}_Y$ in $\phi(p_i)$ $1 \le i \le m$

$$\tilde{\gamma}_{Y,\phi(p_j)} = \mathfrak{m}^{\mu_i}_{\phi(p_i)}$$

Observe that $\mu_i = \mu_j$ if $\phi(p_i) = \phi(p_j)$.

From 3.3, 3.4, 3.5 we obtain the following two lemmas.

LEMMA 3.6. The stalk $\pi_*(\mathcal{R}_{\overline{X}} \otimes_{\mathcal{O}_{\overline{X}}} \gamma_{\overline{X}})_x$ is an ideal in $\overline{\mathcal{O}}_x$ of degree

$$d = \sum_{1 \le i \le r} (e_i + \lambda_i - 1)$$

LEMMA 3.7. $\pi_*\phi^*\tilde{\gamma}_{Y,x}$ is an ideal in $\overline{\mathcal{O}}_x$ of degree $\sum_{1 \le i \le r} \mu_i e_i$.

PROPOSITION 3.8. For each $x \in X$ the following formula holds:

$$\sum_{1 \le i \le r} (e_i - 1) + \lambda_x = \sum_{1 \le i \le r} \mu_i e_i + r_x$$

Proof. It follows from (2.7) (ii). by applying 3.2, 3.6, 3.7. □

PROPOSITION 3.9. In the situation described by 1.1 with X,Y Gorenstein curves, there is the following relation between the arithmetic genus of X and that one of Y:

$$n(2\pi_Y - 2) - (2\pi_X - 2) = \deg(\gamma_X) + \deg(\mathcal{R}_X) + n \deg(\gamma_Y)$$

where π denotes the arithmetic genus and deg of an invertible sheaf means the degree of the associated divisor.

Proof. If $\nu: \overline{C} \longrightarrow C$ is the normalization of a Gorenstein curve C, it is well known that $\deg(\nu^*\omega_C) = 2\pi_C - 2$ (see [S_1] ch.IV,11).

From this equality and the definition of r_X, it follows that:

$$r_X = \deg(\phi^*\nu^*\omega_Y) - \deg(\pi^*\omega_X) = n(2\pi_Y - 2) - (2\pi_X - 2).$$

Then the thesis is proved by 2.7 (i). □

Now we want to illustrate the preceding formula 3.8 with some examples.

EXAMPLE 3.10. In $\mathbb{P}^2_k$ ($k = \mathbb{C}$) let X, Y be the curves of equations $x^3 - y^2z = 0$, $x^5 - y^4z = 0$ respectively and let $O = (0,0,1)$.

We define $f: X \longrightarrow Y$ such that $f(t^2u, t^3, u^3) = (t^4u, t^5, u^5)$;

we have: $\omega_{X,o} = 1/t^2\, k[t^2,t^3]\circ dt$, $f^*\omega_{Y,o} = 1/t^{12}\, k[t^2,t^3]\circ dt$

hence $r_o = -10$, $e_o = 1$ because f is birational, $\lambda_o = 2$, $\mu_o = 12$ and

$$r_o + \mu_o = \lambda_o + e_o - 1$$

EXAMPLE 3.11. In $\mathbb{P}^2_k$ ($k = \mathbb{C}$) let X be as in 3.10 and $Y = \mathbb{P}^1_k$, let $f: X \longrightarrow Y$ be such that $f(t^2u, t^3, u^3) = (t^2, u^2)$ (deg f = 2).

We have $\omega_{X,o} = 1/t^3\, k[t^2,t^3]\circ dt^2$, $\omega_{Y,o} = k[t^2]\circ dt^2$; hence $r_o = 3$, $e_o = 2$, $\lambda_o = 2$, $\mu_o = 0$ and $r_o = \lambda_o + e_o - 1$.

4. *THE COMPLEMENTARY MODULE*

Let $f: X \longrightarrow Y$ be any finite and separable morphism of Gorenstein curves. If U is an open affine of Y, we set $A' = \Gamma(U, O_Y)$ and $B' = \Gamma(U, f_*O_X)$.

DEFINITION 4.1. As in [Z.S.] vol. I, we call complementary module $L(B'/A')$ of B' with respect to A' the following B'-module:

$$L(B'/A') =: \{x \in k(X) \text{ s.t. } Tr(xb) \in A' \text{ for each } b \in B'\}.$$

Note that $L(B'/A')$ is a fractional ideal of B' and if A' is integrally closed in its field of fractions, then $L(B'/A')$ contains B'.

DEFINITION 4.2. Inspired by Kunz (see [K]), we call generalized complementary module of B' with respect to A' the B'-module:

$$\hat{L}(B'/A') =: \{ w \in \Omega(k(X)) \text{ s.t. } Tr(wB') \in \Gamma(U, \omega_Y)\}.$$

Remark 4.3. If S is a multiplicative subset of A', we have:

(i) $S^{-1}L(B'/A') = L(S^{-1}B'/S^{-1}A')$

(ii) $S^{-1}\hat{L}(B'/A') = \hat{L}(S^{-1}B'/S^{-1}A') =: \{w \in \Omega(k(X)) \text{ s.t. } Tr(wS^{-1}B') \subseteq S^{-1}\Gamma(U, \omega_Y)\}$.

For the proof of (i) see [S_2] , ch. III, 4 and the proof (ii) is quite similar.

In view of (4.3) it is natural to give the following

DEFINITION 4.4. We call complementary module (generalized complementary module respectively) the coherent $f_*\mathcal{O}_X$-module $\mathcal{L}_{X/Y}$ ($\hat{\mathcal{L}}_{X/Y}$) defined by the collection $\{ L((f_*\mathcal{O}_X)_y/\mathcal{O}_y)\}_{y\in Y}$ ($\{ \hat{L}((f_*\mathcal{O}_X)_y/\mathcal{O}_y)\}_{y\in Y}$ respectively).

We notice that if U, A', B' are as above, then:

$$\Gamma(U,\mathcal{L}_{X/Y}) = L(B'/A') \text{ and } \Gamma(U,\hat{\mathcal{L}}_{X/Y}) = \hat{L}(B'/A').$$

The next proposition states the connection between $\mathcal{L}_{X/Y}$ and $\hat{\mathcal{L}}_{X/Y}$.

PROPOSITION 4.5. There exists a natural isomorphism of $f_*\mathcal{O}_X$-modules

$$\hat{\mathcal{L}}_{X/Y} \cong \mathcal{L}_{X/Y} \otimes_{\mathcal{O}_Y} \omega_Y$$

Proof. Let y be a point of Y; we put $A = \mathcal{O}_y$, $B = (f_*\mathcal{O}_X)_y$, $\tau\in \Omega(k(Y))$ a fixed basis of $\Omega(k(Y))$ over $k(Y)$ and, consequently, $\omega_y = \alpha A\tau$, $\alpha\in k(Y)$ as in (0.4). Then it is enough to show the equality between the stalks at y: $\hat{L}(B/A) = L(B/A) \otimes_A \omega_y$. Clearly the map $\chi: L(B/A) \otimes_A \omega_y \longrightarrow \hat{L}(B/A)$ defined by $\chi(x \otimes \alpha\tau) = x\alpha\tau$ is an isomorphism of B-modules, because the trace map is k(A)- linear. □

In (1.5) we have seen that the trace of a differential regular at each point of $f^{-1}(y)$ is also regular at y. Now we are able to prove the converse.

PROPOSITION 4.6. The following equality holds:

$$\hat{\mathcal{L}}_{X/Y} = f_*\omega_X$$

Proof. As in (4.5) we have to show that $\hat{L}(B/A) = (f_*\omega_X)_y$. By (1.5) we get: $(f_*\omega_X)_y \subsetneq \hat{L}(B/A)$. To prove the reverse inclusion, we divide the proof in two steps.

Step 1). First we show that $\sum_{p_i\to x_i} \mathrm{Res}_{p_i}(b\omega) = 0$ for each $b\in B$. Let $-k$ the minimum order of ω in the set $(\nu\circ\phi)^{-1}(y)$. Let $\lambda\in B$ be a function such that $\lambda\omega$ is regular in $x_2,\ldots,x_r$ and $\lambda\equiv 1 \bmod \mathfrak{m}_{x_1}^k$. Then for each $b\in B$ we have:

$$0 = \sum_{q_i\to y}\mathrm{Res}_{q_i}\mathrm{Tr}(\lambda b\omega) = \sum_{p_i\to y}\mathrm{Res}_{p_i}(\lambda b\omega) = \sum_{p_i\to x_1}\mathrm{Res}_{p_i}(b\omega).$$

Step 2). Now we show that $\sum_{p_i \to x_1} (h\omega) = 0$ for each $h \in \mathcal{O}_{x_1}$.
If $h \in \mathcal{O}_{x_1}$ then $h = b/s$ where $b \in B$ and $s \in B - \mathfrak{m}_{x_1}$. Let $s' \in B$ be such that $ss' \equiv 1 \mod \mathfrak{m}_{x_1}^k$. Then $\sum_{p_i \to x_1} \mathrm{Res}(bs'/ss')\omega = \sum_{p_i \to x_1} \mathrm{Res}_{p_i}(bs'\omega) = 0$.

COROLLARY 4.7. There is a sequence of isomorphisms of $f_*\mathcal{O}_X$-modules:

$$\mathcal{L}_{X/Y} \cong f_*\omega_X \otimes_{\mathcal{O}_Y} \omega_Y^{-1} \cong f_*\omega_{X/Y} \cong \gamma_{X/Y}$$

Proof. It follows immediately from 4.5, 4.6 and 2.6. □

Finally we state the following relation involving complementary modules and conductor sheaves.

PROPOSITION 4.8. With the same notations of 1.1 and the further hypothesis that X and Y are Gorenstein curves, there is an isomorphism of $f_*\pi_*\mathcal{O}_{\bar{X}}$-modules:

$$f_*\gamma_X \otimes_{f_*\mathcal{O}_X} \mathcal{L}_{X/Y} \cong \nu_*\mathcal{L}_{\bar{X}/\bar{Y}} \otimes_{\nu_*\mathcal{O}_{\bar{Y}}} \gamma_Y$$

Proof. By 2.7 $\tilde{\gamma}_X \otimes_{\mathcal{O}_{\bar{X}}} \pi^*\omega_{X/Y} \cong \mathcal{R}_{\bar{X}}^{-1} \otimes_{\mathcal{O}_{\bar{X}}} \phi^*\tilde{\gamma}_Y$. Then by applying projection formula: $f_*(\gamma_X \otimes_{\mathcal{O}_X} \omega_{X/Y}) \cong \nu_*(\phi_*\mathcal{R}_{\bar{X}}^{-1} \otimes_{\mathcal{O}_{\bar{Y}}} \tilde{\gamma}_Y)$ and the conclusion follows from Corollary 4.7. □

As a consequence of Proposition 4.8, applied to the particular case Y non singular, we recover the following well known result:

COROLLARY 4.9 (see [S_2], ch.III, 4). Let A, B be as in the proof of 4.5. Let $\bar{B}$ be the normalization of B and assume that A is integrally closed. Then we have:

$$\gamma_{\bar{B}/B} = L(\bar{B}/\bar{A}) \cdot L(B/A)^{-1}$$

where $L(B/A)^{-1}$ denotes the inverse of $L(B/A)$ as a B-module.

Acknowledgment. We are grateful to Ragni Piene for a critical and stimulating reading of our work.
We also thank Silvio Greco for useful suggestions.

REFERENCES.

[A.K.] A.ALTMAN, S.KLEIMAN - *Introduction to Grothendieck Duality Theory*- Lecture Notes in Mathematics No. 146 - Springer - 1970.

[C] C.CHEVALLEY - *Introduction to the theory of algebraic functions of one variable* - A.M.S. New York - 1951.

[G.V.] S.GRECO, P.VALABREGA - On the theory of adjoints- Algebraic Geometry- Copenhagen 1978 - Lecture Notes in Mathematics No. 732 - 1979.

[H] R.HARTSHORNE - *Algebraic Geometry* - Springer Berlin - 1978.

[K] E.KUNZ - Holomorphe Differentialformen auf algebraischen Varietaten mit Singularitaten - Manuscripta Mathematica 15; 91-108 - 1975.

[P] R.PIENE - Ideals associated to a desingularization - Algebraic Geometry - Copenhagen 1978 - Lecture Notes in Mathematics No. 732- 1979.

[S_1] J.P.SERRE - *Groupes algebriques et corps de classes* - Hermann Paris- 1959.

[S_2] J.P.SERRE - *Corps locaux* - Hermann Paris - 1962.

[Z.S.] O.ZARISKI, P.SAMUEL - *Commutative Algebra* - Van Nostrand New York - 1952.

NOTE ON LOCAL DOMAINS WHOSE ORDER FUNCTION IS A VALUATION

U. Orbanz

Mathematisches Institut der

Universität zu Köln

West Germany

Introduction. This note is concerned with Zariski's Factorization Theorem for regular surfaces ((9), p.538). Locally this says the following. If R and S are two dimensional regular local rings with the same quotient field and such that S dominates R, then S can be obtained from R by a finite number of quadratic transformations ((1), Theorem 3). The first step in the proof is to show that S dominates an immediate quadratic transform of R. Recently J. Sally ((7)) observed that this property is equivalent to saying that $l(M(R)S) < 2$ if $S \neq R$, where l denotes the analytic spread ((5)). Now let R and S be two regular local rings of dimension d having the same quotient field and such that S dominates R. Although it is known that the Factorization Theorem is no longer true if $d > 2$ (see (6) and (8)), J. Sally proved that still $S \neq R$ implies $l(M(R)S) < d$.

In this note we give a different proof of this result. One advantage of the new proof is that from the assumption $l(M(R)S)=d$ we derive some positive statements (see Prop. 3 and 4) instead of a contradiction. More important however is that our argument applies to a more general situation. The crucial property of R - besides being normal - is that the order function of R should be a valuation. Clearly this assumption is much weaker than regularity. For $d=2$ we obtain again that S dominates a quadratic transform of R. To deduce a Factorization Theorem we need to know that the assumptions made about R are stable under quadratic transformations. An example shows that this is not true in general.

The note closes with the observation that in any regular local ring S, an ideal I such that $l(I) \leq 2$ and not contained in $M(S)^2$ defines a complete intersection (actually a certain hypersurface). This generalizes Prop. (2.5) of (7).

Notation. Any local ring R is assumed to be noetherian. The maximal ideal will be denoted by $M(R)$.

We start by recalling some facts about analytic independence and analytic spread (see (5)). If I is any proper ideal in a local ring R and $x_1,\ldots,x_m$ are elements of I, then $x_1,\ldots,x_m$ are said to be analytically independent in I if the following property is satisfied: Given any form $F(X_1,\ldots,X_m) \in R[X_1,\ldots,X_m]$, of degree s say, such that $F(x_1,\ldots,x_m) \in I^s M(R)$, it follows that all coefficients of F belong to $M(R)$. The elements $x_1,\ldots,x_m$ are said to be analytically independent if they are so in the ideal they generate in R. If $x_1,\ldots,x_m$ are analytically independent in any ideal, then they are analytically independent in the absolute sense.

Let $x_1,\dots,x_m$ be elements of R, contained in $M(R)$, and let $y_2 = x_2/x_1,\dots,y_m = x_m/x_1$ in R_{x_1}. Let $R' = R[y_2,\dots,y_m]$, the subring of R_{x_1} generated by R and $y_2,\dots,y_m$, and let

$$h: R/M(R)[Y_2,\dots,Y_m] \longrightarrow R'/M(R)R'$$

be the canonical R-homomorphism such that $h(Y_i)$ is the image of y_i in $R'/M(R)R'$. The following lemma can be proved by an easy computation.

<u>LEMMA</u>. $x_1,\dots,x_m$ are analytically independent if and only if h is an isomorphism.

In particular we note that if $x_1,\dots,x_m$ are analytically independent then $M(R)R'$ is a prime ideal.

Now we turn to the notion of analytic spread. Let I be a proper ideal in a local ring R again. The analytic spread $l(I)$ is defined to be the dimension of the ring $gr_I(R)\otimes_R R/M(R)$, where $gr_I(R)$ denotes the graded ring of R with respect to I.

<u>THEOREM</u> (Northcott-Rees (<u>5</u>)). Assume $R/M(R)$ to be infinite. Then the following numbers are equal:

a) $l(I)$

b) the least number of elements $x_1,\dots,x_m$ in I such that $I^n = (x_1,\dots,x_m)I^{n-1}$ for some n

c) the maximal number of elements which are analytically independent in I.

Moreover, if $x_1,\dots,x_{l(I)}$ are given as in b), then they are analytically independent in I.

Recall that an element x of R is called integral over some ideal J of R, if x satisfies a relation

$$x^n + a_1x^{n-1} + \dots + a_n = 0 \ , \ a_i \in J^i \ , \ i=1,\dots,n$$

for some n. It is well-known that I is integral over $J \subset I$ if and only if $I^n = JI^{n-1}$ for some n. Therefore b) in the Theorem above signifies that $l(I)$ is the least number of elements generating an ideal over which I is integral. ($x_1,\dots,x_{l(I)}$ as in b) above are said to generate a minimal reduction of I.) Let us note that in general we have

$$ht(I) \leq l(I) \leq \dim R$$

and in particular

$$l(M(R)) = \dim R.$$

For the rest of this paper we fix the following notation. R denotes a local domain with quotient field K. For any ideal I in R, $ord(R,I)$ denotes the order function with respect to I, i.e.

$$ord(R,I)(f) = \sup\{n \mid f \in I^n\} \ , \ f \in R.$$

If $I = M(R)$, we simply write $ord(R)$ instead of $ord(R,I)$.

We say that $ord(R,I)$ is a valuation, if for any $f,g \in R$ we have

$$ord(R,I)(fg) = ord(R,I)(f) + ord(R,I)(g) \ ,$$

i.e. if $ord(R,I)$ can be uniquely extended to a (discrete rank 1) valuation of K. In this case I must be prime. Note that $ord(R,I)$ is a valuation if and only if $gr_I(R)$ is an integral domain.

PROPOSITION 1. Let P be a prime ideal of R such that $ord(R,P)$ is a valuation v with valuation ring V. Let $x_1,\dots,x_m$ be a minimal system of generators of P. Put $t = x_1$, $y_i = x_i/t$ for $i=2,\dots,m$ and $R' = R[y_2,\dots,y_m]$. Then

a) $M(V) = x_iV$ for $i=1,\dots,m$

b) $R' \subset V$ and $M(V) \cap R' = tR'$

c) $V = (R')_{tR'}$.

Proof. a) is trivial since $v(x_i) = 1$, so also $R' \subset V$ is clear. Let $f \in M(V) \cap R'$ and write

$$f = t^{-r} g(x_1, \ldots, x_m) ,$$

where $g(X_1, \ldots, X_m) \in R[X_1, \ldots, X_m]$ is a homogeneous polynomial of degree r. Now $v(f) \geq 1$ implies

$$v(g(x_1, \ldots, x_m)) \geq r + 1 ,$$

and therefore $g(x_1, \ldots, x_m) \in P^{r+1}$. This implies $f \in PR' = tR'$, thereby proving b) and c).

PROPOSITION 2. Let ord(R) be a valuation. Assume that R/M(R) is infinite and let $x_1, \ldots, x_d$ generate a minimal reduction of M(R) (d = dim R). Put $t = x_1$, $y_i = x_i/t$ for i=2,...,d and $R'' = R[y_2, \ldots, y_m]$. If V denotes the valuation ring of ord(R), then we have

a) $R'' \subset V$

b) $M(V) \cap R'' = M(R)R''$

c) V is the unique valuation ring of K dominating $(R'')_{M(R)R''}$.

Proof. $x_1, \ldots, x_d$ can be completed to a minimal system $x_1, \ldots, x_m$ of generators of M(R), so $R'' \subset V$. Since $x_1, \ldots, x_d$ are analytically independent, M(R)R" is a prime ideal. Let $y_i = x_i/t$ for i=d+1,..,m and $R' = R[y_2, \ldots, y_m]$. Then R' is integral over R", and therefore

$$M(R)R'' = P' \cap R'' \quad \text{for some prime ideal } P' \text{ of } R'.$$

Hence b) follows from

$$M(R)R'' \subset M(R)R' \cap R'' \subset P' \cap R'' = M(R)R''$$

together with $M(R)R' = M(V) \cap R'$ (see Prop.1). To prove c), let W be any valuation ring of K dominating $(R'')_{M(R)R''}$. Then $M(W) \cap R'$ contains the prime ideal $tR' = M(V) \cap R'$ and contracts to the same ideal in R". It follows that $M(W) \cap R' = tR'$, and therefore W dominates V by Proposition 1.

REMARK. An analogous statement to Proposition 2 is true if $M(R)$ is replaced by a prime ideal P of R such that $\mathrm{ord}(R,P)$ is a valuation and $\mathrm{ht}(P) = l(P)$. The proof is obtained by localizing.

COROLLARY. If $\mathrm{ord}(R)$ is a valuation with valuation ring V, then V is a prime divisor of R, i.e. the transcendence degree of $V/M(V)$ over $R/M(R)$ is $\dim R - 1$.

Proof. Assume first that $R/M(R)$ is infinite. Using the notation of Proposition 2 we see that the residue classes of $y_2,\ldots,y_d$ in $V/M(V)$ are algebraically independent over $R/M(R)$. This proves the assertion, since the transcendence degree in question is at most d-1. If $R/M(R)$ is finite, we pass to $T = R[X]_{M(R)R[X]}$ to obtain the result. (Note that $\mathrm{ord}(T)$ is the unique extension of $\mathrm{ord}(R)$ to $K(X)$ giving value 0 to X.)

Now in addition to R we consider another local domain S between R and K .

PROPOSITION 3. Assume that $\mathrm{ord}(R)$ is a valuation. If S dominates R and $l(M(R)S) = \dim R$, then

$$\mathrm{ord}(R) = \mathrm{ord}(S)|R.$$

Proof. Passing to $R[X]_{M(R)R[X]}$ and $S[X]_{M(S)S[X]}$ if necessary, we may assume $R/M(R)$ to be infinite. Let $d = \dim R$ and let $x_1,\ldots,x_d$ generate a minimal reduction of $M(R)$. Then, by assumption, $x_1,\ldots,x_d$ generate a minimal reduction of $M(R)S$. Let $t = x_1$, $y_i = x_i/t$ for $i=2,\ldots,d$, and put

$$R'' = R[y_2,\ldots,y_d] \quad \text{and} \quad S'' = S[y_2,\ldots,y_d] .$$

Since $x_1,\dots,x_d$ are analytically independent in R as well as in S, the inclusion $R'' \longrightarrow S''$ induces an inclusion $R''/M(R)R'' \longrightarrow S''/M(S)S''$, which means that

$$M(S)S'' \cap R'' = M(R)R''.$$

Therefore $(S'')_{M(S)S''}$ dominates $(R'')_{M(R)R''}$, and from Proposition 2, c), we see that V is the unique valuation ring of K dominating $(S'')_{M(S)S''}$, where V denotes the valuation ring of ord(R) again. So we get

$$S \subset V \quad \text{and} \quad M(V) \cap S = M(S).$$

It follows that $\mathrm{ord}(R)(x) \geq n$ for all $x \in M(S)^n$, and therefore

$$M(S)^n \cap R = M(R)^n \quad \text{for all } n.$$

<u>REMARK</u>. Let R,S,V be the same as in Proposition 3 above. Then $l(M(R)S) = \dim R \leq \dim S$. But we have the general inequality

$$\dim R \geq \dim S + \mathrm{tr.deg.}_{R/M(R)} S/M(S)$$

((<u>8</u>), Lemma 5.3). This implies that $\dim S = \dim R$ and $S/M(S)$ is algebraic over $R/M(R)$. On the other hand, in the Corollary to Proposition 2 we noted that V is a prime divisor of R, so $V/M(V)$ is a finitely generated field extension of $R/M(R)$ ((<u>1</u>), Prop.2), and since $S \subset V$ we obtain

$$[S/M(S):R/M(R)] < \infty .$$

<u>PROPOSITION 4</u>. Assume that S dominates R, that R and S have the same dimension d and that ord(R) and ord(S) are valuations. Let $x_1,\dots,x_d$ generate a minimal reduction of M(R). If $l(M(R)S) = d$, then the residue classes of $x_1,\dots,x_d$ in $M(S)/M(S)^2$ (as elements of $gr_{M(S)}(S)$) are algebraically independent over $S/M(S)$.

Proof. Let $M_i(X_1,\dots,X_d)$ be monomials in $X_1,\dots,X_d$ of degree s and let $f(X) = \sum a_i M_i(X)$, $a_i \in S$. Assume that

$f(x_1,\ldots,x_d) \in M(S)^{s+1}$. Now from Proposition 3 we know that ord(R) and ord(S) define the same valuation v of K, and consequently

$$v(f(x_1,\ldots,x_d)/x_1^s) > 0,$$

and therefore (see Prop.3)

$$f(x_1,\ldots,x_d)/x_1^s \in M(S)S[x_2/x_1,\ldots,x_d/x_1].$$

Since the residue classes mod M(S) of $x_2/x_1,\ldots,x_d/x_1$ are algebraically independent over S/M(S), we see that $a_i \in M(S)$ for all i.

THEOREM. Let R be a normal local domain whose order function is a valuation, and let K be the quotient field of R. Let S be a regular local ring in K dominating R and assume $l(M(R)S) = \dim R$. Then $S = R$.

Proof. Without loss of generality we may assume that R/M(R) is infinite. Let $x_1,\ldots,x_d$ generate a minimal reduction of M(R). Then by Proposition 4, $x_1,\ldots,x_d$ are a regular system of parameters of S. (Note that $\dim S = d$ by the remark following Prop.3.) Since $[S/M(S):R/M(R)]$ is finite, Zariski's Main Theorem ((4),(37.4)) implies $S = R$.

COROLLARY 1. Let R be a normal local domain with quotient field K and S a regular local overring of R in K. Let $P = M(S) \cap R$ and assume that ord(R,P) is a valuation and $ht(P) = l(P)$. If $l(PS) = l(P)$, then $S = R_P$.

Proof. Apply the Theorem to the local homomorphism $R_P \longrightarrow S$.

COROLLARY 2. Let R be a two dimensional normal local domain having a valuation as order function, and let S be a two dimensional regular local ring dominating R and contained in the quotient field of R. If $S \neq R$, then S dominates a quadratic transform of R.

Proof. By the Theorem we must have $l(M(R)S) = 1$, which implies that $M(R)S$ is principal, since S is normal.

In the situation of Corollary 2, let R' be the quadratic transform of R dominated by S. Then first of all R' is normal, since the powers $M(R)^n$ are integrally closed ((3), Theorem 1). If ord(R') was a valuation again, then we could continue to factor $R \longrightarrow S$ into a sequence of quadratic transformations. But ord(R') need not be a valuation, as we see from the following

EXAMPLE. Take for R the origin of the surface $z^3 + x^4z + xy^2$ (over a field of characteristic zero). Then R is normal, and since the initial form $z^3 + xy^2$ is irreducible, ord(R) is a valuation. But the origin of the surface $z^3 + x^2z + y^2$ is a quadratic transform of R having a non-reduced tangent cone.

It is well-known that Zariski's Factorization Theorem also holds if R is assumed to be a rational singularity ((2), Theorem (4.1)), and actually this was the motivation to look for an extension of J. Sally's result to non-regular R. Rational singularities and normal local domains whose order function is a valuation have in common that their quadratic transforms are normal again, but this seems to be the only link between those two types of rings. The example given above is non-rational, since the multiplicity is 3, and on the other hand rational double points of type F_4 or G_2 have a non-reduced tangent cone ((2), p. 271; take for example $z^2 + y^3 + x^5$).

Here is a class of examples to which our Theorem applies. Let C be an irreducible, non-singular, plane projective curve and take for R the local ring of the vertex of the affine cone over C. Then R is normal and the irreducibility of C implies that ord(R) is a valuation. (We note that R is rational if and only if the degree of C is 2.) It is easy to see that only one quadratic transformation resolves R. Now by Corollary 2 any resolution of R dominates the blowing up of R with center M(R), hence this is the minimal resolution of R.

The following result generalizes Proposition (2.5) of (7).

PROPOSITION 5. Let S be a regular local ring and let I be an ideal of S satisfying

$$I \not\subset M(S)^2 \quad \text{and} \quad l(I) \leq 2.$$

Then I is a complete intersection.

Proof. We may S/M(S) assume to be infinite. There is nothing to show if $l(I) = 1$, so assume $l(I) = 2$. Let x,y generate a minimal reduction of I. Since I is integral over $xS + yS$ and $M(S)^2$ is integrally closed, we have $xS + yS \not\subset M(S)^2$, say $x \notin M(S)^2$. Then in the regular local ring S/xS we have $l(I/xS) = 1$, which implies $I = xS + yS$. (Actually I defines a hypersurface, namely S/I is defined by one element in the regular local ring S/xS.)

COROLLARY. Let R be a 3-dimensional normal local domain whose order function is a valuation, and let S be a 3-dimensional regular local ring dominating R and having the same quotient field. If $M(R) \not\subset M(S)^2$, then M(R)S is a complete intersection.

Proof. If $l(M(R)S) \leq 2$, the result follows from Proposition 5. If $l(M(R)S) = 3$, then R = S is regular by the Theorem.

REFERENCES.

(1) S. Abhyankar, On the valuations centered in a local domain, Amer. J. Math. 78 (1956), 321 - 348

(2) J. Lipman, Rational singularities, with applications to algebraic surfaces and unique factorization, Publ. Math, IHES 36 (1969), 195 - 279

(3) A. Mattuck, Complete ideals and monoidal transforms, Proc. AMS (1970), 555 - 560

(4) M. Nagata, Local rings, Interscience Publishers, New York 1962

(5) D. G. Northcott - D. Rees, Reductions of ideals in local rings, Proc. Camb. Phil. Soc. 50 (1954), 145 - 158

(6) J. Sally, Regular overrings of regular local rings, Transactions AMS 171 (1972), 291 - 300

(7) J. Sally, Fibres over closed points of birational morphisms of nonsingular varieties, to appear in the Amer. J. Math.

(8) D. Shannon, Monoidal transforms of regular local rings, Amer. J. Math. 95 (1973), 294 - 320

(9) O. Zariski, Reductions of the singularities of algebraic three dimensional varieties, Annals of Math. 45 (1944), 472 - 542

MAXIMAL HILBERT FUNCTIONS OF ONE-DIMENSIONAL LOCAL RINGS

Ferruccio Orecchia

Istituto di Matematica
dell'Università di Napoli,
NAPOLI-ITALY

INTRODUCTION. This paper is an expanded version of the lecture given by the author at the Meeting on Commutative Algebra of Trento (June '80).

Let $A=k[X_o,\ldots,X_r]_{(X_o,\ldots,X_r)}/I$ be the local ring of a curve C at a singular point O. Let M and e(A) be the maximal ideal and multiplicity of A. We want to collect under the same general theory various examples of ideals I which need an arbitrary large number of generators (see [Ge], $[G.O.]_2$, [Ma]). The unifying idea of all these examples (and this was not pointed out before) is the maximality of the Hilbert function of A. More precisely this means that the tangent cone Spec(G(A)) of C in O is Cohen-Macaulay and $H_A^o(n)=\dim_k(M^n/M^{n+1})=\mathrm{Min}\{\binom{n+r}{r},e(A)\}$, for any $n\in\mathbf{N}$. Under these assumptions we compute the minimal number v(I) of generators of I and we recover the results on v(I), of [Ge], §4, [G.O.], [Ma] as special cases of the results of this paper. In particular the <u>classical Macaulay examples</u> fit in this theory.

All the rings will be commutative, noetherian and with identity.

1. Let R be a ring and I an ideal of R. With $v(I)$ we denote the least number of generators of I. Let (A,M) be a local ring and $G(A)$ the graded $k=A/M$-algebra $\bigoplus_{n\geq 0}(M^n/M^{n+1})$. Given an element $a\in A$, $a\neq 0$, let n be the largest integer such that $a\in M^n$. The residue class $a^\star$ of a in M^n/M^{n+1} is the initial form of a and n is the degree of $a^\star$ (or of a). If I is an ideal of A we denote by $I^\star$ the homogeneous ideal of $G(A)$ generated by the initials forms of the elements of I. It follows from Nakayama (homogeneous) Lemma that any set of (homogeneous) generators of I (of $I^\star$) contains a minimal generating set.

LEMMA 1. Let n be a positive integer. If $I^\star$ is generated by forms of degree n and n+1 then $v(I^\star)=v(I)$.

PROOF. Let $(f_1^\star,\dots,f_p^\star,f_{p+1}^\star,\dots,f_q^\star)$, $q\geq p$, be a minimal generating set of $I^\star$, where n and n+1 are respectively the degrees of $f_i^\star$ for $i=1,\dots p$ and $i=p+1,\dots,q$. It is easily seen ([H], p. 190) that $I=(f_1,\dots,f_q)$. Thus the claim is proved if we show that the f_i are all necessary to generate I . In fact suppose $f_j=\sum_{i\neq j} a_i f_i$ then, if $f_j^\star$ has degree n, $f_j^\star=\sum_{i\neq 1}^{p} c_i f_i^\star$ where $c_i\in k$, if $f_j^\star$ has degree n+1 then $f_j^\star=\sum_{i=1}^{p} P_i f_i^\star+\sum_{i=p+1}^{q} c_i f_i^\star$ where $c_i\in k$ and P_i are zero or forms of degree 1. This contradicts the minimality of the $f_i^\star$.

REMARK 2. The fact that the forms generating $I^\star$ have two consecutive degrees is crucial in lemma 1, as the following examples show:
if $I=(xy+xz^2,yz,x^3)\subset k[X,Y,Z]_{(X,Y,Z)}$ then $I^\star=(xy,yz,x^3,xz^3)$. If $I=(xz-y^3,yz-x^3)$, then $I^\star=(xz,yz,x^4-y^4)$.

We recall that the Hilbert functions of the maximal ideal M in the local ring A are defined as follows: if $k=A/M$ and $n\in\mathbf{N}$

$$H_A^o(n)=\dim_k(M^n/M^{n+1}) \quad , \quad H_A^j(n)=\sum_{i=o}^{n} H_A^{j-1}(i)$$

<u>From now on</u> A <u>will be a one-dimensional Cohen-Macaulay ring</u>.

It is well known (see[S] and [L]) that if $e=e(A)$ is the multiplicity of A then:

(a) $H_A^o(n)\leq e$ and, for n large enough, $H_A^o(n)=e$. Further if $H_A^o(n)=e$ then $H_A^o(n')=e$, for $n'>n$.

An element x of M is superficial if there is a positive integer c such that : $(M^n:x)\cap M^c=M^{n-1}$, for all $n>c$.

THEOREM 3. (i) Let x be a superficial element and a non zero divisor of A. Then $e(A)=\dim_k(A/xA)$. (ii) Let $x\in M-M^2$, then $x^\star\in G(A)$ is a non zero divisor if and only if $H_A^o(A)=H_{A/xA}^1(n)$ for any $n\in\mathbf{N}$.

PROOF. (i) (See[G], thm 16) . (ii) (See[S], Ch. 2, Note to thm 3.1).

Let emdim A = $H_A^o(n)=r+1$ be the embedding dimension of A . Clearly:

$$H_A^o(n)\leq\binom{n+r}{r}. \text{ Hence } H_A^o(n)\leq\mathrm{Min}\{\binom{n+r}{r},e\}, \text{ if } n\in\mathbf{N}.$$

DEFINITION 4. The ring A has maximal Hilbert function if: $H_A^o(n)=\mathrm{Min}\{\binom{n+r}{r},e\}$.

REMARKS 5. (i) By (a) it is enough to check the maximality of $H_A^o(n)$ for the integers n such that $\binom{n+r}{r}\leq e$. (ii) The notion of extremal graded

Cohen-Macaulay ring , given by Schenzel in [Sc] , is for one-dimensional rings a particular case of Def. 4. In fact it is easily shown that if (B,N) is any extremal graded 1-dimensional Cohen-Macaulay ring, then B_N has maximal Hilbert function, while the rings A of examples 2 and 3 of §2 have G(A) Cohen-Macaulay and not extremal.

COROLLARY 6. Let $m=\mathrm{Min}\{n \mid \binom{n+r}{r}>e\}$ and suppose that A has maximal Hilbert function and x is an element of A . Then:
$x^\star$ is a non zero divisor of G(A) if and only if $H^o_{A/xA}(n)=\binom{n+r-1}{r-1}$ for $n<m$, $H_{A/xA}(m)=e-\binom{m+r-1}{r}$, $H^o_{A/xA}(n)=0$ for $n>m$.

PROOF. Follows easily from thm 3, (ii).

REMARK 7. Let A be the local ring at the origin O of a curve C over an algebraically closed field. Suppose that G(A) is reduced; then the tangent cone Proj(G(A)) of C in O consists of e=e(A) points $\{P_1,\dots,P_e\}\in\mathbb{P}^r$. The ring A has maximal Hilbert function if and only if the points $\{P_1,\dots,P_e\}$ are in generic position (see[O_2] ,§3). It is shown in [G.O.]$_1$,thm 4, that the e-tuples of points of $\mathbb{P}^r$ $(P_1,\dots,P_e)$ considered as points of $\mathbb{P}^r \times \dots \times \mathbb{P}^r$ (e-times) which are in generic position, form a non-empty open subset of $\mathbb{P}^r \times \dots \times \mathbb{P}^r$.

Now we assume that A=R/I is a quotient of a regular local ring R of dimension r+1 modulo an ideal I (of height r). Then :

$$G(A)=k[x_o^\star,\dots,x_r^\star]\simeq k[X_o,\dots,X_r]/I^\star \text{ (where } \{x_o^\star,\dots x_r^\star\} \text{ is a basis of } M/M^2).$$

Let $I^\star=\bigoplus_{n\geq o} I_n$ where I_n is the k=A/M-vector space of forms of degree n of $I^\star$. Clearly:

(b) $H_A^o(n)+\dim_k I_n=\binom{n+r}{r}$

From now on we assume that k=A/M is an infinite field and G(A) is Cohen-Macaulay. In this case we can suppose that $x_o^\star$ is a non zero divisor of G(A). Then $X_o\notin\sqrt{I^\star}$. Let $x_o^\star$ be the class in G(A) of an element x of A. We have:

$$G(A/xA)=G(A)/x_o^\star G(A)\simeq k[X_1,\dots,X_r]/J.$$

If $f=f(X_o,\dots,X_r)\in k[X_o,\dots X_r]$ we set $f(0,X_1,\dots,X_r)=\bar{f}$. Thus:

$J=\{\bar{f}\mid f\in I\}$ and $J=\bigoplus_{n\geq o} J_n$, $J_n=\{\bar{f}\in J\mid f$ has degree $n\}$. Further:

LEMMA 8. If $(\bar{f}_1,\dots,\bar{f}_t)=J$ then $(f_1,\dots,f_t)=I^\star$.

PROOF. Let n be the least degree of a form $f\in I^\star$ such that $f\notin(f_1,\dots,f_t)$. Then $\bar{f}\in J=(\bar{f}_1,\dots,\bar{f}_t)$. Hence $f-\sum_i h_if_i=X_og$ and g has degree n-1. Thus $X_og\in I^\star$, but $X_o\notin\sqrt{I^\star}$, so $g\in I^\star$. Contradiction.

PROPOSITION 9. If $H_A^o(n)=H_A^o(n+1)$ then I is generated by forms of degree less or equal to n+1.

PROOF. By thm. 3,(ii), we have $H^1_{A/xA}(n+1)=H^1_{A/xA}(n)$. Then $H^o_{A/xA}(n+1)=0$ and $\dim_k J_{n+1}=\binom{n+r}{r-1}$ (see (b)). Thus $J_{n+1}=$ degree n+1-piece of $k[X_1,\dots X_n]$. Then J is generated by forms of degree n+1 and the result follows from lemma 8 .

From now on we assume that :

A has maximal Hilbert function and $m=\mathrm{Min}\{n\mid\binom{n+r}{r}>e\}$

COROLLARY 10. The ideal $I^\star$ is generated by forms of degree m and m+1.

PROOF. We have $H_A^o(m)=H_A^o(m+1)=e$ and so $I^\star$ is generated by forms of degree less or equal to m+1 (prop. 9). Further $H_A^o(m)=e<\binom{m+r}{r}$ and $I_m\neq 0$ (see (b)). Hence there is at least a form of degree m in any minimal generating set of $I^\star$.

Let now $W=\{f\in I_{m+1} \mid f=\sum_{i=0}^{r} X_i f_i, f_i\in I_m\}$ and $\bar{W}=\{\bar{f}\mid f\in W\}$. Then , if $\{f_1,\ldots,f_N\}$, $N=\binom{m+r}{r}-e$ is a basis of I_m, a set of generators of W is given by the forms $X_i f_j$, $i=0,\ldots,r$, $j=1,\ldots,N$. Hence a minimal generating set of $I^\star$ is given by $f_1,\ldots,f_N$ plus a set of forms of I_{m+1} which extend $\dim_k W$ linearly independent elements, extracted from $\{X_i f_j\}$, to a basis of I_{m+1}. Thus we have proved:

THEOREM 11. $v(I)=v(I^\star)=\dim_k I_m+\dim_k I_{m+1}-\dim_k W=\binom{m+r}{r}+\binom{m+r+1}{r}-2e-\dim_k W.$

In general $\dim_k W$ (and consequently $v(I)$) does not depend only on $e=e(A)$ and $r+1=\text{emdim } A$, but there are two cases in which this happens:

PROPOSITION 12. If $e=\binom{m+r-1}{r}$, then $v(I)=\binom{m+r-1}{r-1}$. In particular , if $e=\binom{m+1}{2}$, then $v(I)=m+1$.

PROOF. We have $H_A^o(m-1)=H_A^o(m)=e$, then $I^\star$ is generated by forms of degree m (prop. 9) and $v(I)=v(I^\star)=\dim_k I_m=\binom{m+r}{r}-e=\binom{m+r}{r}-\binom{m+r-1}{r}=\binom{m+r-1}{r-1}$.

PROPOSITION 13. If $e=\binom{m+r}{r}-1$, then $v(I)=1+\binom{m+r}{r-1}-r$.

PROOF. In this case $\dim_k I_m=1$ and $\dim_k I_{m+1}=\binom{m+r+1}{r}-e=\binom{m+r}{r-1}+1$. Further if $0\neq g\in I_m$ then $\{X_o g,\ldots,X_r g\}$ is a basis of W and $\dim_k W=r+1$ and the result folllows from thm. 11 .

REMARKS 14. (i) If $I^\star$ is a graded ideal of points $\{P_1,\ldots,P_e\}\in\mathbb{P}^r$ in generic position (remark 7) , in [G.O.]$_2$ and [G.M.] it is shown how to compute $\dim_k W$. Further some examples are given which show that also in this case $\dim_k W$ does not depend only on e and r. (ii) The ideas inspiring the previous results have first pointed out in [O$_1$] and [G.O.]$_2$. Here we have extended the results of those papers from the case of points in generic position to the general case of one-dimensional rings with maximal Hilbert function.

2. EXAMPLES. We show various examples of local rings A of curves at a singular point in $\mathbb{A}^{r+1}$ such that : (a) A has maximal Hilbert function and G(A) is Cohen-Macaulay. In particular , if we set r=2 and $e(A)=\binom{m+1}{2}$ in example 1, we recover the famous Macaulay examples (see [O$_3$] for a detailed discussion) of analytically reducible primes $P_n\subset k[X,Y,Z]$ such that $v(P_n)=n+1$. In example 4 (Maurer) we have analytically irreducible primes $P_n\subset k[X,Y,Z]$ with $v(P_n)=n+1$. We note that proposition 12 gives a class of curves,with arbitrary large number of local generators,of which the examples of Macaulay and of Maurer are special cases. The first examples of analytically irreducible primes having arbitrary large number of generators were given by Moh (see[Mo]), but we have been unable to prove that they satisfy the conditions (a).

EXAMPLE 1.(Singularities with reduced tangent cone). We construct local rings A of curves of $\mathbb{A}^{r+1}$ at a singular point O having reduced (hence Cohen-Macaulay)tangent cone Spec(G(A)) and maximal Hilbert function, for any multiplicity e(A) and embedding dimension r+1.This construction is taken from [G.O.]$_2$ and [O_1]. Let $P_1,\dots,P_e$ be points in generic position (remark 7). We choose coordinates in $\mathbb{P}^r$ so that $P_i=(1,a_{i1},a_{i2},\dots,a_{ir})$ and $a_{i1}\neq a_{j1}$ for $i\neq j$. Let $Q_i=(a_{i1},\dots a_{ir})\in\mathbb{A}^r$ be the corresponding affine points. Let $x_1=t$, $x_2=f_2(t),\dots,x_n=f_n(t)$ a rational curve of $\mathbb{A}^r$ containing the points Q_i. Let $g(t)=\prod_{i=1}^{e}(t-a_{i1})$ and $A'=k[g,tg,f_2g,\dots,f_ng]$. If A is the local ring of A' at the maximal ideal $(g,tg,\dots,f_ng)$ then $e(A)=e$, the ring G(A) is reduced and $H^o_A(n)=\mathrm{Min}\{\binom{n+r}{r},e\}$ (see [O_1],§4 and [G.O.]$_2$ example 13).

In all the following examples we shall work for simplicity with the completion A of the local ring B of the curve at the singular point. For our purposes this is not restrictive because $G(A)\simeq G(B)$ and $H^o_A(n)=H^o_B(n)$.

EXAMPLE 2. Let $A=k[[t^5,t^6,t^7]]$. Then one easily gets that $e(A)=5$, the image of t^5 in G(A) is a non zero divisor and $H^o_A(1)=3$, $H^o_A(2)=5$. Then, by remark 5, (i) and prop. 13, if $I\subset k[[X,Y,Z]]$ is the ideal of the curve we have $v(I)=3$.

EXAMPLE 3. If $A=k[[t^9,t^{10},t^{11},t^{12}]]$, then $(t^9)^*$ is a non zero divisor of G(A) and $H^o_A(1)=4$, $H^o_A(2)=9$. Then $v(I)=8$ (prop. 13).

EXAMPLE 4 (Maurer). Let k be a field of characteristic zero and m be a positive integer. If m is odd we set $a=((m+1)/2)m$, $b=((m+1)/2)(m+1)$, $c=((m+1)/2)(m+2)$. If m is even we set $a=(m/2)(m+1)$, $b=(m/2)(m+2)$, $c=(m/2)(m+3)$. Let $A=k[[t^a,t^b,t^c(t+1)]]$. It is easily checked that $\bar{A}=k[[t]]$ (where $\bar{A}$ is the normalization of A), $e(A)=a=\binom{m+1}{2}$ and emdim $A=H^o_A(1)=3$. Further we will prove that:

(a) G(A) <u>is Cohen-Macaulay and</u> A <u>has maximal Hilbert function</u>.

Then if I is the ideal defining the curve $v(I)=m+1$ (prop. 12). This is proved by Maurer using different methods which do not involve esplicitly the Hilbert functions.

<u>We show</u> (a) <u>in the case</u> m <u>odd</u> . The proof for m even is the same and omitted. We divide the proof in three steps. We need first an easy numerical lemma:

(i) Let m and $h<m$ be positive integers, m odd. The equation: $m(m+1)+h=mu+(m+1)v+(m+2)w$ has at least $p=(m+1)/2$ solutions $(u_i,v_i,w_i)\varepsilon\mathbf{N}^3$ such that $u_i\neq 0$, $i=1,\dots,p$, and $w_i<w_{i+1}$, $i=1,\dots,p-1$. Further if $h=m-1$ then $w_p=p-1=(m-1)/2$ and if $h<m-1$ then $w_p<m$.

PROOF. We can decompose $(m+1)m+h$ in two ways:

$(m+1)m+h=(m+1)(m-h)+(m+2)h=m(h'/2)+(m+1)(m-h-h')+(m+2)(h+h'/2)$

$(m+1)m+h=m(m+1-h)+(m+1)h=m(m+1-h+(h''/2))+(m+1)(h-h'')+(m+2)(h''/2)$

where h' and h'' are even integers such that $o<h'\leq m-h$, $o\leq h''\leq h$. Clearly the number of h' and h'' is $(m-h)/2+(h+1)/2$ if h is odd, $(m-h-1)/2+h/2+1$ if h is even; both numbers are equal to $(m+1)/2$. Then if we set: $(u_i,v_i,w_i)=(m+1-h+h''/2,h-h'',h''/2)$ for $i=h''/2$ and $(u_i,v_i,w_i)=(h'/2,m-h-h',h+h'/2)$ for $i=(h'+h'')/2$ we have the first claim. If $h=m-1$ then $w_p=h''/2=h/2=(m-1)/2$. If $h<m-1$ then $w_p=\mathrm{Max}\{h+h'/2,h''/2\}\leq\mathrm{Max}\{h+(m-h)/2,h/2\}=(m+h)/2\leq(m+m-2)/2=m-1$.

(ii) $(t^b)^m t^s=t^{bm+s}\varepsilon t^a A$ for any $s\varepsilon\mathbf{N}$.

PROOF. Clearly it suffices to show that $t^{bm+s}\varepsilon t^a A$ for $o\leq s<a$ (if $s>a$, then $s=aq+s'$, $o\leq s'<s$). If we set $p=(m+1)/2$ we have $c-b=b-a=p$. Let $s=hp+r$, where $o\leq h\leq m$, $o\leq r<p$; then $t^{bm+s}=t^{bm+hp}t^r=t^{b(m-h)}t^{ch}t^r$. Now $t^{b(m-h)}t^{ch}=(t^a)^{(m+1-h)}t^{bh}\varepsilon t^a A$. Then the claim is proved if we show that if $d=b(m-h)+ch$ then $t^d t^r\varepsilon t^a A$ for $o<r<p$.. Multiplying the equation of (i) by p,

we get $d=b(m-h)+ch=pm(m+1)+ph=pmu+p(m+1)v+p(m+2)w=au+bv+cw$, then $(t^a)^u(t^b)^v(t^c(t+1))^w=t^d(t+1)^w$. By (i) the equation $d=au+bv+cw$ has $p=(m+1)/2$ solutions (u_i,v_i,w_i) with $u_i\neq 0$ and $w_i<w_{i+1}$. So the matrix , with p rows and w_p columns, which has as columns the coefficients of $(t+1)^{w_i}$ (with $a_{ij}=0$ if $w_p\geq j>w_i$) has maximum rank (for this we need ch k =0). Thus to prove that $t^dt^r\in t^aA$, for $o<r<p$, it is enough to prove that $t^dt^r\in t^aA$ for $w_p\geq r\geq p$. But if $h=m-1$ $w_p=(m-1)/2=p-1$ (see (i)). Then $h<m-1$. In this case $w_p<m$ (see (i)) . We can suppose the claim true for h+1 and prove it for h. Now $t^dt^r=t^{d+p}t^{r-p}=t^{b(n-h-1)+c(h+1)}t^{r-p}\in t^aA$ because $o\leq r-p<p$. Thus (ii) is proved.

(iii) Let $x=t^a\in A$. The ring A has maximal Hilbert function and $H^o_{A/xA}(n)=n+1$ for $n<m$, $H^o_{A/xA}(n)=0$ for $n\geq m$. Then, by cor.6, $x^\star$ is not a zero divisor of G(A).

PROOF. It is easily seen that $x^\star$ does not belong to the minimal prime of G(A). Hence x is a superficial element of A (see[Z.S.]Ch 8, §8, Lemma 5). Furthermore A is a domain so $\dim_k(A/xA)=e(A)=a$ (thm. 3,(i)). If m_1 and m_2 are two positive integers such that $m_1+m_2\geq m$. We have $(t^b)^{m_1}(t^c)^{m_2}=(t^b)^{(m_1+m_2)}t^{pm_2}=t^{bm}t^{pm_2}\in t^aA$ by (ii) (recall that c-b=p). Then if we call y and z respectively the classes of t^b and t^c in A/xA we have that $y^{m_1}z^{m_2}$ is zero. Further the number of elements $y^{m_1}z^{m_2}$ is, for $m_1+m_2<m$, equal to $\binom{m+1}{2}=a=e$, so $y^{m_1}z^{m_2}$ are a basis of A/xA. This gives immediately the second part of the claim. Finally we prove that $H(n)=\binom{n+2}{2}$ if $n<m$. This, by remark 5,(i), gives the maximality of the Hilbert function of A. But if $n<m-1$ then $(c+1)n\leq b(n+1)$ and so if f is a homogeneous polynomial of degree n in t^a,t^b,t^c such that $f\in M^{n+1}$ then f=0. Thus $H^o_A(n)=\binom{n+2}{2}$.

REFERENCES

[G] GERAMITA A.V., Some remarks on multiplicity, The curves seminar at Queen's, Queen's Papers in Pure and Applied Math., Vol. 58, 1981, Queen's University, Kingston, Canada,

[Ge] GEYER W.D., On the number of equations which are necessary to describe an algebraic set in n-space, Atas da 3ª Escola de Algebra, Brasilia IMPA Lecture Notes, 9, Rio de Janeiro, 1977, 183-317,

[G.M.] GERAMITA A.V.-MAROSCIA P., The ideal of forms vanishing at a finite set of points in $\mathbb{P}^n$, Queen's Math. Preprint, No. 1981-5,

[G.O.]$_1$ GERAMITA A.V.-ORECCHIA F., On the Cohen-Macaulay type of s-lines in $\mathbb{A}^{n+1}$ Journal of Algebra, Vol.70, No. 1, (1981),

[G.O.]$_2$ GERAMITA A.V.-ORECCHIA F., Minimally generating ideals defining certain tangent cones, (to appear in Journal of Algebra),

[H] HIRONAKA H., Resolutions of singularities of an algebraic variety over a field of characteristic zero, Ann. of Math., 79 (1974), 109-326,

[L] LIPMAN J., Stable ideals and Arf rings, Amer. J. Math., 93 (1971), 649-685

[Ma] MAURER J., Eine Variante der Moh-Kurven (Preprint)

[Mo] MOH T.T., On the unboundedness of generators of prime ideals in power series rings of three variables, J. Math. Soc. Japan 26 (1974), 722-734,

[O$_1$] ORECCHIA F. One-dimensional local rings with reduced associated graded ring and their Hilbert function, Manuscripta Math., 32, (1980), 391-405,

[O$_2$] ORECCHIA F. Points in generic position and conductors of curves with ordinary singularities, J. London Math. Soc. (2), 24 (1981), 85-96,

[O$_3$] ORECCHIA F. Gli esempi di Macaulay, Conference at "Istituto di Alta Matematica, Bologna", To appear in Quaderni del C.N.R.,

[S] SALLY J., Number of generators of ideals in local rings, Lect. Notes in Pure and Applied Math., Marcel Dekker, New York, 1978,

[Sc] SCHENZEL P., Über die freien Auflösungen extremaler Cohen-Macaulay Ringe, Journal of Algebra, Vol 64, No 1, (1980)

[Z.S.] ZARISKI O.-SAMUEL P., Commutative Algebra II, D. Van Nostrand, Princeton, 1960.

ON NORMAL FLATNESS AND SOME RELATED TOPICS

Lorenzo Robbiano

ISTITUTO MATEMATICO DELL' UNIVERSITA' di GENOVA

GENOVA (ITALIA)

This paper is an expansion of the talk which I gave at the Trento Conference on Commutative Algebra (8-13 June 1981). In the last years I have been strongly interested in the theory of the graded rings associated to an ideal, specially in connection with the topic of normal flatness.
This notion was introduced by Hironaka in [4] and since then several authors dealt with it; however, in my opinion, even the best references are too technical and complicated to candidate themselves as natural introduction to the subject. My first purpose is that one of building a simple and accessible frame for the theory of normal flatness; to this end I use the techniques developed in [10] and in the first section I describe a new approach to the so called "Hironaka-Grothendieck isomorphism" (see Theorems 1.8 and 1.9). This approach has already been used for studying new aspects

of the theory (see for instance [7]). The second purpose is that one of describing new results on normal flatness, particularly in the embedded case; to this end I introduce some new terminology, which is useful in the study of some "fiber rings". The main result is Theorem 2.6 , which essentially describes conditions of normal flatness in term of suitable homomorphisms between modules of Koszul homology.

The last section is entirely devoted to the description of some examples and comments on Corollary 3.2, which relates the normal flatness to the equality of suitable integers. Example 6 can also be interesting for another reason: a simple Cohen-Macaulay ring is described, such that the scheme, which is obtained by blowing up its maximal ideal, is not locally Cohen-Macaulay. The rings are supposed to be commutative, with identity and noetherian.

As I said in the introduction, the reader should refer to [10] for the basic definitions and propositions on filtered rings and modules and their associated graded objects. Let me just mention the following simple

Lemma 1.1. *Let R be a multiplicatively filtered ring and* $M \xrightarrow{f} N$ *a surjective omomorphism of filtered R-modules. Assume that the filtration* $\{F_p M\}$ *induces a separated topology on* $K=\mathrm{Ker}(f)$ *and* $F_p N = f(F_p M)$. *If* $G(M) \xrightarrow{G(f)} G(N)$ *is an isomorphism, then f is an isomorphism.*

Proof. Consider the exact sequence $0 \longrightarrow K \xrightarrow{i} M \xrightarrow{f} N \longrightarrow 0$ and put on K the filtration $F_p K = i^{-1}(F_p M)$; then i, f are strict morphisms, hence $0 \longrightarrow G(K) \xrightarrow{G(i)} G(M) \xrightarrow{G(f)} G(N) \longrightarrow 0$ is exact by [10] 1.4.

It follows that $G(K) = 0$, hence $K = 0$ (see [10] 1.3).

Another consequence of the theory, which is described in [10] is the following simple proof of the "local criterion of flatness" (see [5] 0_{III} 10.2.1 [6] and [3]) in the case of finitely generated modules over a local ring.

<u>Proposition</u> 1.2. <u>Let</u> S <u>be a local ring,</u> J <u>an ideal and</u> M <u>a finitely generated</u> S-<u>module. Then the following conditions are equivalent</u>

1) M <u>is free over</u> S.

2) M/JM <u>is free over</u> S/J <u>and</u> $Tor_1^S(S/J, M) = 0$.

3) M/JM <u>is free over</u> S/J <u>and the canonical surjective homomorphism</u>

$$M/JM \otimes_{S/J} G_J(S) \xrightarrow{\pi} G_J(M)$$ <u>is an isomorphism.</u>

<u>Proof.</u> 1)⇒2) and 1)⇒3) are clear.

2)⇒1) Let $r = rk_{S/J}(M/JM)$; the isomorphism $(S/J)^r \cong M/JM$ can be lifted to an exact sequence $0 \to K \to S^r \to M \to 0$.

Tensoring with S/J we get the exact sequence

$0 \to Tor_1^S(S/J, M) \to K/JK \to (S/J)^r \to M/JM \to 0$ hence $K/JK = 0$ and $K = 0$ by Nakayama Lemma.

3)⇒1) As before we get the exact sequence $0 \to K \xrightarrow{i} S^r \xrightarrow{f} M \to 0$

Consider the following filtrations: $F_p S^r = J^p S^r$; $F_p M = J^p M$; $F_p K = K \cap F_p S^r$

With respect to these filtrations, it is clear that i, f are strict morphisms, hence we get (see [10] 1.4) an exact sequence

$0 \to G(K) \xrightarrow{G(i)} G(S^r) \xrightarrow{G(f)} G(M) \to 0$ But $G(S^r) \cong (G_J(S))^r \cong M/JM \otimes G_J(S)$

and $G(M) = G_J(M)$; moreover G(f) is taken to π via these isomorphisms.

Therefore $G(K) = 0$, hence $K = 0$ (see [10] 1.3)

<u>Proposition</u> 1.3. <u>Let</u> A, B <u>be rings such that</u> $A \subset B = A[f_1, \ldots, f_r]$ <u>and</u>

assume that $f_1,\ldots f_r$ is a regular B-sequence and $(f_1,\ldots,f_r)\cap A = 0$.

Then $f_1,\ldots,f_r$ are algebraically independent over A.

Proof. Let us consider the canonical projection $A[T_1,\ldots T_r] \xrightarrow{\pi} B$ given by $\pi(T_i) = f_i$. Let us put the $(T_1,\ldots,T_r)$-adic filtration on $A[T_1,\ldots,T_r]$ and the $(f_1,\ldots,f_r)$-adic filtration on B. Passing to the graded rings we get $A[T_1,\ldots,T_r] \xrightarrow{G(\pi)} A[T_1,\ldots,T_r]$ (see [5] 0_{IV} 15.1.9, [10] 3.5) and $G(\pi)$ is the identity map. The conclusion follows by Lemma 1.1.

Lemma 1.4. Let S be a local ring, $\underline{x} = x_1,\ldots,x_r$ an S-sequence and M a finitely generated S-module. Then the following conditions are equivalent

1) $\underline{x}$ is an M-sequence.

2) $\mathrm{Tor}_1^S(S/\underline{x}S, M) = 0$.

Proof. The Koszul complex $\mathbb{K}(\underline{x}; S)$ is a free resolution of $S/\underline{x}S$, hence $\mathrm{Tor}_1^S(S/\underline{x}S, M) \cong H_1(\underline{x}; M)$ and the conclusion follows (see [11]).

Corollary 1.5. With the same assumption as in 1.4, the following conditions are equivalent

1) M is free over S.

2) $M/\underline{x}M$ is free over $S/\underline{x}S$ and $\underline{x}$ is an M-sequence.

Proof. It is a straightforward consequence of Proposition 1.2 and Lemma 1.4 (see [5] 0_{IV} 15.1.21).

Let now $(R,\mathfrak{m},k)$ be a local ring, I an ideal, $x_1,\ldots,x_r \in R$ and denote by $\underline{x}$ the sequence $x_1,\ldots,x_r$, by $\underline{x}R$ the ideal generated by $\underline{x}$, and by L the ideal $I+\underline{x}R$.

Lemma 1.6. a) There is a canonical surjective homomorphism of graded rings

$$g(I,\underline{x}) : G_I(R) \otimes_R R/L \longrightarrow G_{L/\underline{x}R}(R/\underline{x}R)$$

b) There is a canonical surjective homomorphism of graded rings

$$G(I,\underline{x}) : (G_I(R) \otimes_R R/L)[T_1,\dots,T_r] \longrightarrow G_L(R)$$

such that $G(I,\underline{x})(T_i) = x_i^* = \overline{x}_i \in L/L^2$.

Proof. a) $(G_I(R) \otimes_R R/L)_n = I^n / I^n L = I^n/(I^{n+1} + I^n \cdot \underline{x}R)$

$(G_{L/\underline{x}R}(R/\underline{x}R))_n = ((I+\underline{x}R)/\underline{x}R)^n / ((I+\underline{x}R)/\underline{x}R)^{n+1} \cong I^n/I^{n+1} + I^n \cap \underline{x}R$

and a) is done, while b) is clear.

Lemma 1.7. a) The following conditions are equivalent

1) $g(I,\underline{x})$ is an isomorphism.

2) $I^n \cap \underline{x}R = I^n \cdot \underline{x}R$ for every n.

3) $\mathrm{Tor}_1^R(R/\underline{x}R, R/I^n) = 0$ for every n.

b) If $g(I,\underline{x})$ is an isomorphism, then $G(I,\underline{x})_0$ is injective, where $G(I,\underline{x})_0$ is the restriction of $G(I,\underline{x})$ to the subring $G_I(R) \otimes_R R/L$.

Proof. a) After the proof of Lemma 1.6 a) it is clear that $g(I,\underline{x})$ is an isomorphism if and only if $I^n \cap \underline{x}R \subseteq I^n \cdot \underline{x}R + I^{n+1} \cap \underline{x}R$ for every n, hence if and only if $I^n \cap \underline{x}R \subseteq \bigcap_{k \geq 0} (I^n \cdot \underline{x}R + I^{n+k}) = I^n \cdot \underline{x}R$. The equivalence between 2) and 3) is standard.

b) It is enough to prove that $I^n \cap L^{n+1} \subseteq I^n L$. Now $I^n \cap L^{n+1} \subseteq I^n \cap (I^n L + \underline{x}R) =$ $= I^n L + I^n \cap \underline{x}R = I^n L$ by a).

Theorem 1.8. The following conditions are equivalent

1) $g(I,\underline{x})$ is an isomorphism and $\underline{x}$ is an R-sequence.

2) $\underline{x}$ is a regular sequence on $G_I(R)$.

3) $\underline{x}$ is a regular sequence on R/I^n for every n.

4) $G(I,\underline{x})$ is an isomorphism.

Proof. The equivalence between 2) and 3) follows from the exact sequence

$0 \to I^n/I^{n+1} \to R/I^{n+1} \to R/I^n \to 0$ and the induced exact sequences of the Koszul homology.

1) $\Rightarrow$ 3) follows from Lemma 1.4 and Lemma 1.7 a).

3) $\Rightarrow$ 1) Since 3) $\Rightarrow$ 2), we get that $\underline{x}$ is an R-sequence from [12] and [10] 3.4, moreover $g(I,\underline{x})$ is an isomorphism by Lemma 1.4 and Lemma 1.7 a).

1) $\Rightarrow$ 4) After Lemma 1.7 a) we get that $L^n \cap \underline{x}R = (I^n + L^{n-1} \cdot \underline{x}R) \cap \underline{x}R = L^{n-1} \cdot \underline{x}R$, hence $\{x_1,\ldots,x_r\}$ is an L-standard base of $\underline{x}R$ by [10] 1.8 and since $\underline{x}$ is an R-sequence, it follows that $x_1^*,\ldots x_r^*$ is a $G_L(R)$-sequence by [10] 3.4. After Lemma 1.7 b), $G_I(R) \otimes_R R/L$ is embedded into $G_L(R)$ by $G(I,\underline{x})$ and $G_L(R) = (G_I(R) \otimes_R R/L)[x_1^*,\ldots,x_r^*]$ It is therefore sufficient to use Proposition 1.3 with $A = G_I(R) \otimes_R R/L$, $f_i = x_i^*$, $B = G_L(R)$.

4) $\Rightarrow$ 1) Since $G(I,\underline{x})$ is an isomorphism and in this isomorphism x_i^* correspond to T_i, we know that $x_1^*,\ldots,x_r^*$ is a $G_L(R)$-sequence, hence $\underline{x}$ is an R-sequence by [10] 3.4 and $\{x_1,\ldots,x_r\}$ is an L-standard base of $\underline{x}R$ again by [10] 3.4. Hence $G_L(R)/(x_1^*,\ldots,x_r^*) \xrightarrow{a} G_{L/\underline{x}R}(R/\underline{x}R)$ is an isomorphism.

From the isomorphism $G(I,\underline{x})$ we get the isomorphism $G_I(R) \otimes_R R/L \xrightarrow{b} G_L(R)/(x_1^*,\ldots,x_r^*)$ by passing to the quotients; hence we get the isomorphism $a \circ b$, which clearly turns out to be $g(I,\underline{x})$.

Theorem 1.9. Assume in addition that $\underline{x}$ is a regular sequence on R/I. Then the following conditions are equivalent

1) $G_I(R)$ is free over R/I.

2) $G_I(R) \otimes_R R/L$ is free over R/L and $\underline{x}$ is a regular sequence on $G_I(R)$.

3) $G_I(R) \otimes_R R/L$ is free over R/L and $\underline{x}$ is a regular sequence on R/I^n for every n.

4) $g(I,\underline{x})$ is an isomorphism of free R/L-modules and $\underline{x}$ is an R-sequence.

5) $G(I,\underline{x})$ is an isomorphism of free R/L-modules.

Proof. The equivalence between 1) and 2) follows from Corollary 1.5, while the other equivalences follow from Theorem 1.8.

2. Let now $(R,\mathfrak{m},k)$ be a local ring and I an ideal of R. If f is an element of R and $v = v_I(f)$, i.e. the maximum integer n such that $f \in I^n$, then $\mathrm{in}_I(f)$ (or simply f^*) denotes the residue class of f in I^v/I^{v+1} and it is called the initial form of f with respect to I. If J is another ideal, then $\mathrm{In}_I(J)$ (or simply J^*) denotes the ideal of $G_I(R)$ which is generated by the initial forms of the elements of J and is called the initial ideal of J with respect to I. Then we have an exact sequence

$$(*) \qquad 0 \longrightarrow \mathrm{In}_I(J) \longrightarrow G_I(R) \longrightarrow G_{I+J/J}(R/J) \longrightarrow 0$$

Suppose we are given three ideals J,I,L such that $J \subseteq L$, $I \subseteq L$.

Definition 2.1. a) We denote by $\mathrm{in}_{(I,L)}(f)$ (or simply by f°) the residue class of f in I^v/I^vL and we call it the fiber form of f with respect to (I,L).

b) We denote by $\mathrm{In}_{(I,L)}(J)$ (or simply by J°) the ideal generated by the fiber forms of the elements of J and we call it the fiber ideal of J with respect to (I,L).

By definition we have an exact sequence

$$(**) \qquad 0 \longrightarrow \mathrm{In}_{(I,L)}(J) \longrightarrow G_I(R) \otimes_R R/L \longrightarrow G_{I+J/J}(R/J) \otimes_R R/L \longrightarrow 0$$

Definition 2.2. A set of elements $\{f_1,\ldots,f_r\}$ of J is called an (I,L)-fiber base of J if $\mathrm{In}_{(I,L)}(J) = (\mathrm{in}_{(I,L)}(f_1),\ldots,\mathrm{in}_{(I,L)}(f_r))$.

Lemma 2.3. There is a canonical surjective homomorphism

$In_I(J) \otimes_R R/L \longrightarrow In_{(I,L)}(J)$ *hence an I-standard base of J is also an (I,L)-fiber base of J.*

Proof. The proof simply follows by tensoring (*) with R/L over R and then comparing with (**).

Now we give here a "standard type" application of the results of the first section (see for instance [1])

Proposition 2.4 *Let* $(R,\mathfrak{m},k)$ *be a local ring,* $\underline{x} = x_1,\ldots,x_r$ *a sequence of elements of* R, I *an ideal of* R *and* $L = I + \underline{x}R$. *Let* $J \subseteq I$ *be another ideal of* R *and assume that* $G_I(R)$ *is free over* R/I and $\underline{x}$ *is a regular sequence on* R/I. *Then the following conditions are equivalent*

1) $G_{I/J}(R/J)$ *is free over* R/I.

2) $G_{L/J}(R/J)$ *is free over* R/L *and the canonical morphism* $(In_{(I,L)}(J))[T_1,\ldots,T_r] \xrightarrow{a} In_L(J)$ *induced by* $G(I,\underline{x})$ *is an isomorphism.*

3) $G_{L/J}(R/J)$ *is free over* R/L *and there exists an* L-*standard base* $\{f_1,\ldots,f_s\}$ *of* J *such that* $v_I(f_i) = v_L(f_i)$ *for every* i.

Proof. Let us consider the following commutative diagram

$$\begin{array}{ccccccc}
 & & 0 & & 0 & & \\
 & & \uparrow & & \uparrow & & \\
 & & (G_{I/J}(R/J) \otimes_R R/L)[T_1,\ldots,T_r] & \overset{G(I/J,\,\overline{x})}{\dashrightarrow} & G_{L/J}(R/J) & \longrightarrow & 0 \\
 & & \uparrow & & \uparrow & & \\
0 & \longrightarrow & (G_I(R) \otimes_R R/L)[T_1,\ldots,T_r] & \xrightarrow{G(I,\,\underline{x})} & G_L(R) & \longrightarrow & 0 \\
 & & \uparrow & & \uparrow & & \\
0 & \longrightarrow & (In_{(I,L)}(J))[T_1,\ldots,T_r] & \xrightarrow{a} & In_L(J) & & \\
 & & \uparrow & & \uparrow & & \\
 & & 0 & & 0 & &
\end{array}$$

where $\bar{x}_i$ is the class of x_i mod J, $\underline{\bar{x}} = \bar{x}_1, \ldots, \bar{x}_r$, the arrows are the canonical ones (see (*) and (**)) and $G(I,\underline{x})$ is an isomorphism by Theorem 1.9. By standard arguments we get the exact sequence

$$(\circ)\quad 0 \longrightarrow (\mathrm{In}_{(I,L)}(J))[T_1,\ldots,T_r] \xrightarrow{\ a\ } \mathrm{In}_L(J) \longrightarrow$$

$$\longrightarrow (G_{I/J}(R/J) \otimes_R R/L)\,[T_1,\ldots,T_r] \xrightarrow{G(I/J,\underline{\bar{x}})} G_{L/J}(R/J) \longrightarrow 0$$

1) ⇔ 2) By Theorem 1.9 and the exact sequence (°) we get the equivalence between 1) and 2).

2) ⇒ 3) By assumption $\mathrm{In}_L(J)$ can be generated by non zero homogeneous elements of $\mathrm{In}_{(I,L)}(J)$, hence by the non zero images of homogeneous elements of $\mathrm{In}_I(J) \otimes_R R/L$ (see Lemma 2.3). This means that $\mathrm{In}_L(J)$ can be generated by the residue classes of elements $f_i \in I^{v_i} - I^{v_i}L$ where $v_i = v_I(f_i)$. On the other hand $I^{v_i}L \subseteq I^{v_i} \cap L^{v_i+1} \subseteq I^{v_i} \cap (I^{v_i}L + \underline{x}R) = I^{v_i}L + I^{v_i} \cap \underline{x}R$;

but $G_I(R)$ is free over R/I, hence $g(I,\underline{x})$ is an isomorphism by Theorem 1.9 and $I^{v_i} \cap \underline{x}R = I^{v_i} \cdot \underline{x}R \subseteq I^{v_i}L$. Therefore $\mathrm{In}_L(J)$ can be generated by the residue classes of elements $f_i \in I^{v_i} - L^{v_i+1}$.

3) ⇒ 2) Following the preceding arguments, it is clear that the assumption 3) implies that a is surjective.

Keeping the preceding notations, assume that $J \subseteq I$.

<u>Lemma</u> 2.5. <u>Consider the following conditions</u>

1) $G_{I/J}(R/J)$ <u>is free over</u> R/I.

2) <u>The canonical surjective homomorphism</u> $\mathrm{In}_I(J) \otimes_R R/\mathfrak{m} \longrightarrow \mathrm{In}_{(I,\mathfrak{m})}(J)$ (see Lemma 2.3) <u>is an isomorphism.</u>

3) $\mathfrak{m}I^n \cap J = \mathfrak{m}(I^n \cap J) + I^{n+1} \cap J$ <u>for every</u> n.

Then 1) $\Rightarrow$ 2) $\Leftrightarrow$ 3) and if $G_I(R)$ is free over R/I they are all equivalent.

Proof. Consider the canonical exact sequence

$$0 \longrightarrow In_I(J) = \oplus_n (I^n \cap J)/(I^{n+1} \cap J) \longrightarrow G_I(R) \longrightarrow G_{I/J}(R/J) \longrightarrow 0$$

By tensoring with $k = R/\mathfrak{m}$ over R/I, we get the exact sequence

$$(\alpha) \quad Tor_1^{R/I}(G_I(R), k) \longrightarrow Tor_1^{R/I}(G_{I/J}(R/J), k) \longrightarrow In_I(J) \otimes k \cong$$

$$\cong \oplus_n (I^n \cap J)/(\mathfrak{m}(I^n \cap J) + I^{n+1} \cap J) \longrightarrow G_I(R) \otimes k \longrightarrow G_{I/J}(R/J) \otimes k \longrightarrow 0$$

On the other hand we have the canonical exact sequence

$$(\beta) \quad 0 \longrightarrow In_{(I,\mathfrak{m})}(J) \longrightarrow G_I(R) \otimes k \longrightarrow G_{I/J}(R/J) \otimes k \longrightarrow 0$$

and it is easy to check that $In_{(I,\mathfrak{m})}(J) = \oplus_n (I^n \cap J)/(\mathfrak{m}I^n \cap J)$.

The conclusion easily follows by comparing (α) and (β).

Now let us consider an I-standard base $\{f_1, \ldots, f_r\}$ of J and the corresponding initial forms $f_1^*, \ldots, f_r^*$ in $G_I(R)$, whose degrees are $v_i = v_I(f_i)$; take the free graded module $\oplus_i G_I(R)(-v_i)$ and its graded submodule S of syzygies of $f_1^*, \ldots, f_r^*$. Then consider the $(I,\mathfrak{m})$-fiber base $\{f_1, \ldots f_r\}$ of J (see Lemma 2.3) and the corresponding fiber forms $f_1^\circ, \ldots, f_r^\circ$ in $G_I(R) \otimes k$; take in $\oplus_i (G_I(R) \otimes k)(-v_i)$ the graded submodule S' of syzygies of $f_1^\circ, \ldots, f_r^\circ$. Of course there is a canonical homomorphism $S \longrightarrow S'$ which takes trivial syzygies into trivial syzygies; therefore we get a canonical homomorphism of graded modules $H_1(f_1^*, \ldots, f_r^*; G_I(R)) \xrightarrow{u} H_1(f_1^\circ, \ldots, f_r^\circ; G_I(R) \otimes k)$ between the first modules of the Koszul homology.

Theorem 2.6. Consider the following conditions

1) $G_{I/J}(R/J)$ is free over R/I.

2) For every I-standard base of J, the corresponding canonical homomorphism u is surjective.

3) There exists an I-standard base of J, such that the corresponding canonical homomorphism u is surjective.

Then 1) ⇒ 2) ⇒ 3) and if $G_I(R)$ is free over R/I the conditions are equivalent.

Proof. 1) ⇒ 2) Let $\{f_1,\dots,f_r\}$ be an I-standard base of J and denote by v_i the integers $v_I(f_i)$. Of course it is enough to show that a homogeneous syzygy of $f_1^\circ,\dots,f_r^\circ$ can be lifted to a syzygy of $f_1^*,\dots,f_r^*$. Let $\Sigma\alpha_i f_i^\circ = 0$ where $\deg(\alpha_i) + v_i = t$ (note that $\deg(\alpha_i)$ can also be $\leqslant 0$) and let $a_1,\dots,a_r$ be elements of R such that $\alpha_i = \tilde{a}_i \in (I^{t-v_i}/I^{t-v_i+1})\otimes k$ and such that $a_i = 0$ if $t-v_i < 0$. Then $\Sigma a_i f_i \in \mathfrak{m}I^t \cap J$ hence $\Sigma a_i f_i \in \mathfrak{m}(I^t \cap J) + I^{t+1} \cap J$ by Lemma 2.5. Let $f_1,\dots f_s$ be such that $v_I(f_i) \leqslant t$ $i=1,\dots,s$ and $f_{s+1},\dots,f_r$ be such that $v_I(f_i) > t$ $i=s+1,\dots,r$. Then $\mathfrak{m}(I^t \cap J) + I^{t+1} \cap J =$ $\mathfrak{m}\Sigma_1^s I^{t-v_i} f_i + (f_{s+1},\dots,f_r)$. Therefore $\sum_1^r a_i f_i = \sum_1^s b_i f_i + \sum_{s+1}^r c_i f_i$ where $b_i \in \mathfrak{m}I^{t-v_i}$, $c_i \in R$; then $\sum_1^s (\bar{a}_i - \bar{b}_i) f_i^* = 0$ in I^t/I^{t+1} where $\bar{a}_i$, $\bar{b}_i \in I^{t-v_i}/I^{t-v_i+1}$ and clearly $(\bar{a}_1-\bar{b}_1,\dots,\bar{a}_s-\bar{b}_s,0,\dots,0)$ lifts $(\alpha_1,\dots,\alpha_s,\alpha_{s+1},\dots,\alpha_r)$.

2) ⇒ 3) is obvious

Assume now that $G_I(R)$ is free over R/I

3) ⇒ 1) Let $f \in \mathfrak{m}I^t \cap J$ and let $\{f_1,\dots,f_r\}$ be the I-standard base of J such that the corresponding u is surjective. Since $\mathfrak{m}I^t \cap J \subseteq I^t \cap J =$ $= \sum_1^r I^{t-v_i} f_i$, we may write $f = \sum_1^r a_i f_i$ with $a_i \in I^{t-v_i}$.

Denote by $\bar{a}_i$ the residue class of a_i in I^{t-v_i}/I^{t-v_i+1} (hence $\bar{a}_i = 0$ if $t < v_i$), by $\bar{f}$ the residue class of f in I^t/I^{t+1} and by $\tilde{a}_i$ the residue class of a_i in $(I^{t-v_i}/I^{t-v_i+1})\otimes k$. Then $\bar{f} = \sum_1^r \bar{a}_i f_i^*$ and $0 = \sum_1^r \tilde{a}_i f_i^\circ$.

Let $(\bar{b}_1,\dots,\bar{b}_r)$ be a homogeneous syzygy $(\bar{b}_i \in I^{t-v_i}/I^{t-v_i+1})$ of $f_1^*,\dots,f_r^*$ which lifts $(\tilde{a}_1,\dots,\tilde{a}_r)$; then $a_i - b_i \in I^{t-v_i}\mathfrak{m}$ and $\sum_1^r b_i f_i \in I^{t+1}$.

Hence $f = \sum_1^r a_i f_i = \sum_1^r (a_i - b_i) f_i + \sum_1^r b_i f_i \in \mathfrak{m}(I^t \cap J) + I^{t+1} \cap J$.

The conclusion follows from Lemma 2.5.

Corollary 2.7. The following conditions are equivalent

1) There exists an I-standard base $\{f_1, \ldots, f_r\}$ of J such that $f_1, \ldots, f_r$ is a regular R-sequence and $G_{I/J}(R/J)$ is free over R/I.

2) There exists a system of generators $\{f_1, \ldots, f_r\}$ of J such that their initial forms $f_1^*, \ldots, f_r^*$ with respect to I are a regular $G_I(R)$-sequence and $G_{I/J}(R/J)$ is free over R/I.

3) There exists a system of generators $\{f_1, \ldots, f_r\}$ of J such that their fiber forms $f_1^\circ, \ldots, f_r^\circ$ with respect to $(I, \mathfrak{m})$ are a regular $G_I(R) \otimes k$-sequence and $G_I(R)$ is free over R/I.

Proof. The equivalence between 1) and 2) comes directly from [10] 3.4. The equivalence between 2) and 3) is a consequence of [5] 0_{IV} 15.1.16. However it is easy to see that part of it is a straightforward consequence of Theorem 2.6.

Remark. To see that the hypothesis " $G_I(R)$ is free over R/I" is essential in 3) of the above Corollary, consider the following example:

$R = k[[X,Y]]/(XY)$ $I = J = (y)$. Then $G_I(R) \cong k[X][T]/(XT)$ is not free over R/I. Of course $\{y\}$ is an I- standard base of J, $G_I(R) \otimes k \cong k[T]$ and $y^* = T$ so that y^* divides 0, while $y^\circ = T$ and y° does not divide 0.

Corollary 2.8. If $G_{I/J}(R/J)$ is free over R/I and $\{f_1, \ldots f_r\}$ is a minimal I-standard base of J, then $f_i \in I^{v_I(f_i)} - I^{v_I(f_i)} \cdot \mathfrak{m}$, i. e. $f_i^\circ \neq 0$.

Proof. If $f_i^\circ = 0$, $(0, \ldots, 0, 1, 0, \ldots, 0)$ is a syzygy of $f_1^\circ, \ldots, f_r^\circ$. It can be lifted by Theorem 2.6 and then $f_i^* \in (f_1^*, \ldots, f_{i-1}^*, f_{i+1}^*, \ldots, f_r^*)$, which is a contradiction.

Corollary 2.9. Assume that $J = (f)$ and consider the following conditions

1) $G_{I/J}(R/J)$ is free over R/I.

2) $f^\circ \neq 0$

Then 1) $\Rightarrow$ 2) and if I is generated by a regular sequence they are equivalent.

Proof. 1) $\Rightarrow$ 2) is clear after Corollary 2.8 since clearly f is part of a minimal standard base of J.

2) $\Rightarrow$ 1) By assumption $G_I(R) \cong R/I[T_1,\ldots,T_s]$ and $G_I(R) \otimes k \cong k[T_1,\ldots,T_s]$, hence f° does not divide 0 and this means that $f^* \in R/I[T_1,\ldots,T_s]$ is a homogeneous polynomial with some invertible coefficient; hence f^* does not divide 0 in $G_I(R)$ and $\{f\}$ is an I- standard base of J by [10] 3.4. We may now apply Corollary 2.7.

3. In this section I produce some comments and examples. First of all let $(R,\mathfrak{m},k)$ be, as usual, a local ring, P a prime ideal such that R/P is regular; then $\mathfrak{m}/P$ can be generated by a regular sequence, hence, if we put $L = \mathfrak{m}$ and $I = P$, we deduce from Theorem 1.9 the well-known

Corollary 3.1. The following conditions are equivalent

1) $G_P(R)$ is free over R/P.

2) $G(P,\underline{X}) : (G_P(R) \otimes k)[T_1,\ldots,T_r] \longrightarrow G_{\mathfrak{m}}(R)$ is an isomorphism.

Hereafter we consider the relative case i. e. the case when the local ring is $(R/J, \mathfrak{m}/J, k)$ $J \subseteq P \subseteq \mathfrak{m}$ and R, R/P are regular, $\mathfrak{m} = P+\underline{x}R$ where $\underline{x} = x_1,\ldots,x_r$ is a regular system of parameters on R/P.

Let us denote by $v(\ldots)$ the "minimal number of generators of $(\ldots)$".

Then we get

Corollary 3.2. The following conditions are equivalent

1) $G_{P/J}(R/J)$ is free over R/P.

2) The canonical homomorphism $(In_P(J) \otimes_R k)[T_1,\dots,T_r] \longrightarrow In_{\mathfrak{m}}(J)$ is an isomorphism.

3) There exists an $\mathfrak{m}$-standard base $\{f_1,\dots,f_s\}$ of J of equimultiple elements i.e. such that $v_P(f_i) = v_{\mathfrak{m}}(f_i)$ for every i.

These conditions imply

4) $v(In_P(J)) = v(In_{(P,\mathfrak{m})}(J)) = v(In_{\mathfrak{m}}(J))$.

Proof The equivalence of 1), 2), 3) is an immediate consequence of Lemma 2.3, Proposition 2.4 and Lemma 2.5, after remarking that $G_P(R)$ is free over R/P since P is generated by a regular sequence. Moreover 4) easily follows from 2) and Lemma 2.5.

The following examples give a closer insight into Corollary 3.2. I avoid to write down all the computations, which can be done by using the methods developed in [9] and [10].

Example 1 $R = k[X,Y,Z,W]_{loc}$ $\mathfrak{m} = (X,Y,Z,W)$ $P = (Y,Z,W)R$ $J = (f, g)$ where $f = XY^2 - ZW$ $g = XW^2 - Z^2$

Then $in_{\mathfrak{m}}(f) = -ZW$ $in_{\mathfrak{m}}(g) = -Z^2$ and f, g is a regular R-sequence, hence $\{f, g\}$ cannot be an $\mathfrak{m}$-standard base of J (see [10] 3.4).

By computing, we get that $\{f, g, XY^2Z - XW^3\}$ is an $\mathfrak{m}$-standard base of J.

However $v(In_P(J)) = v(In_{(P,\mathfrak{m})}(J)) = 2$, $v(In_{\mathfrak{m}}(J) = 3$

and $G_{P/J}(R/J)$ is not free over R/P.

Example 2 $R, \mathfrak{m}, P$ as before $J = (f,g)$ where $f = XY^2 - Z^2$ $g = XY - ZW$

Then $\{f,g\}$ is an $\mathfrak{m}$-standard base of J and also a P-standard base of J, but $in_P(f) = XY^2 - Z^2$ $in_P(g) = XY$ hence $in_{(P,\mathfrak{m})}(f) = -Z^2$

$in_{(P,\mathfrak{m})}(g) = 0$. Then $v(\,In_P(J)) = 2$, $v(\,In_{(P,\mathfrak{m})}(J)) = 1$, $v(\,In_{\mathfrak{m}}(J) = 2$ and $G_{P/J}(R/J)$ is not free over R/P.

Example 3 $R, \mathfrak{m}$, P as before $J = (f,g)$ where $f = XY - Z^2$ $g = XZ - W^3$

Then $v(\,In_P(J)) = 3$, $v(\,In_{(P,\mathfrak{m})}(J)) = 1$, $v(In_{\mathfrak{m}}(J)) = 2$ and $G_{P/J}(R/J))$ is not free over R/P.

Example 4 $R, \mathfrak{m}$, P as before $J = (f,g)$ where $f = XY - Z^2$ $g = Y^2 - Z^3$

Then $v(\,In_P(J)) = 3$, $v(\,In_{(P,\mathfrak{m})}(J)) = 2$, $v(\,In_{\mathfrak{m}}(J)) = 2$

Example 5 (see [8] for the explicit computations)

Let S be the affine surface in A_k^4, which is given parametrically by $X = t^2$, $Y = tu^5$, $Z = u^{10}$, $W = u^4$ and let J be the prime ideal associated to S in $k[X,Y,Z,W]$. Then $J = (f,g)$ where $f = Z^2 - W^5$ $g = Y^2 - XZ$.

Let $R, \mathfrak{m}$, P be as before. Then $\{f,g\}$ is an $\mathfrak{m}$-standard base of J, while $\{f, g, XW^5 - Y^2Z, Y^4 - X^2W^5\}$ is a P-standard base of J and $\{f, XW^5 - Y^2Z, Y^4 - X^2W^5\}$ is a $(P,\mathfrak{m})$-fiber base of J. Therefore $v(\,In_P(J)) = 4$, $v(\,In_{(P,\mathfrak{m})}(J)) = 3$, $v(\,In_{\mathfrak{m}}(J)) = 2$ and $G_{P/J}(R/J)$ is not free over R/P although $e(R/J) - e((R/J)_{P/J}) = 4$ (see [2] for a good account on the relation between normal flatness and equimultiplicity.

Example 6. Let S be the affine surface in A_k^4, which is given parametrically by $X = t^2$, $Y = tu^3$, $Z = tu^9$, $W = u^4$ and let J be the prime ideal associated to S in $k[X,Y,Z,W]$. Then it is not difficult to see that J is the ideal generated by the maximal minors of the matrix

$$\begin{pmatrix} X & Y^2 & Z \\ Y & Z & W^3 \end{pmatrix}$$

Denote by f_1, f_2, f_3 respectively $Y^2W^3 - Z^2$, $YZ - XW^3$, $XZ - Y^3$.

Then $\{f_1, f_2, f_3, Y^4 - X^2W^3\}$ is an $\mathfrak{m}$-standard base of J and also a P-standard base of J, while $\{f_1, f_2\}$ is a $(P,\mathfrak{m})$-fiber base of J. Hence $v(\mathrm{In}_P(J)) = 4$, $v(\mathrm{In}_{(P,\mathfrak{m})}(J)) = 2$, $v(\mathrm{In}_{\mathfrak{m}}(J)) = 4$ and $G_{P/J}(R/J)$ is not free over R/P.

Remark 1. It may be interesting to note that, in example 6,

$G_{\mathfrak{m}/J}(R/J) \cong k[X,Y,Z,W]/(Z^2, YZ, XZ, Y^4) \cong k[X,Y,Z,W]/((Z,Y^4)\cap(X,Y,Z)^2)$.

Then this computation provides an example of a <u>local Cohen-Macaulay ring, such that the scheme which is obtained by blowing up the maximal ideal is not locally Cohen-Macaulay</u>.

In fact $\mathrm{Proj}(G_{\mathfrak{m}/J}(R/J))_{W\neq 0} \cong \mathrm{Spec}(k[x,y,z]/((z,y^4)\cap(x,y,z)^2)$

(where $x = \frac{X}{W}$, $y = \frac{Y}{W}$, $z = \frac{Z}{W}$) which is not Cohen-Macaulay at the origin.

Remark 2 The above described examples show that the numbers $v(\mathrm{In}_P(J))$, $v(\mathrm{In}_{(P,\mathfrak{m})}(J))$, $v(\mathrm{In}_{\mathfrak{m}}(J))$ are totally unrelated except for the inequality $v(\mathrm{In}_P(J)) \geq v(\mathrm{In}_{(P,\mathfrak{m})}(J))$, which is a consequence of Lemma 2.3, while Corollary 3.2 says that they are equal if $G_{P/J}(R/J)$ is free over R/P.

So we are led to the following question: is it true that 4) $\Rightarrow$ 1) in Corollary 3.2? This question has an affirmative answer when J is a principal ideal. In fact if $J = (f)$, $\{f\}$ is an $\mathfrak{m}$-standard base of J and a P-standard base of J, therefore, if the three numbers are equal, then $v(\mathrm{In}_{(P,\mathfrak{m})}(J)) = 1$ and we know that a P-standard base is also a $(P,\mathfrak{m})$-fiber base. Hence $f^\circ \neq 0$ and we may apply Corollary 2.9 since P is a regular prime in a regular ring.

References

[1] ACHILLES,R.- SCHMIDT,R. A theorem on Normal Cohen-Macaulay structure Preprint.

[2] ORBANZ,U.- HERRMANN,M. Between equimultiplicity and normal flatness

To appear in "Proceedings of the Conference on Algebraic Geometry" La Rabida Spain (1981).

[3] HERRMANN, M.- SCHMIDT, R.- VOGEL, W. *Theorie der normalen Flachheit Teubner Texte zur Mathematik Leipzig (1977).*

[4] HIRONAKA, H. Resolution of singularities of an algebraic variety over a field of characteristic 0. Ann. of Math. 79 (1964).

[5] GROTHENDIECK, A.- DIEUDONNE' E.G.A. III and IV.

[6] MATSUMURA, H. *Commutative Algebra Benjamin (1970).*

[7] ORBANZ, U.- ROBBIANO, L. Projective normal flatness and Hilbert functions Preprint.

[8] ROBBIANO, L. Remarks on blowing up divisorial ideals. Manuscripta Math. 30 (1980)

[9] ROBBIANO, L.- VALLA, G. On the equations defining tangent cones. Math. Proc. Camb. Phil. Soc. 88 (1980)

[10] ROBBIANO, L.- VALLA, G. Free resolutions for special tangent cones. In this volume.

[11] SERRE, J.P. *Algèbre locale- Multiplicités Springer (1965).*

[12] VALABREGA, P.- VALLA, G. Form rings and regular sequences. Nagoya Math. J. 72 (1978).

FREE RESOLUTIONS FOR SPECIAL TANGENT CONES

Lorenzo Robbiano Giuseppe Valla

ISTITUTO MATEMATICO DELL'UNIVERSITA' DI GENOVA

GENOVA (ITALIA)

The purpose of this paper is first of all that one of producing a suitable introduction to the theory of tangent cones. In algebraic language, tangent cones to points of schemes are described by suitable graded rings; more precisely, if x is a point of a scheme (X,O_X), then the tangent cone to x is $\mathrm{Spec}(G_{\mathfrak{m}_x}(O_{x,X}))$, where $\mathfrak{m}_x$ denotes the maximal ideal of the local ring $O_{x,X}$. Suppose we know that O_X is regularly immersed at x, i.e. that $O_{x,X}$ is the quotient of a suitable regular local ring $(R,\mathfrak{m},k)$ with respect to an ideal J; then $G = G_{\mathfrak{m}/J}(R/J)$ is isomorphic to $G_{\mathfrak{m}}(R)/J^*$, where $G_{\mathfrak{m}}(R)$ is a polynomial ring over the field k, and J^* is the ideal generated by the initial forms of the elements of J. It is clear that knowledge of an explicit description of J^* is an essential point in the study of G and the problem of determinating elements $f_1,\dots,f_r$ of J, such that their initial forms $f_1^*,\ldots,f_r^*$ generate J^* is known as the problem of determinating an $\mathfrak{m}$-standard base of J. In the last years several authors dealt with this topic and an attempt to give general rules for this kind of computations was made in [12] and [14]; but after these papers many progresses were made, also in connection with the study of related topics, such as normal flatness (see for instance [11]). The first section of the present paper deals with the concept of standard base and presents a systematic way of introducing it; Corollary 1.8 and 1.10 describe equivalent conditions and in particular the homological point of view is investigated, since it is, in our opinion, the most natural to look at standard bases. The second section has been inspired by a paper of Herzog (see [6]) and treats the problem of lifting a standard base from a quotient ring to the ring itself and the inverse problem. The third section deals with the main result of [14] and treats it in a new way, namely what is brought into light is the property that "standard base" plus "regular sequence" implies "super-standard base" (see Definition 2.6 and Proposition 2.7) This allows us to get Theorem 3.4, which essentially says that, if J is generated by a regular sequence $f_1,\ldots$ $\ldots,f_r$ such that $\{f_1,\ldots,f_r\}$ is an I-standard base of J (I is an ideal containing J), then the Koszul complex of $f_1^*,\ldots,f_r^*$ is a resolution of $G_{I/J}(R/J)$. While the first sections clarify the fact that giving a standard base is the same thing as giving a first step of a free resolution of $G_{I/J}(R/J)$, Theorem 3.4 shows that in a particular circumstance not only a first step but a full free resolution of $G_{I/J}(R/J)$ can be achieved. It may be interesting to note that an application of that theorem is a simple proof of a well-known theorem of Rees (see Corollary 3.5). However that one described in Theorem 3.4 is not the only possibility of getting a full free resolution; sections 4 and 5 develop some ideas of [12], in particular some methods of computing free resolutions of the tangent cone in the cases of "Cohen-Macaulay codimension 2" and "Gorenstein codimension 3" rings. Also some applications are discussed at the end of the last section and an explicit example is computed. The rings are supposed to be commutative, with identity and noetherian.

1. Let us start by fixing some terminology. By a "multiplicatively filtered ring R" will be meant a ring R with a decreasing family $\{F_pR\}_{p\in Z}$ of ideals such that $F_pR=R$ for $p\leq 0$ and $(F_pR)(F_qR)\subseteq F_{p+q}R$ for every p,q. By a "filtered R-module M" will be meant an R-module M with a decreasing family $\{F_pM\}_{p\in Z}$ of submodules such that $F_pM=M$ for $p\leq 0$ and $(F_pR)(F_qM)\subseteq F_{p+q}M$ for every p,q. If M, N are filtered R-modules and $f:M\to N$ an R-homomorphism, f is said to be a "morphism of filtered modules" if $f(F_pM)\subseteq F_pN$.

<u>Definition-Proposition</u> 1.1. <u>Let</u> $f:M\to N$ <u>be a morphism of filtered modules; f is said to be strict if one of the following equivalent conditions holds</u>:

1) $F_pN\cap f(M)=f(F_pM)$ <u>for every</u> p.

2) $F_pN\cap f(F_{p-1}M)=f(F_pM)$ <u>for every</u> p.

<u>Proof</u>. 1) ⇒ 2) is obvious. As for 2) ⇒ 1) there is nothing to prove if $p\leq 0$, hence we may assume that $F_rN\cap f(M)=f(F_rM)$ if $r<p$. Now $F_pN\cap f(M)=F_pN\cap F_{p-1}N\cap f(M)=$ $=F_pN\cap f(F_{p-1}M)=f(F_pM)$.

Now we consider $G(R)=\oplus_{p\in Z}(F_pR/F_{p+1}R)$ and $G(M)=\oplus_{p\in Z}(F_pM/F_{p+1}M)$; then G(R) has a natural structure as a graded ring and G(M) as a graded G(R)-module and since $G(R)_p=0$ for $p<0$ and $G(M)_p=0$ for $p<0$, they may be also considered as graded objects over $N=\{0,1,2,\dots\}$. It is clear that G(.) is a functor from the category of filtered R-modules into the category of graded G(R)-modules.

If $m \in M$ we denote by $v_F(m)$ the largest integer n such that $m \in F_n M$. If such n does not exist we say $v_F(m)=\infty$ and if $v_F(m)<\infty$ we denote by $in_F(m)$ or simply by m^* the residue class of m in $F_{v_F(m)}M/F_{v_F(m)+1}M$, which is called the "initial form of m".

Let $L \xrightarrow{g} M \xrightarrow{f} N$ be a complex of filtered R-modules whose homology we denote by H, and $G(L) \xrightarrow{G(g)} G(M) \xrightarrow{G(f)} G(N)$ the corresponding complex of graded modules, whose homology we denote by H(G); it is then clear that on H there is a canonical induced filtration $F_pH=(F_p\mathrm{Ker}(f)+g(L))/g(L)$ where $F_p\mathrm{Ker}(f)=\mathrm{Ker}(f)\cap F_pM$.

<u>Lemma</u> 1.2. <u>We have canonical isomorphisms</u>

a) $G(H)_p \cong \mathrm{Ker}(f)\cap F_pM+F_{p+1}M/g(L)\cap F_pM+F_{p+1}M$

b) $H(G)_p \cong f^{-1}(F_{p+1}N)\cap F_pM/g(F_pL)+F_{p+1}M$

<u>Hence</u> G(H) <u>is a subquotient of</u> H(G).

<u>Proof</u>. a). $G(H)_p \cong \mathrm{Ker}(f)\cap F_pM+g(L)/\mathrm{Ker}(f)\cap F_{p+1}M+g(L) \cong$

$\cong \mathrm{Ker}(f)\cap F_pM/\mathrm{Ker}(f)\cap F_{p+1}M+g(L)\cap F_pM \cong$

$\cong \mathrm{Ker}(f)\cap F_pM+F_{p+1}M/\mathrm{Ker}(f)\cap F_{p+1}M+g(L)\cap F_pM+F_{p+1}M \cong$

$\cong \mathrm{Ker}(f)\cap F_pM+F_{p+1}M/g(L)\cap F_pM+F_{p+1}M$

b) comes directly from the definition.

Hereafter we assume that the given filtrations have the property that <u>every submodule is closed in the induced topology</u>. However the reader will note that this hypothesis is often unnecessary.

<u>Lemma</u> 1.3. <u>If</u> $L \xrightarrow{g} M \xrightarrow{f} N$ <u>is a complex of filtered modules</u>, <u>then</u> H=0 <u>if and only if</u> G(H)=0.

<u>Proof</u>. If G(H)=0 then $\mathrm{Ker}(f)=\mathrm{Ker}(f)\cap F_0M \subseteq g(L)+F_1M$, hence $\mathrm{Ker}(f) \subseteq g(L)+$

$+\mathrm{Ker}(f)\cap F_1M\subseteq g(L)+F_2M$ and so on. Eventually we get $\mathrm{Ker}(f)\subseteq\bigcap_{p\geq 0}(g(L)+F_pM)=g(L)$.

Proposition 1.4. If $L\xrightarrow{g}M\xrightarrow{f}N$ is a complex of filtered modules, then the following conditions are equivalent

1) g,f are strict and H=0.

2) H(G)=0.

Proof. 1) $\Rightarrow$ 2). Let $x\in f^{-1}(F_{p+1}N)\cap F_pM$; then $f(x)\in F_{p+1}N\cap f(M)=f(F_{p+1}M)$, hence $f(x)=f(y)$ with $y\in F_{p+1}M$. Therefore $x-y\in \mathrm{Ker}(f)\cap F_pM=g(L)\cap F_pM=g(F_pL)$ and we are done by b) of Lemma 1.2.

2) $\Rightarrow$ 1). From Lemma 1.1 we get G(H)=0 and from Lemma 1.2 we get H=0. Let $y\in F_pN\cap f(M)$; then $y=f(x)$ with x belonging to some F_qM. If $q<p$ we have $x\in f^{-1}(F_{q+1}N)\cap F_qM=g(F_qL)+F_{q+1}M$ by Lemma 1.2; hence $y=f(x')$ with $x'\in F_{q+1}M$. Eventually we get $y\in f(F_pM)$ and this shows that f is strict. Now $F_pM\cap g(L)\subseteq$ $\subseteq F_pM\cap f^{-1}(F_{p+1}N)\subseteq g(F_pL)+F_{p+1}M$ hence $F_pM\cap g(L)\subseteq g(F_pL)+F_{p+1}M\cap g(L)$; in conclusion we get $F_pM\cap g(L)\subseteq\bigcap_{r\geq 0}(g(F_pL)+F_{p+r}M)=g(F_pL)$.

Corollary 1.5. Let $f:M\to N$ be a morphism of filtered modules; then there is a canonical embedding $G(\mathrm{Ker}(f))\to \mathrm{Ker}(G(f))$. The embedding is an isomorphism if and only if f is strict.

Proof. We consider the exact sequence $0\to\mathrm{Ker}(f)\xrightarrow{i}M\xrightarrow{f}N$. Since on Ker(f) there is the induced filtration, i is strict, hence $0\to G(\mathrm{Ker}(f))\xrightarrow{G(i)}G(M)$ is exact by Proposition 1.4. Moreover $0\to G(\mathrm{Ker}(f))\xrightarrow{G(i)}G(M)\xrightarrow{G(f)}G(N)$ is a complex, hence there is a canonical embedding $G(\mathrm{Ker}(f))\to\mathrm{Ker}(G(f))$. The other fact follows again from Proposition 1.4.

Let now $(R,\mathfrak{m})$ be a local ring and $I\supseteq J$ ideals of R. Given an element f of R we denote by $v_I(f)$ or simply by v(f) the number $v_F(f)$ where $F_pR=I^p$ and

$I^p = R$ if $p \leq 0$. We put $0^* = 0$ and we denote by $In_I(J)$ or simply by J^* the homogeneous ideal of $G_I(R) = \oplus_{p \in Z} (I^p/I^{p+1})$ generated by the initials forms of the elements of J; J^* is easily seen to be the kernel of the canonical graded epimorphism $G_I(R) \to G_{I/J}(R/J)$.

Definition 1.6. *A set of elements* $\{f_1,\ldots,f_r\} \subseteq J$ *is called an I-standard base of J if* $(f_1^*,\ldots,f_r^*) = J^*$.

Proposition 1.7. (see [7] Lemma 6). *If* $\{f_1,\ldots,f_r\}$ *is an I-standard base of J, then* $J = (f_1,\ldots,f_r)$.

Proof. Let $J' = (f_1,\ldots,f_r)$; then $J_p^* = I^p \cap J + I^{p+1}/I^{p+1}$ while $(f_1^*,\ldots,f_r^*)_p \subseteq$ $\subseteq J' + I^{p+1}/I^{p+1}$. Hence $I^p \cap J \subseteq J' + I^{p+1}$. Now $J' \subseteq J$ hence $I^p \cap J \subseteq J' + I^{p+1} \cap J$ and we deduce $J' \subseteq J \subseteq \bigcap_{s \geq 0} (J' + I^s) = J'$.

Remark 1. Definition 1.6 and Proposition 1.7 allows us to say that $\{f_1,\ldots,f_r\}$ is an I-standard base with the meaning that $\{f_1,\ldots,f_r\}$ is an I-standard base of $(f_1,\ldots,f_r)$

Remark 2. After [1] p.107, it is easy to see that an upper bound for the set of integers $\{v(f_1),\ldots,v(f_r)\} = \{\deg f_1^*,\ldots,\deg f_r^*\}$ where $\{f_1,\ldots,f_r\}$ is an I-standard base of J, is given by an Artin-Rees number of J with respect to I, i.e. a number k such that $I^p \cap J = I^{p-k}(I^k \cap J)$ for every $p \geq k$ (see also [5])

Let now $J = (f_1,\ldots,f_r)$ and $v_i = v_I(f_i)$; then we may consider the exact sequence $R^r \xrightarrow{f} R \xrightarrow{\pi} R/J \to 0$ where $f(a_1,\ldots,a_r) = \Sigma a_i f_i$ and π is the canonical epimorphism. Let $F_p R^r = \oplus_{i=1}^{r} I^{p-v_i}$, $F_p R = I^p$, $F_p(R/J) = (I/J)^p$. With these filtrations f and π are morphisms of filtered modules and we get the corresponding complex of graded modules

$$(*) \qquad G(R^r) \xrightarrow{G(f)} G(R) \xrightarrow{G(\pi)} G(R/J) \to 0$$

where $G(R^r)=\bigoplus_{i=1}^{r} G_I(R)(-v_i)$, $G(R)=G_I(R)$, $G(R/J)=G_{I/J}(R/J)$, $G(f)(1,0,\ldots,0)=f_1^*$, $\ldots,G(f)(0,\ldots,0,1)=f_r^*$.

Corollary 1.8. The following conditions are equivalent:

1) $\{f_1,\ldots,f_r\}$ is an I-standard base of J.

2) The complex (*) is exact.

3) $G(\mathrm{Ker}(f)) \cong \mathrm{Ker}(G(f))$.

4) f is a strict morphism.

5) $I^p \cap J=\sum_{i=1}^{r} I^{p-v_i} f_i$ for every p.

6) $I^p \cap \sum_{i=1}^{r} I^{p-1-v_i} f_i = \sum_{i=1}^{r} I^{p-v_i} f_i$ for every p.

Proof. It is clear by definitions that 1)⇔ 2) and 4)⇔ 5)⇔ 6) by Definition-Proposition 1.1. The equivalence between 2) and 4) follows from Proposition 1.4 and the equivalence between 3) and 4) follows from Corollary 1.5.

Let us now consider the canonical map $\mathrm{Ker}(f) \to G(\mathrm{Ker}(f))$, whose image is the set of homogeneous elements of $G(\mathrm{Ker}(f))$ and the canonical injective morphism of graded modules $G(\mathrm{Ker}(f)) \to \mathrm{Ker}(G(f))$ (see Corollary 1.5). We get a map h from the module of syzygies of $f_1,\ldots,f_r$ to the module of syzygies of $f_1^*,\ldots,f_r^*$.

Lemma 1.9. Let $(a_1,\ldots,a_r)$ be a non zero element of Ker(f) and $m=\min_i(v(a_i)+v_i)$. Then $h(a_1,\ldots,a_r)=(\bar{a}_1,\ldots,\bar{a}_r)$ where $\bar{a}_i \in I^{m-v_i}/I^{m-v_i+1}$; hence $\bar{a}_i=0$ if $v(a_i)+v_i>m$ and $\bar{a}_i=a_i^*$ if $v(a_i)+v_i=m$.

Proof. It is a strightforward consequence of the above defined filtrations.

Corollary 1.10. The following conditions are equivalent:

1) $\{f_1,\ldots,f_r\}$ is an I-standard base of J.

2) There exists a system of generators of Ker(G(f)), whose elements can be lifted, via h, to elements of Ker(f).

Proof. It follows from the equivalence between 1) and 3) of Corollary 1.8.

2. Let $(R,\mathfrak{m})$ be a local ring, I, J and L ideals of R and denote by "$\bar{\ }$" reduction modulo J and by "$\sim$" reduction modulo L.

Lemma 2.1. The following conditions are equivalent:

1) $I^n \cap (L+J) = I^n \cap L + I^n \cap J$ for all integers $n \geq 0$.

2) For all elements $f \in J$ there exists g in J such that $\tilde{f}=\tilde{g}$ and $v_{\tilde{I}}(\tilde{g}) = v_I(g)$.

3) There exist elements $f_1,\dots,f_s$ in J such that $\{\tilde{f}_1,\dots,\tilde{f}_s\}$ is an $\tilde{I}$-standard base of $\tilde{J}$ and $v_{\tilde{I}}(\tilde{f}_i) = v_I(f_i)$ for $i=1,\dots,s$.

Proof. 1) ⇒ 2). If $\tilde{f}=0$ take g=0. Otherwise let $v = v_{\tilde{I}}(\tilde{f})$; then $f \in I^v + L$, so $f=c+d$ with $c \in I^v$ and $d \in L$. Thus $c \in I^v \cap (L+J)$, hence $c=e+g$ with $e \in I^v \cap L$ and $g \in I^v \cap J$. It follows $\tilde{f}=\tilde{g}$ and $v \leq v_I(g) \leq v_{\tilde{I}}(\tilde{g}) = v$.

2) ⇒3) is obvious.

3) ⇒1). Let $y \in I^n \cap (L+J)$; then $y=c+a$ with $c \in L$ and $a \in J$. If $\tilde{a}=0$ we are done; if not let $p = v_{\tilde{I}}(\tilde{a})$. Then $p \geq n$, hence $a = \sum_{i=1}^{s} c_i f_i + d$ with $c_i \in I^{p-p_i}$ and $d \in L$, where we write p_i for $v_I(f_i) = v_{\tilde{I}}(\tilde{f}_i)$. It follows $y = c+d+\sum_{i=1}^{s} c_i f_i$, where, since $\sum_{i=1}^{s} c_i f_i \in I^n \cap J$, $c+d \in I^n \cap L$. Hence $I^n \cap (L+J) \subseteq I^n \cap L + I^n \cap J$ and, since the other inclusion is clear, the result follows.

As an easy consequence of the above Lemma we get the following criterium of standard base in a relative case (see [6]).

Theorem 2.2. Let $\{x_1,\dots,x_r\}$ be an I-standard base of L and $v_i = v_I(x_i)$. Assume that for all $x_i \notin J$, $v_i = v_{\bar{I}}(\bar{x}_i)$. Then the following conditions are equivalent:

1) $\{\bar{x}_1,\dots,\bar{x}_r\}$ is an $\bar{I}$-standard base of $\bar{L}$.

2) $I^n \cap (L+J) = \sum_{i=1}^{r} I^{n-v_i} x_i + I^n \cap J$, for all integers $n \geq 0$.

3) For all elements $f \in J$ there exists $g \in J$ such that $\tilde{f} = \tilde{g}$ and $v_{\tilde{I}}(\tilde{g}) = v_I(g)$.

4) There exist elements $f_1, \ldots, f_s$ in J such that $\{\tilde{f}_1, \ldots, \tilde{f}_s\}$ is an $\tilde{I}$-standard base of $\tilde{J}$ and $v_{\tilde{I}}(\tilde{f}_i) = v_I(f_i)$ for $i=1,\ldots,s$.

Proof. With our assumptions we have $I^n \cap L = \sum_{i=1}^{r} I^{n-v_i} x_i$, hence 2), 3), and 4) are equivalent by the above lemma. Now let $p_i = v_{\bar{I}}(\bar{x}_i)$; then, since $p_i \geq v_i$, we have $\sum_{i=1}^{r} I^{n-p_i} x_i \supseteq \sum_{i=1}^{r} I^{n-v_i} x_i$ and ,in our case, $\sum_{i=1}^{r} I^{n-p_i} x_i + J = \sum_{i=1}^{r} I^{n-v_i} x_i + J$. Since $\bar{I}^n \cap \bar{L} = (I^n \cap (L+J)+J)/J$ and $\sum_{i=1}^{r} \bar{I}^{n-p_i} \bar{x}_i = (\sum_{i=1}^{r} I^{n-p_i} x_i + J)/J = (\sum_{i=1}^{r} I^{n-v_i} x_i + J)/J$, it follows that $\{\bar{x}_1, \ldots, \bar{x}_r\}$ is an $\bar{I}$-standard base of $\bar{L}$ if and only if $I^n \cap (L+J)+J = \sum_{i=1}^{r} I^{n-v_i} x_i + J$. But this is equivalent to $I^n \cap (L+J) = \sum_{i=1}^{r} I^{n-v_i} x_i + I^n \cap J$; thus 1) $\Leftrightarrow$ 2) and the result follows.

Corollary 2.3. With the same assumptions and notations as in Theorem 2.2, suppose $J \subseteq L$. Then $\{\bar{x}_1, \ldots, \bar{x}_r\}$ is an $\bar{I}$-standard base of $\bar{L}$.

Proof. We have $I^n \cap (L+J) = I^n \cap L = \sum_{i=1}^{r} I^{n-v_i} x_i$.

Corollary 2.4. Let $\{x_1, \ldots, x_r\}$ be an I-standard base of L and $J=(x_1, \ldots, x_{r-1})$. If $x_r \notin J$, then $v_{\bar{I}}(\bar{x}_r) = v_I(x_r)$ and $\{\bar{x}_r\}$ is an $\bar{I}$-standard base of $\bar{L}$.

Proof. Let $v = v_I(x_r)$. If $x_r \in I^{v+1}+J$, then there exist $a_1, \ldots, a_{r-1}$ in R such that $x_r + \sum_{i=1}^{r-1} a_i x_i \in I^{v+1} \cap L$; thus $x_r + \sum_{i=1}^{r-1} a_i x_i = \sum_{i=1}^{r} b_i x_i$ with $b_r \in I$, which contradicts the fact that $x_r \notin J$. It follows that $v = v_{\bar{I}}(\bar{x}_r)$ and this implies, by Corollary 2.3, that $\{\bar{x}_r\}$ is an $\bar{I}$-standard base of $\bar{L}$.

In Theorem 2.2 we found necessary and sufficient conditions for preserving the property of beeing a standard base after reducing modulo an ideal J. Now we want to study the inverse problem.

Lemma 2.5. Let I and $J=(f_1,\dots,f_r)$ be ideals of the local ring R and let $v_i = v_I(f_i)$ for $i=1,\dots,r$. If for some integer $v\geq 0$ there exists $a\in \mathfrak{m}\cap I^v$ such that $I^n\cap J=\sum_{i=1}^r I^{n-v_i}f_i+a(I^{n-v}\cap J)$ for all integers $n\geq 0$, then $\{f_1,\dots,f_r\}$ is an I-standard base of J.

Proof. We have to prove $I^n\cap J=\sum_{i=1}^r I^{n-v_i}f_i$ for all integers $n\geq 0$, hence if $v=0$ the conclusion follows by Nakayama. If $v>0$, making induction on n, we get $I^n\cap J=\sum_{i=1}^r I^{n-v_i}f_i+a(\sum_{i=1}^r I^{n-v-v_i}f_i)=\sum_{i=1}^r I^{n-v_i}f_i$ as required.

Definition 2.6. Let I be an ideal of the local ring R. A sequence $\{f_1,\dots,f_r\}$ is said to be a super I-standard base if $\{f_{i_1},\dots,f_{i_t}\}$ is an I-standard base of $(f_{i_1},\dots,f_{i_t})$ for every subset $\{i_1,\dots,i_t\}$ of $\{1,\dots,r\}$.

The next proposition gives a very important example of a super I-standard base.

Proposition 2.7. Let I and $J=(f_1,\dots,f_r)$ be ideals of the local ring R such that $f_1,\dots,f_r$ is a regular sequence and an I-standard base of J. Then $\{f_1,\dots,f_r\}$ is a super I-standard base.

Proof. It is enough to show that $I^n\cap(f_1,\dots,f_{r-1})=\sum_{i=1}^{r-1} I^{n-v_i}f_i$ where $v_i=v_I(f_i)$ for all $i=1,\dots,r$. Let $L=(f_1,\dots,f_{r-1})$; then we get

$$I^n\cap L = I^n\cap J\cap L = \sum_{i=1}^{r-1} I^{n-v_i}f_i + (I^{n-v_r}f_r\cap L) = \sum_{i=1}^{r-1} I^{n-v_i}f_i + f_r(I^{n-v_r}\cap(L:f_r)) =$$

$=\sum_{i=1}^{r-1} I^{n-v_i}f_i+f_r(I^{n-v_r}\cap L)$. The conclusion follows by Lemma 2.5.

Theorem 2.8. Let $L=(x_1,\dots,x_r)$ and $v_i=v_I(x_i)$. Assume that for all $i=1,\dots,r$ $v_i=v_{\bar{I}}(\bar{x}_i)$ and $\{\bar{x}_1,\dots,\bar{x}_r\}$ is an $\bar{I}$-standard base of $\bar{L}$. If there exists a minimal base $f_1,\dots,f_s$ of J such that

1) $\{f_1,\dots,f_s\}$ is a super I-standard base of J

2) $\{\tilde{f}_1,\ldots,\tilde{f}_s\}$ <u>is a regular sequence in</u> $\tilde{R}=R/L$.

<u>then</u> $\{x_1,\ldots,x_r\}$ <u>is an I-standard base of L.</u>

<u>Proof.</u> We use induction on s. Let s=1, $f=f_1$ and $v=v_I(f)$. We have $\bar{I}^n\cap\bar{L}=$ $=\sum_{i=1}^{r}\bar{I}^{n-v_i}\bar{x}_i$, hence $I^n\cap L=\sum_{i=1}^{r}I^{n-v_i}x_i+J\cap I^n\cap L$. But $J\cap I^n\cap L=I^{n-v}f\cap L=$ $=f(I^{n-v}\cap(L:f))=f(I^{n-v}\cap L)$; the conclusion follows by Lemma 2.5.

Now assume $s>1$ and let $N=(f_1,\ldots,f_{s-1})$; further let "'" denote reduction modulo N. Then in the ring R' we have ideals I', J' and $L'=(x_1',\ldots,x_r')$; we get for every $i=1,\ldots,r$: $v_{I'}(x_i')\leqslant v_{\bar{I}'}(\bar{x}_i')=v_{\bar{I}}(\bar{x}_i)=v_I(x_i)\leqslant v_{I'}(x_i')$. Further $\{\bar{x}_1',\ldots,\bar{x}_r'\}$ is an $\bar{I}'$-standard base of $\bar{L}'$ and, by the minimality of $\{f_1,\ldots,f_s\}$ and Corollary 2.4., $\{f_s'\}$ is an I'-standard base of J'; it is also clear that f_s' is a non zerodivisor modulo L'. So by the first step of the induction we get that $\{x_1',\ldots,x_r'\}$ is an I'-standard base of L'. Since we have seen that $v_i=v_{I'}(x_i')$ for all $i=1,\ldots,r$, the ideal N has the same property as the ideal J, so we can conclude by the inductive assumption.

3. Let now consider the Koszul complex $\mathbb{K}(f_1,\ldots,f_r;R)\equiv\mathbb{K}$; it is a complex of finitely generated free R-modules where $\mathbb{K}_i=\Lambda^i R^r$. We denote by $\Sigma(i,r)$ the set of i-uples $j=(j_1,\ldots,j_i)$, $1\leqslant j_1<j_2<\ldots<j_i\leqslant r$; then the set $\{e_{j_1}\wedge\ldots\wedge e_{j_i}\mid j\in\Sigma(i,r)\}$ is a canonical base of $\mathbb{K}_i$. The differentials d_i are defined by $d_i(e_{j_1}\wedge\ldots\wedge e_{j_i})=\sum_{k=1}^{i}(-1)^k f_{j_k}.e_{j_1}\wedge\ldots\wedge e_{j_{k-1}}\wedge e_{j_{k+1}}\wedge\ldots\wedge e_{j_i}$.

On $\mathbb{K}$ we consider the following filtration: $F_p\mathbb{K}_i=\bigoplus_{j\in\Sigma(i,r)}I^{p-v(j)}$ where $v(j)=\sum_{k=1}^{i}v_{j_k}$.

<u>Lemma</u> 3.1. <u>The two complexes</u> $G_F(\mathbb{K})$ <u>and</u> $\mathbb{K}(f_1^*,\ldots,f_r^*;G_I(R))$ <u>are canonically isomorphic</u>.

Proof. We have $G_F(\mathbb{K}_i)=\oplus_{p\geq 0} F_p\mathbb{K}_i/F_{p+1}\mathbb{K}_i \cong (\oplus_{p\geq 0})(\oplus_{j\varepsilon\Sigma(i,r)})(I^{p-v(j)}/I^{p-v(j)+1})\cong$

$\cong \oplus_{j\varepsilon\Sigma(i,r)} G_I(R)(-v(j))\cong \Lambda^i(\oplus_{s=1}^{r} G_I(R)(-v_s))$.

Corollary 3.2. 1) If $J=(f_1,\ldots,f_r)$ then $gr(f_1^*,\ldots,f_r^*)\leq gr(J)$.

2) $gr(J)\geq gr(J^*)$.

Proof. 1) follows from Lemma 3.1., Proposition 1.4. and the grade-sensitivity of the Koszul complex (see for example [10]), while 2) from 1) and Proposition 1.7.

Lemma 3.3. The following conditions are equivalent:

1) $f_1^*,\ldots,f_r^*$ is a regular $G_I(R)$-sequence.

2) $f_1,\ldots,f_r$ is a regular R-sequence, $\{f_1,\ldots,f_r\}$ is an I-standard base and $gr(f_1^*,\ldots,f_r^*)\geq r-1$.

Proof. 1) ⇒ 2). Let us consider the Koszul complex of $f_1,\ldots,f_r$

$$\mathbb{K}:\quad 0\to\Lambda^r R^r\to\ldots\to\Lambda^2 R^r\xrightarrow{d_2}\Lambda^1 R^r\xrightarrow{d_1}\Lambda^0 R^r\cong R\to R/J\to 0$$

and the Koszul complex $\mathbb{K}^*$ of $f_1^*,\ldots,f_r^*$ which is canonically isomorphic to $G_F(\mathbb{K})$ after Lemma 3.1:

$$\mathbb{K}^*:\quad 0\to\Lambda^r(\oplus_{s=1}^{r} G_I(R)(-v_s))\to\ldots\to\Lambda^2(\oplus_{s=1}^{r} G_I(R)(-v_s))\xrightarrow{G(d_2)}\Lambda^1(\oplus_{s=1}^{r} G_I(R)(-v_s))\to$$

$$\xrightarrow{G(d_1)}\Lambda^0(\oplus_{s=1}^{r} G_I(R)(-v_s))\cong G_I(R)\to G_I(R)/(f_1^*,\ldots,f_r^*)\to 0$$

After Corollary 3.2 the only thing to be proved is that $\{f_1,\ldots,f_r\}$ is an I-standard base of J,but this follows from the equivalence between 1) and 2) of Corollary 1.8.

2) ⇒ 1). $gr(f_1^*,\ldots,f_r^*)\geq r-1$ implies that $\mathbb{K}^*$ is exact up to Λ^2 hence $d_r,\ldots,d_2$ are strict morphism after Proposition 1.4. Also d_1 is strict since $\{f_1,\ldots,f_r\}$ is an I-standard base of J (see 1) ⇔ 4) in Corollary 1.8.).

Now the homology of $\mathbb{K}$ is zero since $f_1,\ldots,f_r$ is a regular sequence, hence

Proposition 1.4 implies that $\mathbb{K}^*$ is exact.

Theorem 3.4. (see [14]) The following conditions are equivalent:

1) $f_1^*,\dots,f_r^*$ is a regular $G_I(R)$-sequence.

2) $f_1,\dots,f_r$ is a regular sequence and $\{f_1,\dots,f_r\}$ is an I-standard base.

Proof. 1) ⇒ 2) by Lemma 3.3.

2) ⇒ 1). Since there is nothing to be proved for r=0, we may assume that the statement is true for r-1. Now $f_1,\dots,f_{r-1}$ is again a regular sequence and $\{f_1,\dots,f_{r-1}\}$ is an I-standard base by Proposition 2.7, hence $f_1^*,\dots,f_{r-1}^*$ is a regular $G_I(R)$-sequence. Thus we can conclude by applying 2) ⇒ 1) of Lemma 3.3.

Corollary 3.5 (Rees). Let $x_1,\dots,x_r$ be elements of R and denote by $\underline{x}R$ the ideal they generate. The following conditions are equivalent:

1) $x_1,\dots,x_r$ is an R-sequence.

2) The canonical epimorphism $(R/\underline{x}R)[T_1,\dots,T_r] \xrightarrow{\pi} G_{\underline{x}R}(R)$ is an isomorphism.

Proof. Let us denote by x_i^* the initial form of x_i in $G_{\underline{x}R}(R)$ i.e. the residue class of x_i in $\underline{x}R/(\underline{x}R)^2$. We know that π is defined by $\pi(T_i)=x_i^*$. Hence 2) ⇒ 1) is a straightforward consequence of Theorem 3.4. As to the proof of 1) ⇒ 2) assume that r=1 and let $\bar{a}T^n \in \mathrm{Ker}(\pi)$. Then $ax^n \in x^{n+1}R$ whence $ax^n = bx^{n+1}$ and $x^n(a-bx)=0$. But x is a non zerodivisor, hence a=bx and $\bar{a}=0$. We proceed by induction on r. Since $\{x_1,\dots,x_r\}$ is an $\underline{x}R$-standard base of $\underline{x}R$ it follows from Proposition 2.7. that it is a super $\underline{x}R$-standard base; in particular x_1^* generates the ideal $(x_1)^*$ i.e. the sequence

$$0 \longrightarrow G_{\underline{x}R}(R) \xrightarrow{x_1^*} G_{\underline{x}R}(R) \longrightarrow G_{\underline{x}R/(x_1)}(R/(x_1)) \longrightarrow 0$$

is exact. Hence we get a commutative diagram

$$\begin{array}{ccccccccc} 0 & \to & (R/\underline{x}R)[T_1,\dots,T_r] & \xrightarrow{T_1} & (R/\underline{x}R)[T_1,\dots,T_r] & \longrightarrow & (R/\underline{x}R)[T_2,\dots,T_r] & \to & 0 \\ & & \downarrow \pi & & \downarrow \pi & & \downarrow \bar{\pi} & & \\ 0 & \longrightarrow & G_{\underline{x}R}(R) & \xrightarrow{x_1^*} & G_{\underline{x}R}(R) & \longrightarrow & G_{\underline{x}R/(x_1)}(R/(x_1)) & \to & 0 \end{array}$$

Now $\bar{\pi}$ is an isomorphism by induction, and we deduce an exact sequence

$$0 \to \mathrm{Ker}(\pi) \xrightarrow{T_1} \mathrm{Ker}(\pi) \longrightarrow 0$$

whence $\mathrm{Ker}(\pi)=0$ by Nakayama. This concludes the proof.

4. Let R be a local ring and $A=(a_{ij})$ an $r\times(r+1)$ matrix with entries in R. Denote by A_i the determinant of the matrix which is obtained by deleting the ith column of A and by (A) the ideal $(A_1,\dots,A_{r+1})$; let I be an ideal of R such that $I \supseteq (A)$. We consider the following complex

$$\mathbb{K}: \quad 0 \longrightarrow R^r \xrightarrow{g} R^{r+1} \xrightarrow{f} R \xrightarrow{\pi} R/(A) \longrightarrow 0$$

where g is defined by A with respect to the canonical bases, $f(a_1,\dots,a_{r+1}) = \sum_{i=1}^{r+1} a_i A_i$ and π is the canonical projection. It is well known that $\mathrm{gr}(A) \leqslant 2$ and $\mathrm{gr}(A)=2$ if and only if $\mathbb{K}$ is exact. Let $F_p(R/(A))=(I/(A))^p$, $F_pR=I^p$, $F_pR^{r+1}=\bigoplus_{i=1}^{r+1} I^{p-v_i}$ where $v_i=v(A_i)$ and $F_pR^r=\bigoplus_{i=1}^{r} I^{p-p_i}$ where $p_i=\min_j(v(a_{ij})+v_j)$. With these filtrations g, f and π are morphism of filtered modules and we get the corresponding complex of graded modules

$$G(\mathbb{K}): \quad 0 \longrightarrow \bigoplus_{i=1}^{r} G_I(R)(-p_i) \xrightarrow{G(g)} \bigoplus_{i=1}^{r+1} G_I(R)(-v_i) \xrightarrow{G(f)} G_I(R) \xrightarrow{G(\pi)} G_{I/(A)}(R/(A)) \to 0$$

where $G(f)(1,0,\dots,0)=A_1^*,\dots,G(f)(0,\dots,0,1)=A_{r+1}^*$ and $G(g)(1,0,\dots,0)= =(\bar{a}_{11},\dots,\bar{a}_{1,r+1}),\dots,G(g)(0,\dots,0,1)=(\bar{a}_{r1},\dots,\bar{a}_{r,r+1})$ with

$\bar{a}_{ij} \varepsilon I^{p_i-v_j}/I^{p_i-v_j+1}$. Hence G(g) is represented by the matrix $\bar{A}=(\bar{a}_{ij})$ where $\bar{a}_{ij} \varepsilon G_I(R)_{p_i-v_j}$. Let $V=(v(a_{ij}))$ and $H=(p_i-v_j)$; since $p_i-v_j+p_k-v_m = =p_i-v_m+p_k-v_j$, the matrix H is homogeneous. We recall that if $M=(m_{ij})$ is an r×r matrix with entries in Z, the integers $\sum_{t=1}^{r} m_{t,\tau(t)}$ where τ is a permutation of {1,...,r} are called M-sums and σ(M) is defined to be the minimum of the M-sums. If M is an r×(r+1) matrix with entries in Z, we denote by $\sigma_i(M)$ the integer $\sigma(M_i)$, where M_i is the square matrix which is obtained by deleting the ith column of M.

Lemma 4.1. With the above notations we have:

a) $v_i \geqslant \sigma_i(V) \geqslant \sigma_i(H) = \sum_{j=1}^{r} p_j - \sum_{j\neq i} v_j$ for every i=1,...,r+1

If moreover $v_i=\sigma_i(V)$ for every i=1,...,r+1 then

b) $\sum_{i=1}^{r} p_i = \sum_{i=1}^{r+1} \sigma_i(V)$.

Proof. It is an easy consequence of the preceding assumptions that a) holds while b) is Proposition 2.7 in [12].

In the following we denote by $\bar{A}_i$ the determinant of the matrix which is obtained by deleting the ith column of $\bar{A}$ and by $(\bar{A})$ the ideal $(\bar{A}_1,\ldots,\bar{A}_{r+1})$

Corollary 4.2. The following conditions are equivalent:

1) $\bar{A}_i=A_i^*$ for every i=1,...,r+1.

2) $v_i = \sigma_i(V)$ for every i=1,...,r+1.

Proof. It is easy to show that $\bar{A}_i=\overline{A_i} \varepsilon I^s/I^{s+1}$ where $s=\sum_{j=1}^{r} p_j - \sum_{j\neq i} v_j$; hence $\bar{A}_i=A_i^*$ if and only if $v_i=s$. Therefore the conclusion follows by Lemma 4.1.

Theorem 4.3. Assume that $v_i = \sigma_i(V)$ for every i=1,...,r+1. Then $gr(\bar{A})=2$ if and only if $G(\mathbb{K})$ is exact.

Proof. We have seen that G(g) is represented by the matrix $\bar{A}$, while G(f) is represented by the vector-matrix $(A_1^*,\dots,A_{r+1}^*)$. Since $v_i=\sigma_i(V)$, $\bar{A}_i=A_i^*$ by Corollary 4.2, hence we get a complex

$$G'(\mathbb{K}):\quad 0 \longrightarrow \bigoplus_{i=1}^{r} G_I(R)(-p_i) \xrightarrow{G(g)} \bigoplus_{i=1}^{r+1} G_I(R)(-v_i) \xrightarrow{G(f)} G_I(R) \longrightarrow G_I(R)/(\bar{A}) \rightarrow 0$$

which is the Northcott-Eagon complex associated to the matrix $\bar{A}$. If $gr(\bar{A})=2$ $G'(\mathbb{K})$ is exact; this implies that f is strict by Proposition 1.4 applied to the complex $R^r \xrightarrow{g} R^{r+1} \xrightarrow{f} R$. Hence $G'(\mathbb{K})$ coincides with $G(\mathbb{K})$ by Corollary 1.8. Conversely if $G(\mathbb{K})$ is exact, f is a strict morphism, hence $G(\mathbb{K})$ coincides with $G'(\mathbb{K})$, which implies $gr(\bar{A})=2$.

Corollary 4.4. *Assume that* $v_i=\sigma_i(V)$ *for every* $i=1,\dots,r+1$, $G_I(R)$ *is Cohen-Macaulay and* $gr(A)=2$. *Then the following conditions are equivalent:*

1) $gr(\bar{A})=2$

2) $G(\mathbb{K})$ *is exact.*

3) $\{A_1,\dots,A_{r+1}\}$ *is an I-standard base of* (A).

Proof. With respect to the preceding theorem, the only point to be proved is 3) ⇒ 1). Since $G_I(R)$ is Cohen-Macaulay, R is Cohen-Macaulay (see [8]) and they have the same dimension by [13]; moreover $gr((A)) \geq gr(A_1^*,\dots,A_{r+1}^*)$ (see 3.2), hence $\dim(R/(A))=\dim(R)-ht((A))=\dim(R)-gr((A)) \leq \dim(G_I(R))+$ $-gr(A_1^*,\dots,A_{r+1}^*)=\dim(G_I(R))-ht(A_1^*,\dots,A_{r+1}^*)=\dim(G_I(R)/(A^*))=\dim(G_{I/(A)}(R/(A))=$ $=\dim(R/(A))$. From $gr((A))=2$ we get $gr((\bar{A}))=gr(A_1^*,\dots,A_{r+1}^*)=2$.

Remark. It is interesting to note that if I is primary for the maximal ideal of R and $G(\mathbb{K})$ is exact, we may compute the Hilbert-Samuel function $H_{I/(A)}(R/(A)$. This function turns out to be only dependent on dim(R) and the integers $p_1,\dots,p_r,v_1,\dots,v_{r+1}$.

5. Let R be a local ring, n an odd integer and $A = (a_{ij})$ an $n\times n$ skew-symmetric matrix ($a_{ij} = -a_{ij}$ and $a_{ii} = 0$) with entries in R. Denote by Pf_i the pfaffian of the matrix which is obtained by deleting the i-th row and column of A and by Pf(A) the ideal $(Pf_1,\dots,Pf_n)$. Let I be an ideal of R such that $I \supseteq Pf(A)$. We consider the following complex

$$\mathbb{L}\ :\quad 0 \longrightarrow R \xrightarrow{h} R^n \xrightarrow{g} R^n \xrightarrow{f} R \xrightarrow{\pi} R/Pf(A) \longrightarrow 0$$

where h is defined by $h(1) = (Pf_1,\dots,Pf_n)$, g is defined by A with respect to the canonical bases, $f(a_1,\dots,a_n) = \sum_{i=1}^{n} a_i Pf_i$ and π is the canonical projection. It is well-known (see for instance [4]) that $gr(Pf(A)) \leqslant 3$ and $gr(Pf(A)) = 3$ if and only if $\mathbb{L}$ is exact. Let $F_p(R/Pf(A) = (I/Pf(A))^p$; $F_pR = I^p$; $F_pR^n = \bigoplus_{i=1}^{n} I^{p-v_i}$ where $v_i = v(Pf_i)$; $F'_pR^n = \bigoplus_{i=1}^{n} I^{p-p_i}$ where $p_i = \min_j (v(a_{ij})+v_j)$ and $F'_pR = I^{p-v}$ where $v = \min_j (v_j+p_j)$. With these filtrations h, g, f and π are morphisms of filtered modules and we get the corresponding complex of graded modules

$$G(\mathbb{L})\ :\ 0 \to G_I(R)(-v) \xrightarrow{G(h)} \bigoplus_{i=1}^{n} G_I(R)(-p_i) \xrightarrow{G(g)} \bigoplus_{i=1}^{n} G_I(R)(-v_i) \xrightarrow{G(f)}$$

$$\xrightarrow{G(f)} G_I(R) \xrightarrow{G(\pi)} G_{I/Pf(A)}(R/Pf(A)) \longrightarrow 0$$

where $G(f)(1,\dots,0) = Pf^*_1,\dots,$ $G(f)(0,\dots,0,1) = Pf^*_n$, $G(g)(1,\dots,0) = (\bar{a}_{11},\dots,\bar{a}_{1n}),\dots,$ $G(g)(0,\dots,0,1) = (\bar{a}_{n1},\dots,\bar{a}_{nn})$ with $\bar{a}_{ij} \in I^{p_i-v_j}/I^{p_i-v_j+1}$ and $G(h)(1) = (\overline{Pf}_1,\dots,\overline{Pf}_n)$ with $\overline{Pf}_i \in I^{v-p_i}/I^{v-p_i+1}$. Hence G(g) is represented by the matrix $\bar{A} = (\bar{a}_{ij})$ where $\bar{a}_{ij} \in G_I(R)_{p_i-v_j}$.

Let $V = (v(a_{ij}))$ and $H = (p_i-v_j)$; of course the matrix V is symmetric and H is homogeneous. If $M = (m_{ij})$ is a symmetric $2r \times 2r$ matrix with entries in $\mathbb{Z}$, the integers $m_{\sigma(1)\sigma(2)} + m_{\sigma(3)\sigma(4)} + \dots + m_{\sigma(2r-1)\sigma(2r)}$

where σ is a permutation of $\{1,\dots,2r\}$ are called M-pfaffian sums and $\psi(M)$ is defined to be the minimum of the M-pfaffian sums.

Lemma 5.1. If $M = (m_{ij})$ is a homogeneous symmetric $2r\times 2r$ matrix with entries in $\mathbb{Z}$, $\psi(M) = \frac{1}{2}Tr(M)$.

Proof. We have $\psi(M) = m_{\sigma(1)\sigma(2)} + m_{\sigma(3)\sigma(4)} +\dots+ m_{\sigma(2r-1)\sigma(2r)}$ for a suitable σ; hence $\psi(M) = m_{\sigma(2)\sigma(1)} +\dots+ m_{\sigma(2r)\sigma(2r-1)}$. It follows that

$$\psi(M) = \tfrac{1}{2}(m_{\sigma(1)\sigma(2)} + m_{\sigma(2)\sigma(1)} +\dots+ m_{\sigma(2r-1)\sigma(2r)} + m_{\sigma(2r)\sigma(2r-1)}) = $$

$= \frac{1}{2}\, Tr(M)$.

If M is an $n\times n$ matrix (n odd) with entries in $\mathbb{Z}$, we denote by $\psi_i(M)$ the integers $\psi(M_i)$ where M_i is the matrix which is obtained by deleting the i-th row and column of M.

Lemma 5.2. With the above notations we have:

a) $v_i \geqslant \psi_i(V) \geqslant \psi_i(H)$ for every $i = 1,\dots,n$.

If moreover $p_i+v_i = p_j+v_j$ for every i and j, i.e. if H is symmetric, then

b) $\psi_i(H) = \frac{1}{2} \sum_{j\neq i}(p_j-v_j)$.

Proof. a) follows directly from the definitions and b) is a consequence of Lemma 5.1.

Corollary 5.3. The following conditions are equivalent:

1) $\bar{A}$ is skew-symmetric and $Pf_i(\bar{A}) = Pf_i^*$ for every $i = 1,\dots,n$.

2) $p_i+ v_i = Tr(H)$ for every $i = 1,\dots,n$.

Proof. It is clear that $\bar{A}$ is skew-symmetric if and only if H is symmetric, if and only if $p_i+ v_i = p_j+ v_j$ for every i and j and it is easy to show that if $\bar{A}$ is skew-symmetric, $Pf_i(\bar{A}) = \overline{Pf_i} \in I^s/I^{s+1}$ where $s = \frac{1}{2} \sum_{j\neq i}(p_j-v_j)$.

Now if 1) holds, $v_i = \frac{1}{2} \sum_{j\neq i}(p_j-v_j)$, hence $p_i+ v_i = \Sigma(p_j-v_j) = Tr(H)$.

Conversely $p_i + v_i = \mathrm{Tr}(H)$ implies that $\bar{A}$ is skew-symmetric and also

$v_i = \frac{1}{2} \sum_{j \neq i} (p_j - v_j)$, hence $\mathrm{Pf}_i(\bar{A}) = \mathrm{Pf}_i^*$ for every $i=1,\dots,n$.

<u>Theorem</u> 5.4. <u>Assume that</u> $v_i + p_i = \mathrm{Tr}(H)$ <u>for every</u> $i=1,\dots,n$.

<u>Then</u> $\mathrm{gr}(\mathrm{Pf}(\bar{A})) = 3$ <u>if and only if</u> $G(\mathbb{L})$ <u>is exact</u>.

<u>Proof</u>. Since $v_i + p_i = \mathrm{Tr}(H)$ for every i, we get $v = \mathrm{Tr}(H)$, hence $v_i = v - p_i$; moreover $\bar{A}$ is skew-symmetric and $\mathrm{Pf}_i(\bar{A}) = \mathrm{Pf}_i^*$ by Corollary 5.3. Accordingly $G(h)(1) = (\mathrm{Pf}_1(\bar{A}),\dots,\mathrm{Pf}_n(\bar{A}))$ and $G(f)(1,0,\dots,0) = \mathrm{Pf}_1(\bar{A}),\dots$ $\dots, G(f)(0,\dots,0,1) = \mathrm{Pf}_n(\bar{A})$. We get a complex

$$G'(\mathbb{L}):\quad 0 \to G_I(R)(-v) \xrightarrow{G(h)} \bigoplus_{i=1}^{n} G_I(R)(-p_i) \xrightarrow{G(g)} \bigoplus_{i=1}^{n} G_I(R)(-v_i) \xrightarrow{G(f)}$$

$$\xrightarrow{G(f)} G_I(R) \longrightarrow G_I(R)/\mathrm{Pf}(\bar{A}) \to 0$$

which is the Buchsbaum-Eisenbud complex associated to the skew-symmetric matrix $\bar{A}$ (see [4]). If $\mathrm{gr}(\bar{A}) = 3$, $G'(\mathbb{L})$ is exact hence f is strict by Proposition 1.4 applied to the complex $R^n \xrightarrow{g} R^n \xrightarrow{f} R$; hence $G'(\mathbb{L})$ coincides with $G(\mathbb{L})$ by Corollary 1.8. Conversely if $G(\mathbb{L})$ is exact, f is a strict morphism hence $G(\mathbb{L})$ coincides with $G'(\mathbb{L})$, which implies that $\mathrm{gr}(\mathrm{Pf}(\bar{A}) = 3$.

<u>Corollary</u> 5.5. <u>Assume that</u> $v_i + p_i = \mathrm{Tr}(H)$ <u>for every</u> $i=1,\dots,n$, $G_I(R)$ <u>is Cohen-Macaulay and</u> $\mathrm{gr}(\mathrm{Pf}(A)) = 3$. <u>Then the following conditions are equivalent</u>:

1) $\mathrm{gr}(\mathrm{Pf}(\bar{A}) = 3$.

2) $G(\mathbb{L})$ <u>is exact</u>.

3) $\{\mathrm{Pf}_1,\dots,\mathrm{Pf}_n\}$ <u>is an</u> I-<u>standard base of</u> $\mathrm{Pf}(A)$.

<u>Proof</u>. We may use the same arguments as in Corollary 4.4.

The results of this section can be used, for example, to study the tangent cone at the origin of those monomial Gorenstein curves in A^4 with parametric equations $X = t^{n_1}$, $Y = t^{n_2}$, $Z = t^{n_3}$, $W = t^{n_4}$. Unfortunately the skew-symmetric matrix, whose pfaffians generate the ideal of these curves, not always satisfies the condition 2 of Corollary 5.3 (for example take $n_1 = 28$, $n_2 = 20$, $n_3 = 58$, $n_4 = 13$). However, when this condition is fulfilled, we can easily recognize whether the tangent cone is still Gorenstein. An analogous investigation, in the codimension 2 case was completely developed in [12], so we restrict now ourselves to show how the results of this paper can be used, by producing an explicit calculation.

Let $R = k[\![X,Y,Z,W]\!]$ and P the prime ideal, which is the kernel of the morphism $f: R \longrightarrow k[t]$ defined by $f(X) = t^{22}$, $f(Y) = t^{49}$, $f(Z) = t^{31}$, $f(W) = t^{16}$. Using the results of [3] one can show that $P = (Y^2 - X^3W^2, Z^3 - X^2Y, YW^3 - X^3Z, X^5 - Z^2W^3, YZ - W^5)$ so that P is the ideal generated by the pfaffians of the skew-symmetric matrix

$$A = \begin{pmatrix} 0 & W^3 & X^2 & Z & 0 \\ -W^3 & 0 & Y & 0 & X^3 \\ -X^2 & -Y & 0 & -W^2 & Z^2 \\ -Z & 0 & W^2 & 0 & Y \\ 0 & -X^3 & -Z^2 & -Y & 0 \end{pmatrix}$$

and R/P is Gorenstein. We have $v_1 = 2$, $v_2 = 3$, $v_3 = 4$, $v_4 = 5$, $v_5 = 2$ and $p_1 = 6$, $p_2 = 5$, $p_3 = 4$, $p_4 = 3$, $p_5 = 6$ hence $p_i + v_i = 8$ for $i = 1,\dots,5$. Now it is clear that $\{Y^2, Z^3 - X^2Y, X^3Z, X^5, YZ\}$ is an (X,Y,Z)-standard base in the ring $k[X,Y,Z]$ (one can apply Corollary 1.10, after remarking that these elements "coincide" with their initial forms); hence we can use

Theorem 2.8 with $J = (W)$, $I = (X,Y,Z,W)$, $L = P$ and we get that

$\{Y^2 - X^3W^2,\ Z^3 - X^2Y,\ YW^3 - X^3Z,\ X^5 - Z^2W^3,\ YZ - W^5\}$ is an I- standard base of P.

Then Corollary 5.5 proves that $G_{I/P}(R/P)$ is Gorenstein.

References

[1] ATIYAH, M.F. - MACDONALD, I.G. *Introduction to Commutative Algebra. Addison - Wesley London (1969).*

[2] AUSLANDER,M. - BUCHSBAUM,D. Codimension and multiplicity. Ann. of Math. 68 (1958)

[3] BRESINSKY,H. Symmetric semigroups of integers generated by 4 elements. Manuscripta Math. 17 (1975).

[4] BUCHSBAUM,D. - EISENBUD,D. Algebra structure for resolutions and some structure theorems for ideals of codimension 3. Amer. J. Math. 99 (1977).

[5] CHIANTINI,L. On form ideals and Artin-Rees condition. Manuscripta Math. 36 (1981).

[6] HERZOG,J. When is a regular sequence super regular? Nagoya Math. J. 83 (1981).

[7] HIRONAKA,H. Resolutions of singularities of an algebraic variety over a field of characteristic 0. Ann. of Math. 79 (1964).

[8] HOCHSTER,M.- RATLIFF,L.J.Jr. Five theorems on Macaulay rings. Pacific J. Math. 44 (1973).

[9] NASTASESCU,C.- VAN OYSTAEYEN,C. *Graded and filtered rings and modules Springer 758 (1979).*

[10] NORTHCOTT, D.G. *Lessons on rings modules and multiplicities. Cambridge University Press (1968).*

[11] ROBBIANO, L. On normal flatness and some related topics. In this volume.

[12] ROBBIANO, L.- VALLA, G. On the equations defining tangent cones. Math. Proc. Cambridge Philos. Soc. 88 (1980).

[13] ROSSI, M.E. Altezza e dimensione nell'anello graduato associato ad un ideale. Rend. Sem. Mat. Univ. Politec. Torino 36 (1978).

[14] VALABREGA, P.- VALLA, G. Form rings and regular sequences. Nagoya Math. J. 72 (1978).

ON THE STRUCTURE OF CERTAIN EXACT COUSIN COMPLEXES

R.Y. SHARP

Department of Pure Mathematics,

University of Sheffield,

Sheffield, United Kingdom

1. Introduction

Throughout this paper, A denotes a commutative Noetherian ring (with non-zero identity). We shall be concerned with the Cousin complex $C(\mathcal{F},M)$ for an A-module M with respect to a filtration $\mathcal{F}$ of Spec(A) which admits M: this concept was introduced in [4] and is more general than the concept of Cousin complex discussed in [1]. A filtration $\mathcal{F}$ of Spec(A) is a descending sequence $(F_i)_{i\geq 0}$ of subsets of Spec(A), so that

$$\mathrm{Spec}(A) \supseteq F_0 \supseteq F_1 \supseteq \ldots \supseteq F_i \supseteq \ldots,$$

with the property that $F_i \setminus F_{i+1}$ is low [1; (2.1)] with respect to F_i (that is each member of $F_i \setminus F_{i+1}$ is a minimal member of F_i with respect to inclusion) for each $i \geq 0$. We then write $\partial F_i = F_i \setminus F_{i+1}$. We say that the filtration $\mathcal{F}$ admits the A-module M if $\mathrm{Supp}(M) \subseteq F_0$.

In those circumstances, we may construct [4; 1.3] a complex

$$0 \longrightarrow M \xrightarrow{d^{-1}} M^0 \xrightarrow{d^0} M^1 \longrightarrow \ldots \longrightarrow M^i \xrightarrow{d^i} \ldots,$$

denoted by $C(\mathscr{F},M)$ and called the Cousin complex for M with respect to $\mathscr{F}$, such that $M^0 = \bigoplus_{\underline{p}\in\partial F_0} M_{\underline{p}}$;

$$M^i = \bigoplus_{\underline{p}\in\partial F_i} (\operatorname{coker} d^{i-2})_{\underline{p}}$$

for each $i > 0$; the component, for $m \in M$ and $\underline{p} \in \partial F_0$, of $d^{-1}(m)$ in $M_{\underline{p}}$ is $m/1$; and, for $i > 0$, $x \in M^{i-1}$ and $\underline{q} \in \partial F_i$, the component of $d^{i-1}(x)$ in $(\operatorname{coker} d^{i-2})_{\underline{q}}$ is $\pi(x)/1$, where $\pi: M^{i-1} \longrightarrow \operatorname{coker} d^{i-2}$ is the canonical epimorphism.

The "Cousin complex for M" of (2.6) of [1] is just the complex denoted in this paper by $C(\mathscr{H}(M),M)$, where $\mathscr{H}(M) = (K_i)_{i\geq 0}$ is the M-height filtration of Spec(A), defined by

$$K_i = \{\underline{p} \in \operatorname{Supp}(M): \operatorname{ht}_M \underline{p} \geq i\}$$

(for each $i \geq 0$). (See (1.7) of [1] for the definition of the M-height of a prime ideal in the support of M.)

In this paper we shall be particularly concerned with exact Cousin complexes: these are of interest because certain Cohen-Macaulay properties can be characterized in terms of them. For example, a non-zero finitely generated A-module M is Cohen-Macaulay if and only if $C(\mathscr{H}(M),M)$ is exact [2; (2.4)]; in particular, A is a Cohen-Macaulay ring if and only if $C(\mathscr{H}(A),A)$ is exact. Also, if, when A is local, we define the dimension filtration $\mathscr{D}(A) = (D_i)_{i\geq 0}$ of Spec(A) by

$$D_i = \{\underline{p} \in \operatorname{Spec}(A): \dim A/\underline{p} \leq \dim A - i\}$$

(for each $i \geq 0$), then an A-module M for which $M \neq \underline{m}M$ (where $\underline{m}$ is the maximal ideal of A) is a balanced big Cohen-Macaulay A-module [4;§0] if and only if $C(\mathscr{D}(A),M)$ is exact [4; 3.7].

The purpose of this paper is to throw some light on the structure of the terms in certain exact Cousin complexes. We describe a new notion of repeated division in Cousin complexes which is particularly satisfactory when these complexes are exact: if M is such that $C(\mathcal{F},M)$ is (defined and) exact, then, whenever $x \in M$ and $a_1,\ldots,a_i$ is an M-sequence (with $i \geq 1$), the process of repeated division produces an element $x \div a_1,\ldots,a_i$ of the (i-1)-st term of $C(\mathcal{F},M)$ which has a useful relationship to the natural image of x in $M/(a_1,\ldots,a_i)M$; moreover, in certain situations, including those described in the preceding paragraph, every element of the (i-1)-st term of $C(\mathcal{F},M)$ can be expressed as $x \div a_1,\ldots,a_i$ (for a suitable choice of $x \in M$ and (poor) M-sequence $a_1,\ldots,a_i$).

The results described in this paper led to the discovery of the modules of generalized fractions described in [5].

2. Repeated division in Cousin complexes

We begin with some basic lemmas which are concerned with the situation used in [1;§2] to describe the fundamental ideas behind the Cousin complex.

(2.1) LEMMA. *Let* U,U' *be subsets of* Spec(A) *such that* $U' \subseteq U$ *and* $U \setminus U'$ *is low with respect to* U. *Let* M *be an* A-*module such that* $\mathrm{Supp}(M) \subseteq U$, *and set*

$$H = \bigoplus_{\underline{p}\in U\setminus U'} M_{\underline{p}}.$$

Let $\xi: M \to H$ *be the mapping defined in* [1;(2.3)].

Let $x \in M$ *and* $a \in A$ *satisfy the following condition: for all* $\underline{q} \in U \setminus U'$,

$$\mathrm{ann}(x) + Aa \not\subseteq \underline{q}. \qquad (*)$$

(i) *If* $\underline{p} \in U \setminus U'$ *and* $\mathrm{ann}(x) \not\subseteq \underline{p}$, *then clearly* $x/s = 0$ *in* $M_{\underline{p}}$ *for all* $s \in A \setminus \underline{p}$.

(ii) *If* $\underline{p} \in U \setminus U'$ *and* $\mathrm{ann}(x) \subseteq \underline{p}$, *then it follows from* (*) *that* $a \in A \setminus \underline{p}$, *so that we may consider the element* x/a *of* $M_{\underline{p}}$.

(iii) *There is an element* $\alpha = (\alpha_{\underline{p}})_{\underline{p}\in U\setminus U'}$ *of* H *for which*

$$\alpha_{\underline{p}} = \begin{cases} 0 & \text{for } \underline{p} \in U \setminus U' \text{ with } \mathrm{ann}(x) \not\subseteq \underline{p}, \\ x/a & \text{for } \underline{p} \in U \setminus U' \text{ with } \mathrm{ann}(x) \subseteq \underline{p}. \end{cases}$$

We denote the element α *by* $x \div a$.

(iv) $a(x \div a) = (x \div 1) = \xi(x)$.

(v) *Let* $\pi : H \to \operatorname{coker} \xi$ *denote the natural epimorphism. Then*

$$\operatorname{ann}(x) + Aa \subseteq \operatorname{ann}(\pi(x \div a)).$$

Proof. Parts (i) and (ii) are clear, and (iii) is immediate once one recalls from [1; (2.2) and (2.3)] that ann(x) can only be contained in a finite number of prime ideals in $U \setminus U'$.

(iv) Let $a(x \div a) = \beta = (\beta_{\underline{p}})_{\underline{p} \in U \setminus U'}$. Clearly, for $\underline{p} \in U \setminus U'$, we have $\beta_{\underline{p}} = 0$ when $\operatorname{ann}(x) \not\subseteq \underline{p}$ and $\beta_{\underline{p}} = x/1$ when $\operatorname{ann}(x) \subseteq \underline{p}$; however, if $\operatorname{ann}(x) \not\subseteq \underline{p}$, then $x/1 = 0$ in $M_{\underline{p}}$. The result follows.

(v) This is now immediate from (iv), since ann(x) clearly annihilates $x \div a$.

The most interesting applications of (2.1) are to situations in which a is a non-zerodivisor on M. Let us show now that (2.1) is then applicable.

(2.2) LEMMA. *Let the situation be as in* (2.1).

(i) *Let* $b \in A$ *be a non-zerodivisor on* M. *Then for all* $x \in M$ *and for all* $\underline{q} \in U \setminus U'$, *we have* $\operatorname{ann}(x) + Ab \not\subseteq \underline{q}$, *and so we may form the element* $x \div b$ *of* H.

(ii) *Let* $a \in A$ *be such that, for all* $x \in M$ *and for all* $\underline{q} \in U \setminus U'$, *it is the case that* $\operatorname{ann}(x) + Aa \not\subseteq \underline{q}$. *Then*

$$\operatorname{ann}(x) + Aa \subseteq (aM:x) \subseteq \operatorname{ann}(\pi(x \div a)).$$

If, in addition, ξ *is a monomorphism, then* a *must be a non-zerodivisor on* M, *and*

$$(aM:x) = \operatorname{ann}(\pi(x \div a)).$$

Proof. (i) Suppose that, for some $x \in M$ and some $\underline{q} \in U \setminus U'$, we have $\operatorname{ann}(x) \subseteq \underline{q}$. Since $\operatorname{Supp}(M) \subseteq U$ and $U \setminus U'$ is low with respect to U, it follows that $\underline{q} \in \operatorname{Ass}(Ax)$; therefore $b \notin \underline{q}$ since b is a non-zerodivisor on M.

(ii) The first inclusion relation is clear, so let $s \in (aM:x)$. Thus $sx = am$ for some $m \in M$. Let $\underline{p}$ be a prime ideal in $U \setminus U'$. If $\operatorname{ann}(x) \subseteq \underline{p}$, so that $a \notin \underline{p}$, then $s(x/a) = m/1$ in $M_{\underline{p}}$. However, if $\operatorname{ann}(x) \not\subseteq \underline{p}$, then it follows that $\mathrm{ann}(m) \not\subseteq \underline{p}$ and $s.0 = 0 = m/1$ in $M_{\underline{p}}$. Thus, in H,

$$s(x \div a) = \xi(m),$$

and so s annihilates $\pi(x \div a)$.

Now suppose that, in addition, ξ is a monomorphism. Let $x \in M$ be such that $ax = 0$, so that $a\xi(x) = 0$ in H. Thus $a(x/1) = 0$ in $M_{\underline{q}}$ for all $\underline{q} \in U \setminus U'$. Let $\underline{p} \in U \setminus U'$: we claim that $x/1 = 0$ in $M_{\underline{p}}$. This is clear if $a \notin \underline{p}$; on the other hand, if $a \in \underline{p}$, then the hypothesis ensures that $\mathrm{ann}(x) \not\subseteq \underline{p}$, and hence $x/1 = 0$ in $M_{\underline{p}}$, as claimed. Thus $\xi(x) = 0$, so that $x = 0$ since ξ is monomorphic. Hence a is a non-zerodivisor on M.

Now let $c \in \mathrm{ann}(\pi(x \div a))$; thus there exists $y \in M$ such that $c(x \div a) = \xi(y) = (y \div 1)$. Hence $ac(x \div a) = a(y \div 1)$, so that, by (2.1)(iv), $\xi(cx) = \xi(ay)$. But ξ is a monomorphism, and so $cx = ay$. Hence $c \in (aM:x)$, as required.

We now apply these ideas to Cousin complexes.

(2.3) DEFINITION. Suppose the filtration $\mathcal{F} = (F_i)_{i\geq 0}$ of Spec(A) admits the A-module M, and the Cousin complex $C(\mathcal{F},M)$ is

$$M^{-2} \xrightarrow{d^{-2}} M^{-1} \xrightarrow{d^{-1}} M^{0} \xrightarrow{d^{0}} M^{1} \longrightarrow \dots \longrightarrow M^{i} \xrightarrow{d^{i}} \dots,$$

where $M^{-2} = 0$ and $M^{-1} = M$. For each $i \geq -1$, let

$$\pi^{i}: M^{i} \longrightarrow \mathrm{coker}\, d^{i-1}$$

denote the canonical epimorphism.

Let $x \in M$ and let $a_1, a_2, \dots, a_t \in A$. We say that x _is divisible by the sequence_ $a_1, a_2, \dots, a_t$, or that $a_1, a_2, \dots, a_t$ _divides_ x, _with respect to_ $\mathcal{F}$, if the following conditions are satisfied:

(i) $\mathrm{ann}(x) + Aa_1 \not\subseteq \underline{p}$ for each $\underline{p} \in \partial F_0$, so that the element $x \div a_1$ of M^0 may be constructed;

(ii) $\mathrm{ann}(\pi^0(x \div a_1)) + Aa_2 \not\subseteq \underline{p}$ for each $\underline{p} \in \partial F_1$, so that, by (2.1), the element $\pi^0(x \div a_1) \div a_2$ of M^1 may be formed; this element is denoted by $x \div a_1, a_2$;

(iii) $\mathrm{ann}(\pi^1(x \div a_1, a_2)) + Aa_3 \not\subseteq \underline{p}$ for each $\underline{p} \in \partial F_2$, so that, by (2.1) again, the element $\pi^1(x \div a_1, a_2) \div a_3$ of M^2 may be formed; this element is denoted by $x \div a_1, a_2, a_3$;

.

.

.

(t) $\operatorname{ann}(\pi^{t-2}(x \div a_1, a_2, \ldots, a_{t-1})) + Aa_t \nsubseteq \underline{p}$ for each $\underline{p} \in \partial F_{t-1}$, so that the element $\pi^{t-2}(x \div a_1, a_2, \ldots, a_{t-1}) \div a_t$ of M^{t-1} may be formed; this element is denoted by $x \div a_1, a_2, a_3, \ldots, a_t$ and is referred to as the result of dividing x by the sequence $a_1, a_2, \ldots, a_t$.

(2.4) REMARKS. Let the situation be as in (2.3).

(i) If $x \in M$ is divisible, with respect to $\mathcal{F}$, by the sequence $a_1, \ldots, a_t$ of elements of A, then it follows from (2.1)(v) and a simple inductive argument that

$$\operatorname{ann}(x) + \sum_{i=1}^{r} Aa_i \subseteq \operatorname{ann}(\pi^{r-1}(x \div a_1, \ldots, a_r)) \text{ for each } r = 1, \ldots, t.$$

(ii) Let $y \in M$ and $b_1, \ldots, b_t \in A$ be such that, for all $r = 1, \ldots, t$,

$$\operatorname{ann}(y) + \sum_{i=1}^{r} Ab_i \nsubseteq \underline{p} \text{ for each } \underline{p} \in \partial F_{r-1}.$$

A simple inductive argument shows that y is divisible, with respect to $\mathcal{F}$, by the sequence $b_1, \ldots, b_t$.

3. A reduction technique

A small hint about the helpful role which repeated division can play in the study of regular sequences for A-modules with exact Cousin complexes is provided by (2.2). In order to exploit this to the full, we need the reduction technique of [4;§2] to use in inductive arguments. For the reader's convenience we state, without proof, the relevant results from [4;§2], and we also include in this section some other preparatory results which will be needed in Section 4.

The following notation will be in force throughout Section 3. We assume that $\mathscr{F} = (F_i)_{i\geq 0}$ is a filtration of Spec(A) which admits an A-module M, and we shall denote the Cousin complex $C(\mathscr{F},M)$ by

$$M^{-2} \xrightarrow{d^{-2}} M^{-1} \xrightarrow{d^{-1}} M^{0} \longrightarrow \dots \longrightarrow M^{i} \xrightarrow{d^{i}} M^{i+1} \longrightarrow \dots,$$

where $M^{-2} = 0$ and $M^{-1} = M$. We shall denote by $\mathscr{F}[1]$ the filtration $(G_i)_{i\geq 0}$ of Spec(A) defined by $G_i = F_{i+1}$ for all $i \geq 0$. We shall use a to denote a fixed element of A, and * will denote the functor $(0:_{(-)}a)$ (from the category of all A-modules and A-homomorphisms to itself).

Let $d^{0'}: \operatorname{coker} d^{-1} \to M^1$ be the homomorphism induced by d^0. We shall use $C(\mathscr{F},M)^1$ to denote the complex

$$0 \longrightarrow \operatorname{coker} d^{-1} \xrightarrow{d^{0'}} M^1 \xrightarrow{d^1} M^2 \longrightarrow \dots \longrightarrow M^i \longrightarrow \dots,$$

and $C(\mathscr{F},M)^{1*}$ will denote the complex

$$0 \longrightarrow (\operatorname{coker} d^{-1})^* \xrightarrow{d^{0'*}} M^{1*} \xrightarrow{d^{1*}} M^{2*} \longrightarrow \dots \longrightarrow M^{i*} \longrightarrow \dots$$

induced from $C(\mathscr{F},M)^1$ by application of the functor *. Our aim is to compare, when $C(\mathscr{F},M)$ is exact and a is a non-zerodivisor on M, the complexes $C(\mathscr{F},M)^{1*}$ and $C(\mathscr{F}[1],M/aM)$.

(3.1) LEMMA. *Suppose that* $C(\mathscr{F},M)$ *is exact at* M *and that* a $(\in A)$ *is a non-zerodivisor on* M, *so that, by* (2.2)(i), a *divides each element of* M *with respect to* $\mathscr{F}$.

(i) *There is an isomorphism* $\psi^{-1}: M/aM \to (\operatorname{coker} d^{-1})^*$ *which is such that, if* $\tau: M \to M/aM$ *and* $\pi^0: M^0 \to \operatorname{coker} d^{-1}$ *denote the natural epimorphisms, then* $\psi^{-1}(\tau(x)) = \pi^0(x \div a)$ *for all* $x \in M$.

(ii) *The filtration* $\mathscr{F}[1]$ *admits* M/aM.

(iii) *Multiplication by* a *provides an automorphism of* M^0.

The next lemma is a straightforward adaptation of [2;(4.5)], and so its proof is left to the reader.

(3.2) LEMMA. *Suppose that* $C(\mathcal{F},M)$ *is exact and that* a *is a non-zerodivisor on* M. *Then the natural induced homomorphisms*

$$\nu^{0}:\operatorname{coker}(d^{0'*}) \to (\operatorname{coker} d^{0})^{*} \quad (= (\operatorname{coker} d^{0'})^{*})$$

and, for $n > 0$,

$$\nu^{n}:\operatorname{coker}(d^{n*}) \to (\operatorname{coker} d^{n})^{*}$$

are all isomorphisms.

It follows from (3.1)(ii) that, when $C(\mathcal{F},M)$ is exact and a is a non-zerodivisor on M, we may form the Cousin complex $C(\mathcal{F}[1],M/aM)$.

(3.3) THEOREM. *Suppose that* $C(\mathcal{F},M)$ *is exact and that* a *is a non-zerodivisor on* M. *Let* $C(\mathcal{F}[1],M/aM)$ *be*

$$N^{-2} \xrightarrow{e^{-2}} N^{-1} \xrightarrow{e^{-1}} N^{0} \longrightarrow \cdots \longrightarrow N^{i} \xrightarrow{e^{i}} N^{i+1} \longrightarrow \cdots,$$

where $N^{-2} = 0$ *and* $N^{-1} = M/aM$, *and let* $C(\mathcal{F},M)^{1*}$ *be*

$$L^{-2} \xrightarrow{f^{-2}} L^{-1} \xrightarrow{f^{-1}} L^{0} \longrightarrow \cdots \longrightarrow L^{i} \xrightarrow{f^{i}} L^{i+1} \longrightarrow \cdots,$$

where $L^{-2} = 0$, $L^{-1} = (\operatorname{coker} d^{-1})^{*}$, $L^{i} = M^{i+1*}$ *for each* $i \geq 0$, $f^{-1} = d^{0'*}$, *and* $f^{i} = d^{i+1*}$ *for each* $i \geq 0$.

There is an isomorphism of complexes

$$\Psi = (\psi^{i})_{i \geq -2}: C(\mathcal{F}[1],M/aM) \longrightarrow C(\mathcal{F},M)^{1*}$$

which has the following properties.

(i) *The map* ψ^{-1} *is the isomorphism of Lemma* (3.1)(i).

(ii) *Suppose that* $i \geq 0$ *and that*

$$\rho^{i-1}:N^{i-1} \longrightarrow \operatorname{coker} e^{i-2}, \quad \pi^{i}:M^{i} \longrightarrow \operatorname{coker} d^{i-1}$$

denote the natural epimorphisms. If $\beta^{i} = (\rho^{i-1}(x_{\underline{p}})/s_{\underline{p}})_{\underline{p} \in \partial F_{i+1}} \in N^{i}$ *(so that* $x_{\underline{p}} \in N^{i-1}$ *and* $s_{\underline{p}} \in A \setminus \underline{p}$ *for all* $\underline{p} \in \partial F_{i+1}$ *and all but finitely many of the* $\rho^{i-1}(x_{\underline{p}})/s_{\underline{p}}$ *are zero), then*

$$\psi^i(\beta^i) = \begin{cases} (\pi^i(\psi^{i-1}(x_{\underline{p}}))/s_{\underline{p}})_{\underline{p}\in\partial F_{i+1}} & \underline{\text{for}}\ i > 0, \\ (\psi^{-1}(x_{\underline{p}})/s_{\underline{p}})_{\underline{p}\in\partial F_1} & \underline{\text{for}}\ i = 0. \end{cases}$$

Consequently, $C(\mathscr{F}[1],M/aM)$ is also exact.

4. Repeated division in exact Cousin complexes

We are now in a position to make some interesting developments of the work in Section 2 in the special case when the Cousin complex concerned is exact.

Throughout this section, $\mathscr{F} = (F_i)_{i\geq 0}$ will denote a filtration of Spec(A) which admits an A-module M, and we shall denote $C(\mathscr{F},M)$ by

$$M^{-2} \xrightarrow{d^{-2}} M^{-1} \xrightarrow{d^{-1}} M^0 \longrightarrow \cdots \longrightarrow M^i \xrightarrow{d^i} M^{i+1} \longrightarrow \cdots,$$

where $M^{-2} = 0$ and $M^{-1} = M$.

We say that a sequence $a_1,a_2,\ldots,a_n$ of elements of A is a poor M-sequence if a_i is a non-zerodivisor on $M/(a_1,\ldots,a_{i-1})M$ for each $i = 1,\ldots,n$; an M-sequence is a poor M-sequence $a_1,a_2,\ldots,a_n$ such that $M \neq (a_1,\ldots,a_n)M$.

(4.1) REMARK. Let $a \in A$, and suppose that $C(\mathscr{F},M)$ is exact at M. Then it follows from (2.2) that a divides each element of M with respect to $\mathscr{F}$ if and only if a is a non-zerodivisor on M.

(4.2) PROPOSITION. *Suppose that $C(\mathscr{F},M)$ is exact and that $a \in A$ is a non-zerodivisor on M. Let $\tau:M \to M/aM$ be the natural epimorphism. Let, with the notation of Section 3,*

$$\Psi = (\psi^i)_{i\geq -2}:C(\mathscr{F}[1],M/aM) \longrightarrow C(\mathscr{F},M)^{1*}$$

be the isomorphism of Theorem (3.3). Let $x \in M$ and let $a_2,a_3,\ldots,a_t \in A$. Then x is divisible by $a,a_2,a_3,\ldots,a_t$ with respect to $\mathscr{F}$ if and only if $\tau(x)$ is divisible by $a_2,a_3,\ldots,a_t$ with respect to $\mathscr{F}[1]$. When this is the case,

$$\psi^{i-2}(\tau(x) \div a_2,a_3,\ldots,a_i) = x \div a,a_2,a_3,\ldots,a_i$$

for each $i = 2,\ldots,t$.

Proof. Note that, by (3.1), the filtration $\mathscr{F}[1]$ admits M/aM. Also, by (2.4)(i), if x is divisible with respect to $\mathscr{F}$ by $a, a_2, \ldots, a_i$ for some $i \geq 2$, then a annihilates $x \div a, a_2, \ldots, a_i$, and so $x \div a, a_2, \ldots, a_i \in M^{i-1*}$.

Note also that, by (3.1)(i), if $\pi^0 : M^0 \longrightarrow \operatorname{coker} d^{-1}$ is the natural epimorphism, then $\psi^{-1}(\tau(x)) = \pi^0(x \div a) \in (\operatorname{coker} d^{-1})^*$. The proposition can now be established by means of a straightforward inductive argument based on the explicit description of the isomorphism Ψ given in (3.3).

Proposition (4.2) is useful in inductive arguments. For example, an induction involving (4.2), the basis for the induction being provided by (4.1), will establish the following corollary.

(4.3) COROLLARY. Suppose that $C(\mathscr{F},M)$ is exact and let $a_1, a_2, \ldots, a_t \in A$. Then $a_1, a_2, \ldots, a_t$ divides each element of M with respect to $\mathscr{F}$ if and only if $a_1, a_2, \ldots, a_t$ is a poor M-sequence.

For the next result, we need the following extension of notation employed in Section 3. Let q be a positive integer: then $\mathscr{F}[q]$ denotes the filtration $(H_i)_{i\geq 0}$ of $\operatorname{Spec}(A)$ for which $H_i = F_{i+q}$ for all $i \geq 0$. We shall also use $\pi^i : M^i \longrightarrow \operatorname{coker} d^{i-1}$ to denote the natural epimorphism for each $i \geq -1$.

(4.4) THEOREM. Suppose that $C(\mathscr{F},M)$ is exact, and that $a_1, \ldots, a_q, a_{q+1}, \ldots, a_{q+t}$ is a poor M-sequence, where q and t are such that $q > 0$ and $t \geq 0$. Then the filtration $\mathscr{F}[q]$ admits $M/(a_1, \ldots, a_q)M$, and the Cousin comlex $C(\mathscr{F}[q], M/(a_1, \ldots, a_q)M)$ is exact. Let this complex be

$$Q^{-2} \xrightarrow{g^{-2}} Q^{-1} \xrightarrow{g^{-1}} Q^0 \longrightarrow \cdots \longrightarrow Q^i \xrightarrow{g^i} Q^{i+1} \longrightarrow \cdots,$$

where $Q^{-2} = 0$ and $Q^{-1} = M/(a_1, \ldots, a_q)M$.

Let $d^{q-1'} : \operatorname{coker} d^{q-2} \longrightarrow M^q$ be the homomorphism induced by d^{q-1}. We shall use $C(\mathscr{F},M)^q$ to denote the complex

$$0 \longrightarrow \operatorname{coker} d^{q-2} \xrightarrow{d^{q-1'}} M^q \xrightarrow{d^q} M^{q+1} \longrightarrow \cdots \longrightarrow M^{q+i} \longrightarrow \cdots .$$

Let $\dagger$ denote the functor $(0:_{(-)} \sum_{i=1}^{q} Aa_i)$ (from the category of all A-modules and A-homomorphisms to itself). The complex

$$0 \longrightarrow (\operatorname{coker} d^{q-2})^\dagger \xrightarrow{d^{q-1'\dagger}} M^{q\dagger} \xrightarrow{d^{q\dagger}} M^{q+1\dagger} \longrightarrow \cdots \longrightarrow M^{q+i\dagger} \longrightarrow \cdots,$$

induced from $C(\mathcal{F},M)^q$ by the application of the functor $\dagger$, will be denoted by $C(\mathcal{F},M)^{q\dagger}$; it is to be understood that it is indexed so that $M^{q\dagger}$ is its 0th term.

There is an isomorphism of complexes

$$\Gamma = (\gamma^i)_{i\geq -2} : C(\mathcal{F}[q], M/(a_1,\ldots,a_q)M) \longrightarrow C(\mathcal{F},M)^{q\dagger}$$

which has the following properties.

(i) Let $\tau' : M \longrightarrow M/(a_1,\ldots,a_q)M$ denote the natural epimorphism. Let $x \in M$. By (4.3), x is divisible by $a_1,\ldots,a_q$ with respect to $\mathcal{F}$, and the resulting $\pi^{q-1}(x \div a_1,\ldots,a_q)$ is annihilated by $\sum_{i=1}^{q} Aa_i$, and so belongs to $(\operatorname{coker} d^{q-2})^\dagger$, by (2.4)(i). In fact,

$$\gamma^{-1}(\tau'(x)) = \pi^{q-1}(x \div a_1,\ldots,a_q).$$

(ii) Suppose that $i \geq 0$ and that $\xi^{i-1} : Q^{i-1} \longrightarrow \operatorname{coker} g^{i-2}$ denotes the natural epimorphism. If $\alpha^i = (\xi^{i-1}(x_{\underline{p}})/s_{\underline{p}})_{\underline{p}\in\partial F_{i+q}} \in Q^i$ (so that $x_{\underline{p}} \in Q^{i-1}$ and $s_{\underline{p}} \in A \setminus \underline{p}$ for all $\underline{p} \in \partial F_{i+q}$ and all but finitely many of the $\xi^{i-1}(x_{\underline{p}})/s_{\underline{p}}$ are zero), then

$$\gamma^i(\alpha^i) = \begin{cases} (\pi^{i+q-1}(\gamma^{i-1}(x_{\underline{p}}))/s_{\underline{p}})_{\underline{p}\in\partial F_{i+q}} & \text{for } i > 0, \\ (\gamma^{-1}(x_{\underline{p}})/s_{\underline{p}})_{\underline{p}\in\partial F_q} & \text{for } i = 0. \end{cases}$$

(iii) Let $x \in M$. Then x is divisible by $a_1,\ldots,a_q,a_{q+1},\ldots,a_{q+t}$ with respect to $\mathcal{F}$, and the element $\tau'(x)$ of $M/(a_1,\ldots,a_q)M$ is divisible by $a_{q+1},\ldots,a_{q+t}$ with respect to $\mathcal{F}[q]$. In fact

$$\gamma^{i-1}(\tau'(x) \div a_{q+1},\ldots,a_{q+i}) = x \div a_1,\ldots,a_q,a_{q+1},\ldots,a_{q+i}$$

for each $i = 1,\ldots,t$.

Proof. This theorem is proved by induction on q: in the particular case in which $q = 1$, all the claims have already been established (in (3.1), (3.3), (4.2) and (4.3)). So assume that $q = n > 1$ and the result has been established when $q = n - 1$.

Let $X = M/a_1M$. Then $a_2,\ldots,a_{n+t}$ is a poor X-sequence, and, by (4.3), (3.1) and (3.3), the filtration $\mathcal{F}[1]$ admits X and the resulting Cousin complex $C(\mathcal{F}[1],X)$ is exact. We may thus apply the inductive hypothesis to X, $\mathcal{F}[1]$ and $a_2,\ldots,a_n,a_{n+1},\ldots,a_{n+t}$. The conclusions are that $\mathcal{F}[1][n-1] = \mathcal{F}[n]$ admits $X/(a_2,\ldots,a_n)X$, that $C(\mathcal{F}[n],X/(a_2,\ldots,a_n)X)$ is exact and that, if ** denotes the functor $(0:_{(-)} \sum_{i=2}^{n} Aa_i)$, then there is an isomorphism of complexes

$$\Gamma_1 : C(\mathcal{F}[n],X/(a_2,\ldots,a_n)X) \longrightarrow C(\mathcal{F}[1],X)^{n-1**}$$

having the appropriate analogues of properties (i), (ii), (iii).

Now the natural isomorphism $M/(a_1,\ldots,a_n)M \longrightarrow X/(a_2,\ldots,a_n)X$ shows that $\mathcal{F}[n]$ admits $M/(a_1,\ldots,a_n)M$ and also gives rise to an isomorphism of complexes

$$\Gamma_0 : C(\mathcal{F}[n],M/(a_1,\ldots,a_n)M) \longrightarrow C(\mathcal{F}[n],X/(a_2,\ldots,a_n)X).$$

Also, (3.3) provides us with an isomorphism of complexes

$$\Psi : C(\mathcal{F}[1],X) \longrightarrow C(\mathcal{F},M)^{1*},$$

where * denotes the functor $(0:_{(-)} Aa_1)$. Bearing in mind the natural isomorphism $(\nu^0 : \mathrm{coker}(d^{0'*}) \longrightarrow (\mathrm{coker}\, d^0)^*$ or, when $n > 2$,)

$$\nu^{n-2} : \mathrm{coker}(d^{n-2*}) \longrightarrow (\mathrm{coker}\, d^{n-2})^*$$

whose existence was established in (3.2), we see that Ψ induces an isomorphism of complexes

$$\Gamma_2 : C(\mathcal{F}[1],X)^{n-1**} \longrightarrow C(\mathcal{F},M)^{n\dagger}.$$

We now set

$$\Gamma = \Gamma_2 \circ \Gamma_1 \circ \Gamma_0 : C(\mathcal{F}[n],M/(a_1,\ldots,a_n)M) \longrightarrow C(\mathcal{F},M)^{n\dagger}.$$

The isomorphism of complexes Γ_0 shows that $C(\mathcal{F}[n],M/(a_1,\ldots,a_n)M)$ is exact. It is straightforward to check that the isomorphism of complexes Γ has (the appropriate analogues of) properties (i),(ii),(iii), and so the proof of the theorem can be completed by induction.

(4.5) <u>COROLLARY.</u> <u>Suppose that</u> $C(\mathcal{F},M)$ <u>is exact, that</u> $x \in M$ <u>and that</u> $a_1,\ldots,a_q$ <u>is a poor</u> M-<u>sequence, so that, by</u> (4.3), x <u>is divisible by</u> $a_1,\ldots,a_q$ <u>with respect to</u> $\mathcal{F}$. <u>Let</u> $\tau' : M \longrightarrow M/(a_1,\ldots,a_q)M$ <u>and</u> $\pi^{q-1} : M^{q-1} \longrightarrow \operatorname{coker} d^{q-2}$ <u>denote the natural epimorphisms.</u> <u>Then</u>

$$\operatorname{ann}(\tau'(x)) = \operatorname{ann}(\pi^{q-1}(x \div a_1,\ldots,a_q)).$$

<u>Proof.</u> We use the notation of (4.4). It follows from part (i) of that theorem that the isomorphism of complexes

$$\Gamma = (\gamma^i)_{i\geq -2} : C(\mathcal{F}[q],M/(a_1,\ldots,a_q)M) \longrightarrow C(\mathcal{F},M)^{q\dagger}$$

has the property that $\gamma^{-1}(\tau'(x)) = \pi^{q-1}(x \div a_1,\ldots,a_q)$, from which the result is obvious.

(4.6) <u>COROLLARY.</u> <u>Suppose that</u> $C(\mathcal{F},M)$ <u>is exact and that</u> $a_1,\ldots,a_q$ <u>is a poor</u> M-<u>sequence.</u> <u>Then</u>

$$\operatorname{Ass}(M/(a_1,\ldots,a_q)M) \subseteq \partial F_q.$$

<u>Proof.</u> We use the notation of (4.4) again. Let $\underline{p} \in \operatorname{Ass}(M/(a_1,\ldots,a_q)M)$, so that there exists $x \in M$ for which $\operatorname{ann}(\tau'(x)) = \underline{p}$. By (4.5), $\underline{p}$ is the annihilator of an element of $\operatorname{coker} d^{q-2}$, so that $\underline{p} \in \operatorname{Ass}(\operatorname{coker} d^{q-2})$. Since $C(\mathcal{F},M)$ is exact it follows from [4;1.4(i)] that $\underline{p} \in \partial F_q$.

The rather technical (4.4) has another striking consequence which shows that, for each $i \geq 0$, every element of the i-th term in certain exact Cousin complexes (including all those for finitely generated A-modules) can be expressed as a result of a repeated division. We need one preliminary lemma.

(4.7) LEMMA. *Suppose that* $C(\mathcal{F},M)$ *is exact at* M *and that* $\mathrm{Ass}(M)$ *is finite. Let* $\alpha \in M^{0}$. *Then there exist* $x \in M$ *and* $a \in A$ *such that* a *is a non-zerodivisor on* M, *and so, by* (4.1), a *divides each element of* M *with respect to* $\mathcal{F}$, *and*

$$\alpha = x \div a.$$

Proof. We know from [4;1.4(i)] that $\mathrm{Ass}(M) \subseteq \partial F_0$ and that

$$\mathrm{Supp}(A\pi^{0}(\alpha)) \subseteq F_1.$$

Since $\mathrm{Ass}(M)$ is finite, it follows that there exists $a \in \mathrm{ann}(\pi^{0}(\alpha))$ such that a is a non-zerodivisor on M. Thus there exists $x \in M$ such that

$$a\alpha = d^{-1}(x) = x \div 1 = a(x \div a).$$

Since, by (3.1)(iii), multiplication by a provides an automorphism of M^{0}, we deduce that $\alpha = x \div a$.

(4.8) THEOREM. *Suppose that* $C(\mathcal{F},M)$ *is exact and that, for all* $i \geq 0$ *and every* M-*sequence* $a_1,\ldots,a_i$, *the set* $\mathrm{Ass}(M/(a_1,\ldots,a_i)M)$ *is finite. Suppose that* $q \geq 0$ *and that* $\beta \in M^{q}$. *Then there exist* $y \in M$ *and a poor* M-*sequence* $a_1,\ldots,a_q,a_{q+1}$ *(so that, by* (4.3), $a_1,\ldots,a_q,a_{q+1}$ *divides each element of* M *with respect to* $\mathcal{F}$) *such that*

$$\beta = y \div a_1,\ldots,a_q,a_{q+1}.$$

Moreover, we can even demand that $a_1,\ldots,a_q$ *be any prescribed poor* M-*sequence contained in* $\mathrm{ann}(\beta)$.

Proof. By the obvious analogue of [1;(2.7)(vi)], $\mathrm{Supp}(A\beta) \subseteq F_q$. By (4.6) and the hypothesis, for all $i \geq 0$ and every poor M-sequence $a_1,\ldots,a_i$, the set $\mathrm{Ass}(M/(a_1,\ldots,a_i)M)$ is finite and contained in ∂F_i. It is therefore a simple matter to produce a poor M-sequence of length q composed of elements of $\mathrm{ann}(\beta)$, and, indeed, to extend any poor M-sequence of elements of $\mathrm{ann}(\beta)$ of length less than q to one of length q.

Now let $a_1,\ldots,a_q$ be a poor M-sequence contained in ann(β). If $q = 0$, the result follows from (4.7), and so we assume that $q > 0$. We are now going to use Theorem (4.4), and we employ the notation of that theorem, so that, in particular, † denotes the functor $(0:_{(-)} \sum_{i=1}^{q} Aa_i)$. Let

$$\Gamma = (\gamma^i)_{i\geq -2}: C(\mathcal{F}[q], M/(a_1,\ldots,a_q)M) \longrightarrow C(\mathcal{F},M)^{q\dagger}$$

be the isomorphism of (4.4). Now $\beta \in M^{q\dagger}$, the set $\mathrm{Ass}(M/(a_1,\ldots,a_q)M)$ is finite, and, by (4.4), the complex $C(\mathcal{F}[q], M/(a_1,\ldots,a_q)M)$ is exact. Hence, by (4.7), there exist $a_{q+1} \in A$ which is a non-zerodivisor on $M/(a_1,\ldots,a_q)M$ and $y \in M$ such that

$$(\gamma^0)^{-1}(\beta) = \tau'(y) \div a_{q+1},$$

where $\tau': M \longrightarrow M/(a_1,\ldots,a_q)M$ is the natural epimorphism. Thus $a_1,\ldots,a_q,a_{q+1}$ is a poor M-sequence, and, by (4.4)(iii),

$$\beta = \gamma^0(\tau'(y) \div a_{q+1}) = y \div a_1,\ldots,a_q,a_{q+1}.$$

(4.9) REMARK. The hypothesis "for all $i \geq 0$ and every M-sequence $a_1,\ldots,a_i$, the set $\mathrm{Ass}(M/(a_1,\ldots,a_i)M)$ is finite" of (4.8) is, of course, satisfied when M is a finitely generated A-module, and it is also satisfied when A is local and M is a balanced big Cohen-Macaulay A-module: see [3;(2.4)]. Thus, in view of [2;(2.4)] and [4;3.7], Theorem (4.8) may be applied both to (finitely-generated) Cohen-Macaulay modules and to balanced big Cohen-Macaulay modules over local rings.

References

1. R.Y. Sharp, "The Cousin complex for a module over a commutative Noetherian ring", Math. Z., 112 (1969), 340-356.

2. R.Y. Sharp, "Gorenstein modules", Math. Z., 115 (1970), 117-139.

3. R.Y. Sharp, "Cohen-Macaulay properties for balanced big Cohen-Macaulay modules", Math.Proc.Cambridge Philos.Soc., 90 (1981), 229-238.

4. R.Y. Sharp, "A Cousin complex characterization of balanced big Cohen-Macaulay modules", Quart. J.Math.Oxford (2), to appear.

5. R.Y. Sharp and H. Zakeri, "Modules of generalized fractions", Mathematika, to appear.

ON THE BASSET'S LIMITATIONS FOR THE MAXIMUM NUMBER OF ISOLATED SINGULARITIES ON AN ALGEBRAIC SURFACE

Ezio Stagnaro

Istituto di Matematica Applicata
Facoltà di Ingegneria
Università
ITALY

Introduction

Let F_n be an irreducible (reduced) algebraic surface of degree n in the projective space P^3_K , K algebraically closed field of characteristic zero. Suppose F_n has D_n nodes (= conical double points), and no other singularities.

In order to get a limitation for the number D_n , in the years 1906,1907 A.B. Basset (cfr.[5], [6]) made the following reasoning. Projecting F_n from a generic point P in $P^3_{\mathbb{C}}$, $\mathbb{C}$ complex field, over a plane π and calling δ the branch locus, we have:

1) $N = \deg \delta = n(n-1)$

2) $M = \text{class } \delta = n(n-1)^2 - 2D_n$

3) k = number of (ordinary) cusps on δ $= n(n-1)(n-2)$.

Now we know three Plücker's characters of δ and therefore we are able to deduce the others with the Plücker's formulas:

4) i = number of (ordinary) flexes on δ $= 3(M-N)$

5) b = number of bitangents to δ $= M(M-10) + 8N - 3k$.

Since $i \geq 0$, from 4) we get

$$\text{I)} \qquad D_n \leq \frac{2}{3} n(n-1)(n-2) .$$

We have, for example, $D_3 \leq 4$, $D_4 \leq 16$, $D_5 \leq 40$, $D_6 \leq 80$, $D_7 \leq 140$, $D_8 \leq 224$,...

Next, since $b \geq 0$, from 5) and using I), we have

$$\text{II)} \qquad D_n \leq \frac{1}{2}[n(n-1)^2-5-\sqrt{n(n-1)(3n-14)+25}\ ] , \ n \geq 5 .$$

We have, for example, $D_5 \leq 34$, $D_6 \leq 66$, $D_7 \leq 114$, $D_8 \leq 224$,...

We call respectively I) , II) *the first, second limitation of A.B. Basset* for the maximum number of nodes on F_n .
Note that for $n \geq 5$ the limitation II) is better than I).

We make some remarks:

a) Basset supposes that δ is irreducible and reduced (maybe this was well-known in those times) ; moreover , he supposes δ has only Plückerian singularities (punctual or tangential) i.e. δ has only ordinary cusps , nodes, ordinary flexes and bitangents. So Basset's limitations I) and II) work if one proves that it is possible to project F_n from a point P in $P^3_{\mathbb{C}}$ such that the corresponding branch locus is irreducible (reduced) and has only Plückerian singularities; such a projection is called *generic projection* of F_n on π . Unfortunately nobody has proved that such F_n has a generic projection; also,as far as I know nobody has proved that a fixed non singular surface has a generic projection. So this way to prove Basset's limitations I) and II) gives some difficulties.

b) Basset does not use essentially Plücker's formulas, but sometimes he uses only inequalities among Plückerian characters of δ . Hence, in order to prove I) and II), we may try to prove such inequalities, even if δ has complicated singularities, using the generalized Plücker's formulas. Obviously, first we must prove that δ is irreducible and reduced for some P in $P^3_{\mathbb{C}}$.

Following the second way we shall able to prove I) and II) in full, which is the *main purpose* of the paper. More precisely in this paper we prove the analogues of I) and II) for q-fold points with tangent cone without singular lines; for simplicity we call such points *ordinary q-fold points* . In the last part of the paper, we prove that certain surfaces, considered by G. Castelnuovo for other purposes (cfr. [4]), have the maximum number of ordinary q-fold points, for particular values of the integers q and n.

1. Let F_n be an algebraic surface of degree n in P^3_K , K algebraically closed field of caracteristic zero. Let us suppose F_n has Q_n ordinary q-fold points and no other singularities, $q < n$.

Projecting F_n from a point P in P^3_K on a plane π , we denote by P'_{n-1} the first polar of P with respect to F_n, by $C = F_n \cdot P'_{n-1}$ the complete intersection of F_n with P'_{n-1}, by δ the branch locus on π .

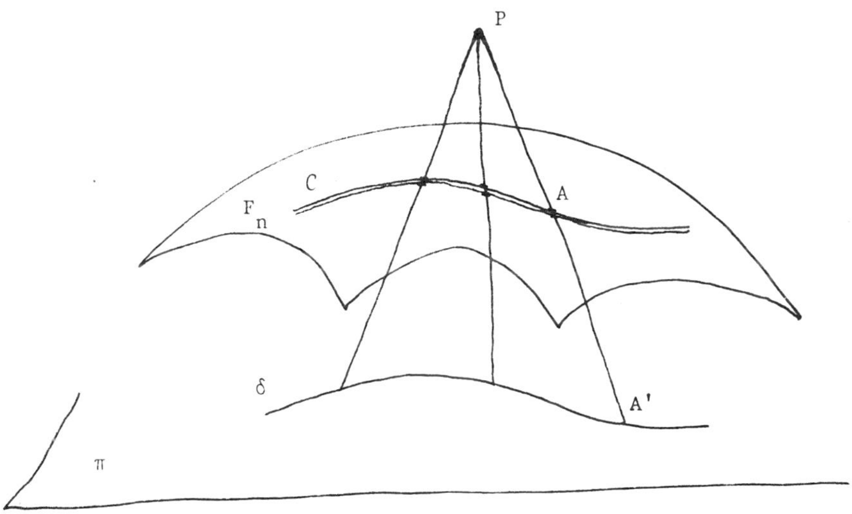

C is called "curve of the coincidences" (curva delle coincidenze); we remark that projecting C from P over π we get the branch locus δ, moreover

$$C = \left\{ \begin{array}{l} A \in F_n \text{ such that there exists a line } \ell \text{ tangent to} \\ F_n \text{ in A and passing through P} \end{array} \right\} .$$

Lemma 1. *If P is a generic point in P^3_K (i.e. P does not belong to an algebraic surface in P^3_K fixed from the context), we have:*

i) C is irreducible (and reduced);

ii) the projection (from P) $C \to \delta$ is birational.

Proof. A very short proof of this lemma can be found in [22], Ch.VI, § 4, Pag. 131, where Zariski says $\mathcal{E}$ "is a simple branch curve. ... The consideration of the surface dual to F shows that the generic tangent cone of F is irreducible." This proves completely the lemma, even if in [22] the surface has a double line and no other singularities, since the proof remains valid also in our case. Anyhow, for reasons of completeness and convenience of the reader, we prefer to give here a more detailed proof. In proving ii) we use a computation of constants, which seems to me the same reasoning (very short) of Zariski to prove δ "is a simple branch curve".

Proof of i). If C is reducible or non reduced for a generic point $P(\lambda_1,\lambda_2,\lambda_3,\lambda_4)$ in P^3_K, then the linear system S cut out on F_n by

$$S : \lambda_1\frac{\partial F_n}{\partial x_1} + \lambda_2\frac{\partial F_n}{\partial x_2} + \lambda_3\frac{\partial F_n}{\partial x_3} + \lambda_4\frac{\partial F_n}{\partial x_4} , \quad \lambda_i \in K ,$$

is reducible.

By Bertini's second theorem, we have only two possibilities:

A) S has a fixed component D;

B) S is composite with a pencil.

In the case A) $\frac{\partial F_n}{\partial x_i}$, for any i, vanishes on D, so D is a singular curve on F_n: contradiction.

In the case B) any two curves of S have no common points outside the base points of the pencil. So F_n has class zero, because any two curves of S are cut out on F_n by two first polars $P'_{n-1}(P_1)$, $P'_{n-1}(P_2)$ of two points $P_1,P_2 \in P^3_K$; this means there is no tangent plane to F_n passing through the generic line P_1P_2. Therefore F_n is a ruled developable (rigata sviluppabile), or a cone (cfr. [1]). Since a ruled developpable, different from a cone, has a singular curve (loc. cit.), we get in any case a contradiction.

Proof of ii). It is enough to prove that δ is reduced. If it is not then $C \to \delta$ is a map $m \leftrightarrow 1$, $m > 1$. In particular from any point in P^3_K pass ∞^1 bitangents if $m = 2$, tritangents if $m = 3$, , ... to F_n (by bitangent, tritangents,... we mean in simple points on F_n). The points in P^3_K are ∞^3 and a bitangent (tritangent ...) is counted ∞^1 times (as many times as the points of a bitangent); it follows that the bitangents to F_n are $\infty^{1+3-1} = \infty^3$. On the other hand we consider a (generic) plane section; the section curve has a finite number of bitangents. We remark that a bitangent to a section plane curve is bitangent to F_n. Next, the planes in P_K^3 are ∞^3, a bitangent is counted ∞^1 times (as many times as the planes passing through the bitangent). So we have now $\infty^{3-1} = \infty^2$, against ∞^3. This proves ii) and the lemma.

Remark. In the proof of lemma 1 we have only used the facts that F_n has no singularities in codimension 1 and that F_n is different from a cone.

2. *Proposition* 1. *If* $P \in P_K^3$ *is generic, the following statements hold:*

i) if B *is an ordinary* q*-fold point on* F_n*, then the curve* C *has in* B *a* $q(q-1)$*-fold point with distinct tangent lines;*

ii) C *has no singularities outside the singular points of* F_n ;

iii) $\deg \delta = \deg C = n(n-1)$;

iv) $\text{class}\ \delta = \text{class}\ F_n = n(n-1)^2 - Q_n q(q-1)^2$.

Proof. i) follows from [3], pag. 143 ; ii) follows applying the first Bertini theorem to the system S considered in the proof of lemma 1; $\deg C = n(n-1)$ it is obvious, and $\deg C = \deg \delta$ follows from lemma 1; finally class $F_n = n(n-1)^2 - Q_n q(q-1)^2$ (cfr. [3], n.4, pag. 143). Moreover class F_n = class δ, namely class F_n = number of tangent planes to F_n passing through a generic line ℓ; intersecting with π, calling $P' = \ell \cap \pi$, we have class δ = number of tangent lines to δ passing through the generic point P' on π: so class F_n = class δ. This proves the proposition .

Remark. If $Q_n = D_n$ we have proved the formulas 1) and 2) given in the introduction.

3. We give first a justification of the formula 3) in the introduction, under the hypothesis that F_n has only nodes.

Let A be a non singular point on C (curve of the coincidences). Since P is generic, it is not difficult to see that the projection A' of A from P is a cusp (ordinary or not) if the tangent line ℓ, in A to C, passes through P. And conversely if A' is a cusp, distinct from the images of singular points on C, then there exists a point A on C, simple on C, such that the projection of A is A' and the line AA' is tangent to C on A and passes through P.

Since the line ℓ = AA' is tangent at A to C, hence tangent at A to the first polar P'_{n-1}, then the second polar P''_{n-2}, of P with respect to F_n, passes through A, and conversely.

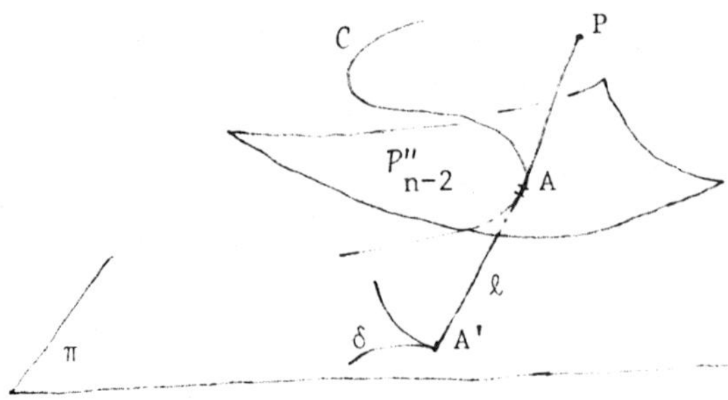

Therefore such points A, distinct from the singular points of F_n, are the points of the complete intersection: $P''_{n-2} \cdot P'_{n-1} \cdot F_n = P''_{n-2} \cdot C$. Now we suppose the points A all distinct, we get:

number of points A = n(n-1)(n-2) ;

hence

number of points A = number of points A' = number cusps δ.

Therefore number cusps δ = n(n-1)(n-2).

Under the hypothesis : points A' all distict, we have proved the formula 3) in the introduction.

Remark. In the case $Q_n = D_n$ the secon polar P''_{n-2} does not pass through the nodes on F_n (if P is generic).

Now we consider the general case

Lemma 2. *Assume* P *is generic. Then:*

(a) Any non ordinary point on δ *(singular point with non distinct tangents) is the projection from* P *of a non singular point on* C .

(b) If A' ∈ δ *is singular and non ordinary, the the points* $A_1, \ldots, A_s \in C$ *which project to* A', *are non singular and correspond to the branch* $B'_1, \ldots, B'_s$ *of* δ *at* A'.

(c) If B' *is the branch at* A' *corresponding to the point* A ∈ C, ν = *ord* B', *then* ν − 1 *is the intersection multiplicity of* C *with* P''_{n-2} *at* A.

Proof. To prove (a) and (b) it is sufficient to choose P outside the tangent cones to F_n in the q-fold points, consider the i),ii) of the proposition 1, n. 2 and use the statements at the beginning of this section.

Now we prove (c).

Without loss of generality we may assume the following statements:

- F_n is an affine closed set in the affine space A^3_K ;

- P is at infinity on the z axis;
- $A = A' = 0 = (0,0,0)$ = origin;
- π the plane $z = 0$.

From these assumptions, F_n, P'_{n-1}, P''_{n-2} are given respectively by

$$F_n \; : \; z^n + \alpha_1 z^{n-1} + \dots + \alpha_{n-1} z + \alpha_n \quad , \qquad \alpha_i \in K[x,y] \; ,$$

$$P'_{n-1} = \frac{\partial F_n}{\partial z} \; : \; nz^{n-1} + (n-1)\alpha_1 z^{n-2} + \dots + 2\alpha_{n-2} z + \alpha_{n-1}$$

$$P''_{n-2} = \frac{\partial^2 F_n}{\partial z^2} \; : \; n(n-1)z^{n-2} + (n-1)(n-2)\alpha_1 z^{n-3} + \dots + 6\alpha_{n-3} z + 2\alpha_{n-2} \; .$$

Assume

$$\alpha_n(0,0) = \alpha_{n-1}(0,0) = \dots = \alpha_{n-\mu-1}(0,0) = 0 \; , \quad \alpha_{n-\mu}(0,0) \neq 0,$$

where the integer μ is characterized by such a property. We remark that $\mu \geq 3$ because $0 = (0,0,0) \in F_n \cap P'_{n-1} \cap P''_{n-2}$.

Now we consider the system

$$6) \quad \begin{cases} F_n : xA + yB + z^{\mu}C & A,B,C,A',\dots, C'' \in K[x,y,z] \\ P'_{n-1} : xA' + yB' + z^{\mu-1}C' & A(0,0,0) \neq 0 \text{ or } B(0,0,0) \neq 0 \\ P''_{n-2} : xA'' + yB'' + z^{\mu-2}C'' & C(0,0,0)\neq 0, C'(0,0,0)\neq 0, C''(0,0,0)\neq 0. \end{cases}$$

The point 0 is non singular on C, and the tangent planes at 0 to F_n and to P'_{n-1} respectively, are distinct. Putting $A(0,0,0) = \bar{A}$, etc. this means

$$\det \begin{pmatrix} \bar{A} & \bar{B} \\ \bar{A}' & \bar{B}' \end{pmatrix} \neq 0 \; .$$

So , from the first two in 6), we can 'resolve' the system with respect to x and y and 'substituting' in P''_{n-2} , we have that the multiplicity of intersection of F_n, P'_{n-1}, P''_{n-2} in 0 is $\mu-2$. (If one prefers the canonical image of z in the discrete valuation ring of C in 0 is an uniformizing parameter,etc.). Hence from the hypothesis in the lemma, we have

$$\mu - 2 = \nu - 1, \quad \mu = \nu + 1 \; .$$

Next we consider a generic plane π' through the z axis

$$\pi' \; : \; \lambda_1 x + \lambda_2 y \quad , \; \lambda_i \text{ parameters} \; ,$$

and we consider the system

$$\begin{cases} F_n : xA + yB + z^{\nu+1}C \\ P'_{n-1}: xA' + yB' + z^{\nu}C' \\ P''_{n-2}: \lambda_1 x + \lambda_2 y \end{cases} \qquad C(0,0,0) \neq 0, \quad C'(0,0,0) \neq 0 \qquad \det\begin{pmatrix} \bar{A} & \bar{B} \\ \bar{A}' & \bar{B}' \end{pmatrix} \neq 0 .$$

As before , we see that,for generic values of λ_1 , λ_2, the multiplicity of intersection of F_n, P'_{n-1}, π', in $0 = (0,0,0)$ is ν .

It follows that the projection of the branch B of C with center 0 is a branch B' of δ of order ν . Namely, if we come back to the projective situation, then the generic plane π' cuts C in $n(n-1) - \nu$ points outside 0, hence the line $\ell' = \pi' \cap \pi$ (ℓ' passes through 0) cuts δ in $n(n-1)$ points , among them ν coincide in 0 along the branch B' of δ ; that is,the multiplicity of intersection of ℓ' with the branch B' is ν. By definition this means that B' has order ν. This proves c) and the lemma.

Corollary. We have the formula

7) $$\sum (\nu - 1) = n(n-1)(n-2) - \sum Q_n q(q-1)(q-2),$$

where the summation on the left is extended to all branches of δ , and ν denotes the order of the branches.

Proof. From c) of lemma 2,it is sufficient to prove that,if P is generic, then the multiplicity of intersection of F_n, P'_{n-1}, P''_{n-2} in an ordinary q-fold point, is $q(q-1)(q-2)$; and this follows from [3], pag. 143, with rather easy modifications.

Remark. The formula 7) is the anlogous of the formula 3) in the introduction.

4. *In this section we prove the limitation 1) of Basset generalized to ordinary q-fld points, $q \geq 2$.*

We consider the generalized Plücker formula(cfr., for instance [2], pag. 424)

8) $$\sum (\nu' - \nu) = 3(\text{class } \delta - \deg \delta),$$

where the summation is extended to any branch of δ of order ν and class ν' (if 0 is the center of a branch D, the class ν' of D is the multiplicity of intersection in 0 of the branch D with its tangent, minus ν)

(The formula 8) is analogous to the formula 4) in the introduction).

We write 8) in the form

9) $$\sum (\nu'-1) = 3(\text{class } \delta - \deg \delta) + \sum (\nu - 1).$$

Substituting $\deg \delta$ and class δ from iii) and iv) of the proposition 1, n. 2 and substituting the 7) of corollary 2, n.3 and considering $\sum (\nu'-1) \geq 0$, as in the introduction, we get for the maximum number of ordinary q-fold points the following limitation

I') $$Q_n \leq \frac{4n(n-1)(n-2)}{q(q-1)(4q-5)} .$$

For instance for $q = 3$, we have

$Q_5 \leq 5$, $Q_6 \leq 11$, $Q_7 \leq 20$, $Q_8 \leq 32$, $Q_9 \leq 48$, $Q_{10} \leq 68, \ldots$.

Remark. In the case $n = 5$ we have that Q_5 gives the maximum number of ordinary triple points for an F_5, namely D. Gallarati in [11] finds an F_5 with five ordinary triple points.

5. *In this section we prove the limitation II) of Basset generalized to q-fold points, $q \geq 2$.*

The proof of the following lemma belongs to S. Greco. Our proof uses the same ideas as the original proof, in part due to Greco (cfr. [16]), but it is simpler.

Lemma 3. Let δ *our branch locus of the projection from the generic point* P . *Then*

10) $$M(M - 10) + 8N - 3\sum(\nu - 1) \geq 0 \quad (N = \deg \delta \ , \ M = \text{class } \delta),$$

where the summation is extended to all branches of δ of order ν .

Proof. Let δ' be the dual curve of δ and $P_1, \ldots, P_h$ singular points on δ' with multiplicity $d_1, \ldots, d_h$ respectively. Let us suppose that the actual points on δ' are $P_1, \ldots, P_\ell$. Applying the first Plücker generalised formula to δ' , we have

$$N = M(M - 1) - \sum_{i=1}^{h} d_i(d_i-1) - \sum(\nu'-1) ,$$

where ν' is the order of the branches of δ', i.e. the class of the branches of δ .

It follows that

$$N \leq M(M - 1) - \sum_{i=1}^{\ell} d_i(d_i-1) - \sum(\nu'-1) .$$

Let B_{i_α} be the branches of δ' in P_i and $\nu'_{i_\alpha} = \text{ord } B_{i_\alpha}$. Then

$d_i = \sum_\alpha \nu'_{i_\alpha}$, hence

$$d_i(d_i-1) = (\sum_\alpha \nu'_{i_\alpha})(\sum_\alpha(\nu'_{i_\alpha} - 1)) \geq \sum_\alpha \nu'_{i_\alpha}(\nu'_{i_\alpha} - 1)$$

Considering the summation with to respect to i , we have $\sum_{i=1}^{\ell} d_i(d_i-1) \geq \sum \nu'(\nu'-1)$, and then

$$N \leq M(M-1) - \sum \nu'(\nu'-1) - \sum(\nu'-1)$$

$$N \leq M(M-1) - \sum(\nu'+1)(\nu'-1)$$

$$N \leq M(M-1) - 3\sum(\nu'-1) .$$

Substituting $\sum(\nu'-1)$,given by 9), we find the inequality 10) and the lemma is proved.

The formula 10) is analogous to formula 5) in the introduction.

Finally sustituting in the inequality 10) of lemma 3 the values of N and M given by iii) and iv) of proposition 1, n.2, and moreover substituting in 10) the value of $\sum(\nu - 1)$ given by 7) in the corollary of n. 3, we get the second limitation of Basset generalized to ordinary q-fold points:

$$\text{II}')\quad Q_n \leq \frac{2n(n-1)^2(q-1)-13q+16-\sqrt{4n(n-1)(3n-11q+8)(q-1)+(13q-16)^2}}{2q(q-1)^3}$$

if $\dfrac{2n(n-1)^2(q-1)-13q+16+\sqrt{\ldots}}{2q(q-1)^3} > \dfrac{4n(n-1)(n-2)}{q(q-1)(4q-5)}$, $M \geq 10$, $n>q\geq 2$.

Obviously II') has meaning when the number under the radical is ≥ 0; for example , when $q = 3$, such a number is positive for $n \geq 8$,and we have:$Q_8 \leq 32$, $Q_9 \leq 46$, $Q_{10} \leq 65$, $Q_{11} \leq 89$, $Q_{12} \leq 118$,... .

We remark that the limitation II') is better than limitation I') for n=9, 10,... .

6. G. Castelnuovo in [4] considered the following surfaces . Let $F(X,Y,Z)$ be a form of degree Q and G_1,G_2,G_3 three quadrics meeting in 8 distinct points.The surface in the projective space P^3_K defined by $F(G_1,G_2,G_3)$ has degree 2q and has 8 q-fold points in

G_1 G_2 G_3. If $F(X,Y,Z)$ is general enough the q-fold points are ordinary.

More generally we consider surfaces of degree s: G_1, G_2, G_3 meeting in s^3 distinct points. Now $F(G_1,G_2,G_3)$ defines in P_K^3 a surface of degree sq with s^3 ordinary q-fold points if $F(X,Y,Z)$ is sufficiently general.

Proposition 2. The number s^3 *is the maximum number of ordinry q-fold points for the surfaces of degree* sq *, if*

$$q \geq \frac{1}{8}\left[3(3s^3-4s^2+3)+\sqrt{9(3s^3-4s^283)^2-16(5s^3-8s+5)}\right] . \tag{11)}$$

Proof. From the LImitation I'), n.4, s^3 is the maximum number of q-fold points for a surface of degree sq,if we have

$$\frac{4sq(sq-1)sq-2)}{q(q-1)(4q-5)} < s^3+1 .$$

That is

$$\frac{9s^3q-12s^2q-5s^3+8s}{4q^2-9q+5} < 1 .$$

Since $4q^2-9q+5 > 0$, for any q, eliminating the denominator we deduce the inequality 11), proving the proposition.

For s=2,the inequality 11) of the proposition gives $q > 7$; so for a surface F_{2q} of degree 2q , 8 is the maximum number of ordinary q-fold points: for instance 8 is the maximum number of ordinary 8-fold points for an F_{16}, of ordinary 9-fold points for an F_{18},... .

For s=3,the inequality 11) gives $q > 32$; so for an F_{3q}, 27 is the maximum number of ordinary q-fold points: for instance 27 is the maximum number of ordinary 33-fold points for an F_{99}.

For s=4,the inequality 11) gives $q > 97$; for instance 64 is the maximum number of ordinary 100-fold points for an F_{400} .

7. The limitations I') and II') for the q-fold points on F_n, can be extended to the case the surface F_n has ordinary q_i-fold points, i=1,2,3,... . If F_n has Q_i ordinry q_i-fold points, we have:

$$\deg \delta = n(n-1);$$

$$\text{class } \delta = n(n-1)^2 - \sum_i Q_i q_i (q_i-1)^2$$

$$\sum(\nu-1) = n(n-1)(n-2) - \sum_i Q_i q_i (q_i-1)(q_i-2) .$$

From these three equalities , we can deduce, as before,limitations similar to I') and II'), but we omit this .

8. We list here examples of surfaces having the highest number respectively of nodes, ordinary triple points,...,ordinary q-fold points. We denote $\mu(n)_q$ the maximum number of ordinary q-fold points for F_n.

$q = 2$ $\mu(3)_2 = 4$
" $\mu(4)_2 = 16$ (Kummer surface)
" $\mu(5)_2 = 31$ (cfr.([8] and [17])+ [20]) (Basset's limit is 34)
" $\mu(6)_2 \geq 64$ (cfr.[15] and [19]) (Basset's limit is 66)
" $\mu(7)_2 \geq 90$ (cfr.[15]) (Basset's limit is 114)
" $\mu(8)_2 \geq 180$ (cfr.[14] and [13]) (Basset's limit is 224)
⋮
" $\mu(n)_2 \geq \ldots$ (cfr.[12] and [14])

$q = 3$ $\mu(5)_3 = 5$ (cfr.[11],see also the remark in the n.4)
" $\mu(6)_3 \geq 8$ (Castelnuovo surfaces,cfr. n. 6; Basset's generalized limit is 11)
" $\mu(7)_3 \geq \ldots$
" $\mu(8)_3 \geq \ldots$
" $\mu(9)_3 \geq 36$ (cfr.[9]) (Basset's generalized limit is 46)
....

In [7] E. Togliatti proves that an F_5, having an ordinary triple point, can have 24 nodes more, and this number is the maximum (among the F_5 with an ordinary triple point).

In [11] D. Gallarati generalizes the method of Togliatti and finds the maximum number of nodes for a surface F_5 having 2,3,4,5 ordinary triple points respectively. The above mentioned in [10] proves that if F_6 is a surface of degree 6 with an ordinary 4-fold point, then the maximum number, for F_6, of nodes is 45.

In [18] N. Chiarli considers an F_6 having an ordinary 4-fold point and she finds the maximum number of ordinary triple points and nodes for F_6.

In these last cases, if we want a limitation when F_n has singular points of different multiplicity, we must use the formulas in n.7.

Finally, we point out that recently in [21] J.W. Bruce has found a limitation which for nodes and for n odd, is better than Basset's limitation.

Acknowledgement : I want to thank S. Greco for his help, his stimulation and his helpful criticism.

References

1. F. Enriques, *Lezioni di Geometria Descrittiva*, N. Zanichelli, Bologna (1925)

2. F. Enriques-O. Chisini, *Teoria geometrica delle equazioni e delle funzioni algebriche*, Vol.II, N. Zanichelli, Bologna (1926).

3. E. Bertini, *Complementi di Geometria Proiettiva*, N. Zanichelli, Bologna (1927).

4. G. Castelnuovo, *Osservazioni intorno alla Geometria sopra una superficie*, Rend. Ist. Lombardo, 24 (1891), 127-137.

5. A.B. Basset, *The maximum number of double points on a surface*, Nature, 73 (1906), pag.246.

6. ———————, *On the singularities of surfaces*, Quart. Journal, 38 (1907), 63-83.

7. E.G. Togliatti, *Alcune osservazioni intorno ad una particolare superficie di ordine 5°*, Volume unico: Stdi in onore di Salvatore Ortu Carboni, Genova (1935), 253-259.

8. ———————, *Una notevole superficie del 5° ordine con solo punti doppi isolati*, Festschrift R. Fueter, Zürich (1940), 127-132.

9. D. Gallarati, *Alcune riflessioni intorno ad una nota del Prof. B. Segre*, Atti Acc. Ligure di Sc. e Lettere, 9 (1952), 1-7.

10. ———————, *Sopra una notevole superficie del 6° ordine*, Boll. Un. Mat. Italiana, 3 (1951), 213-215.

11. ———————, *Sulle superficie del quinto ordine dotate di punti tripli*, Rend. Acc. Naz. dei Lincei, 12 (1950), 70-75.

12. ———————, *Intorno a certe superficie algebriche aventi un elevato numero di punti singolari isolati*, Rend. Acc. Naz. dei Lincei, 11 (1951), 1-7.

13. D. Gallarati, *Una superficie dell'ottavo ordine con 160 nodi*, Atti Acc. Ligure di Sc. e Lettere, 14 (1957), 1-7.

14. H.O. Kreiss, *Uber syzygetische Flachen*, Annali di Mat., 41 (1955), 105-111.

15. E. Stagnaro, *Sul massimo numero di punti doppi isolati di una superficie algebrica di* P^3, Rend. Sem. Mat. di Padova, 59 (1978) 179-198.

16. ———————, *On the maximum number of isolated singularities of algebraic curves and surfaces and on Basset's limitation*, Seminar Notes, Bhaskaracharya Pratishthana Institute, Pune (India), (1979-1980).

17. ——————, *A new construction of a surface of degree 5 with 31 nodes*, Rend Sem. Mat. di Padova (to appear)

18. N. Chiarli, *Sul massimo numero di punti singolari di una superficie del sesto ordine con un punto quadruplo*, Le Matematiche, Catania, 26 (1971), 297-306.

19. F. Catanese-G. Ceresa, *Costructing sextic srfaces with a given number d of nodes*, (preprint).

20. A. Beauville, *Sur le nombre maximum de points double d'une surface dans* P^3 $(\mu(5) = 31)$, Algebraic Geometry, Angers 1979, Sijthoff and Noordhoff (1980), 207-215.

21. J.W. Bruce, *An upper bound for the number of singularities on a hypersurface*, (preprint).

22. O. Zariski, *Algebraic Surfaces (second supplemented edition)*, Springer-Verlag, 61 (1971).

ON THE AFFINENESS OF HENSEL SCHEMES

Rosario Strano

Department of Mathematics
University of Catania
Catania, Italy

Abstract. We prove that, under some finiteness assumptions, if $X \to Y$ is an affine morphism of Hensel schemes with Y affine, then X is affine. As a corollary we prove an analoghe of Serre's criterion for affineness of Hensel schemes.

Introduction. In [3] on the analogy of the theory of formal schemes, we posed the following conjectures about Hensel schemes.

1) If X is an Hensel scheme and I is an ideal of definition, then if $(X, O_X/I)$ is affine, X is affine.

2) If $X \to Y$ is an affine morphism of Hensel schemes and Y is affine, X is affine.

3) (Serre's criterion). If X is a quasi-compact, separated Hensel scheme or an Hensel scheme whose underlying space is noetherian, then if $H^1(X, F) = 0$ for every quasi-coherent

ideal F in O_X , X is affine.

In [3] Prop. 1.22, we proved 2) in the case of an integral morphism.

In this paper we prove that 1) $\Rightarrow$ 2) $\Rightarrow$ 3) and that 2) is true when $\varphi : X \to Y$ is locally of finite type and X is noetherian. As a corollary we get 3) when X is noetherian and of finite type over an affine Hensel scheme Y . In particular 2) and 3) are true for Hensel schemes of finite type over a base field e.g. Hensel schemes which are the henselization of an algebraic variety along a closed subvariety.

In section 1 we prove that 1) $\Rightarrow$ 2) $\Rightarrow$ 3) together with some general facts about Hensel schemes. In section 2 we give the definition of morphism locally of finite type between Hensel schemes and we prove some properties which will be used in section 3 in order to prove the main results of the paper. In section 4 we deduce some corollaries. As for the notion of Hensel schemes, morphisms, fiber products etc. we refer to [3], [4], [5].

1. In this section we prove that 1) $\Rightarrow$ 2) $\Rightarrow$ 3). First we recall a definition.

1.1. Definition. Let $\varphi : X \to Y$ be a morphism of Hensel schemes. We say that φ is *affine* if there exists an affine open covering $\{U_i\}_{i \in I}$ of Y s.t. the induced Hensel open subscheme $\varphi^{-1}(U_i) \subset X$ is affine for all $i \in I$.

The following facts are easily deduced from the definition.

1.2. Proposition. Let $\varphi : X \to Y$ be an affine morphism. Then

a) if $U \subset Y$ is open, the induced morphism $\varphi' : \varphi^{-1}(U) \to U$ is affine

b) if $y \in Y$, the fiber $\varphi^{-1}(y)$ is an affine Hensel scheme

c) if

$$\begin{array}{ccc} X & \rightarrow & Y \\ & \searrow \quad \swarrow & \\ & S & \end{array}$$

is a commutative diagram of Hensel schemes and $S' \to S$ is any morphism, the induced morphism $\varphi_{S'} : X \times_S S' \to Y \times_S S'$ is affine.

1.3. Theorem. Conjecture 1) implies conjecture 2).

Now we recall the definition of closed immersion.

1.4. Definition. Let $\varphi : X \to Y$ be a morphism of Hensel schemes. We say that φ is a *closed immersion* if the following equivalent conditions are verified.

a) There is an affine open covering $\{U_i\}_{i \in I}$ of Y, $U_i = Sph\,(A_i, \underline{a}_i)$ s.t. $\varphi^{-1}(U_i) = Sph\,(A_i/\underline{c}_i, \underline{a}_i + \underline{c}_i/\underline{c}_i)$ for some ideal $\underline{c}_i \subset A_i$, all $i \in I$.

b) For every affine open covering $\{U_i\}_{i \in I}$ of Y, the same as above holds.

The equivalence of the two conditions follows from [3] Th. 1.11. In particular a closed immersion is a finite morphism ([3] 1.21 and 1.22) hence for every quasi-coherent $\mathcal{O}_X$-module F, $\varphi_* F$ is a quasi-coherent $\mathcal{O}_Y$-module.

A closed immersion determines a quasi-coherent ideal of $\mathcal{O}_Y$, i.e. the kernel of $\mathcal{O}_Y \to \varphi_* \mathcal{O}_X$; conversely given a quasi-coherent ideal F of $\mathcal{O}_Y$ its support is a closed subset X of Y and we have a closed immersion $(X, \mathcal{O}_Y/F) \to Y$. The Hensel scheme $(X, \mathcal{O}_Y/F)$ is denoted by Y/F.

1.5. Proposition. Given a closed subset X of Y, there is a uniquely determined quasi-coherent ideal F of $\mathcal{O}_Y$ such that Y/F is a reduced ordinary scheme and its underlying space is X.

Proof. Let J be the canonical ideal of Y and let E be the quasi-coherent ideal of O_Y/J which determines in X the structure of reduced subscheme of Y/J. Let F be the inverse image of $\varphi_* E$ under the canonical map of O_Y-modules $O_Y \to \varphi_*(O_Y/J)$, where $\varphi : Y/J \to Y$ is the canonical immersion. From the exact sequence of O_Y-modules

$$0 \to J \to F \to \varphi_* E \to 0$$

since J, $\varphi_* E$ are quasi-coherent, also F is quasi-coherent by [3] Cor. 1.15. F is unique because the immersion $Y/F \to Y$ factors through φ, since Y/F is a reduced scheme.

1.6. Definition. A morphism $\varphi : X \to Y$ of Hensel schemes is said *separated* if the following equivalent conditions are verified.

a) $\Delta : X \to X \times_Y X$ is a closed immersion

b) $Im(\Delta)$ is a closed subset of $X \times_Y X$

c) $\bar{\varphi} : \bar{X} \to \bar{Y}$ is separated, where $\bar{X} = X/I$, $\bar{Y} = Y/J$ and I, J are the canonical ideals of X, Y respectively.

In fact let us prove the equivalence of the three conditions.

a) $\Rightarrow$ b) is clear. b) $\Leftrightarrow$ c) since the underlying spaces of X, $X \times_Y X$ are the same as those of $\bar{X}$, $\bar{X} \times_Y \bar{X}$ and Δ, $\bar{\Delta}$ coincide as maps of topological spaces. b) $\Rightarrow$ a) let $y \in X \times_Y X$; if $y \notin \Delta(X)$, since $\Delta(X)$ is closed, there is an affine open $y \in U \subset X \times_Y X - \Delta(X)$ and $\Delta^{-1}(U) = \emptyset$. If $y \in \Delta(X)$ then $y = \Delta(x)$ for some $x \in X$; we easily reduce to the case when X, Y are affine i.e. $X = Sph\,(B, \underline{b})$, $Y = Sph\,(A, \underline{a})$ and φ is induced by a morphism of couples $(A, \underline{a}) \to (B, \underline{b})$; then $X \times_Y X = Sph\,({}^h(B \otimes_A B, \underline{c}))$ where $\underline{c} = Im(\underline{b} \otimes B) + Im(B \otimes \underline{b})$ and Δ is induced by ${}^h(B \otimes_A B, \underline{c}) \to (B, \underline{b})$ hence Δ is a closed immersion.

An Hensel sheme is said *separated* if it is separated over *Spec* $\mathbb{Z}$.

1.7. Lemma. Let X be an Hensel scheme, F , G quasi-coherent O_X-submodules of a quasi-coherent O_X-module M . Then $F+G$, $F\cap G$ are quasi-coherent O_X-modules.

Proof. The same as in [1] 2.2.2 using [3] Th. 1.11.

Let X be an Hensel scheme, I an ideal of definition, $A=\Gamma(X,O_X)$, $\underline{a}=\Gamma(X,I)$; the it is easy to see that $(A,\underline{a})$ is an H-couple and there is a canonical morphism $X\to Y=Sph(A,\underline{a})$.

1.8. Theorem. Suppose conjecture 2) true. Let X be an Hensel scheme separated and quasi-compact or such that its underlying space is noetherian. Then the following conditions are equivalent.

a) X is an affine Hensel scheme

b) there are elements $f_i\in A=\Gamma(X,O_X)$ s.t. $X_{f_i}=\{x\in X \mid f_i(x)=0\}$ is affine for every i and the f_i's generate A as an ideal

c) the canonical morphism $\varphi : X\to Y=Sph(A,\underline{a})$ is affine

d) $H^i(X,F)=0$ for every quasi-coherent O_X-module F , all $i>0$

e) $H^1(X,F)=0$ for every quasi-coherent ideal F in O_X .

Proof. a) $\Rightarrow$ c), c) $\Rightarrow$ a), d) $\Rightarrow$ e) are clear. a) $\Rightarrow$ d) follows from [3] Th. 1.11. b) $\Rightarrow$ c) it is enough to prove that $\varphi^{-1}(Y_f)=X_f$ for every $f\in A$. In fact if $x\in X$, $y=\varphi(x)$, then y corresponds to the prime $\underline{q}\subset A/\underline{a}$ which is the contraction of the prime $\underline{p}\subset B/\underline{b}$, where $Sph(B,\underline{b})$ is an open affine of X containing x . Then if $f(x)\neq 0$ we have $f\notin \underline{p}$, hence $f\notin \underline{q}$ i.e. $f(y)\neq 0$ and conversely. e) $\Rightarrow$ b) the proof is similar to the one used in the case of schemes

([2] 5.2.1) using [3] Cor. 1.15, Prop. 1.5 and Lemma 1.7.

2. In this section we give the definition of morphism locally of finite type between Hensel schemes and we prove some properties which will be used later.

2.1. Definition. Let B an A-algebra, $\underline{b} \subset B$ an ideal such that $(B,\underline{b})$ is an H-couple. We say that $(B,\underline{b})$ is of *finite type* over A if there exists an A-algebra C, $C \subset B$, C of finite type over A, such that, if we put $D = {}^{h}(C,\underline{b} \cap C)$, the induced homomorphism $D \to B$ is surjective.

This definition is the same as the one given in [6].

The following proposition can be easily deduced from the definition.

2.2. Proposition. Let $(B,\underline{b})$ be an H-couple with B an A-algebra. Then the following are equivalent.

a) $(B,\underline{b})$ is of finite type over A

b) there exists an A-algebra homomorphism $\varphi : C \to B$, C of finite type over A, s.t., if we put $\underline{c} = \varphi^{-1}(\underline{b})$, the induced homomorphism ${}^{h}(C,\underline{c}) \to B$ is surjective

c) there exist an A-algebra of finite type C, an ideal $\underline{c} \subset C$ and a surjective A-algebra homomorphism ${}^{h}(C,\underline{c}) \to B$ s.t. $\underline{c}B = \underline{b}$

d) same as in b) with C of the form $A[X_1,\ldots,X_n]$

c) same as in c) with C of the form $A[X_1,\ldots,X_n]$.

2.3. Definition. Let $\varphi : X \to Y$ be a morphism of Hensel schemes. We say that φ is *locally of finite type* if for every $x \in X$ there exist affine open subsets $x \in U \subset X$, $\varphi(x) \in V \subset Y$ s.t. $\varphi(U) \subset V$ and, if $U = Sph(B,\underline{b})$, $V = Sph(A,\underline{a})$, $(B,\underline{b})$ is of finite type over A.

2.4. Theorem. The following facts are true.

1) Let φ be a morphism of schemes; then φ is locally of finite type as a morphism of Hensel schemes $\Leftrightarrow$ φ is locally of finite type.

2) Let φ be a morphism of Hensel schemes which is locally an immersion; then φ is locally of finite type.

3) Let $\varphi : X \to Y$, $\psi : Y \to Z$ be morphisms of Hensel schemes locally of finite type; then $\psi \circ \varphi$ is locally of finite type.

4) Let $\varphi : X \to Y$, $\psi : Y \to Z$ be morphisms of Hensel schemes s.t. $\psi \circ \varphi$ is locally of finite type; then φ is locally of finite type.

5) Let

$$\begin{array}{ccc} X & \xrightarrow{\varphi} & Y \\ & \searrow \quad \swarrow & \\ & S & \end{array}$$

be a commutative diagram of Hensel schemes with φ locally of finite type and let $S' \to S$ be any morphism; then the induced morphism $\varphi_{S'} : X \times_S S' \to Y \times_S S'$ is locally of finite type.

6) Let $\varphi : X \to Y$ be a morphism of schemes locally of finite type and let C, D be closed subschemes of X, Y respectively s.t. $\varphi(C) \subset D$. Then the induced morphism ${}^h\varphi : {}^hX_C \to {}^hY_D$ is locally of finite type.

7) Let $\varphi : X \to Y$ be a morphism of Hensel schemes locally of finite type and let I, J be quasi-coherent ideal of O_X, O_Y respectively s.t. $\varphi^*(J)O_X \subset I$. Then the induced morphism $\bar{\varphi} : X/I \to Y/J$ is locally of finite type.

Proof. 1) is easy. 2) follows from the fact that a local immersion is a morphism $\varphi : X \to Y$ such that for every $x \in X$ there exist open subsets $x \in U \subset X$, $\varphi(x) \in V \subset Y$ s.t. $\varphi(U) \subset V$ and $\varphi/U : U \to V$ is a closed immersion.

3) We can suppose $X = Sph(B,\underline{b})$, $Y = Sph(A,\underline{a})$, $Z = Sph(A',\underline{a}')$

such that $(B,\underline{b})$ is of finite type over A and $(A,\underline{a})$ is of finite type over A'.

Then there exist homomorphisms $f : A[X_1,\dots,X_n] \to B$ and $g : A'[Y_1,\dots,Y_m] \to A$ such that $^h(A[X_1,\dots,X_n],f^{-1}(\underline{b})) \to B$ and $^h(A'[Y_1,\dots,Y_m],g^{-1}(\underline{a})) \to A$ are surjective. Consider the induced homomorphisms $h : A'[X_1,\dots,X_n,Y_1,\dots,Y_m] \to B$ and $^h(A'[X_1,\dots,X_n,Y_1,\dots,Y_m],h^{-1}(\underline{b})) \to B$.

Put $D = {}^h(A'[Y_1,\dots,Y_m],g^{-1}(\underline{a}))$ and prove that $^h(D[X_1,\dots,X_n],\underline{d}) \simeq {}^h(A'[X_1,\dots,X_n,Y_1,\dots,Y_m],h^{-1}(\underline{b}))$ where $\underline{d}$ is the extended ideal of $h^{-1}(\underline{b})$ in $D[X_1,\dots,X_n]$: in fact D is a direct limit of local etale neighborhoods of $(A'[Y_1,\dots,Y_m],g^{-1}(\underline{a}))$ and hence $D[X_1,\dots,X_n]_{1+\underline{d}}$ is a direct limit of local etale neighborhoods of $(A'[X_1,\dots,X_n,Y_1,\dots,Y_m],h^{-1}(\underline{b}))$ since $g^{-1}(\underline{a})A'[X_1,\dots,X_n,Y_1,\dots,Y_m] \subset h^{-1}(\underline{b})$; applying [3] Prop. 2.2 we have our isomorphism.

The canonical homomorphism $D[X_1,\dots,X_n] \to A[X_1,\dots,X_n]$ is surjective and we prove that $\underline{d}$ is the inverse image of $f^{-1}(\underline{b})$: in fact both ideals have the same contraction $h^{-1}(\underline{b})$ in $A'[X_1,\dots,X_n,Y_1,\dots,Y_m]$ and both contain $g^{-1}(\underline{a})D[X_1,\dots,X_n]$, hence they are equal because of the isomorphism

$$A'[X_1,\dots,X_n,Y_1,\dots,Y_m]/g^{-1}(\underline{a})A'[X_1,\dots,X_n,Y_1,\dots,Y_m] \simeq$$
$$\simeq D[X_1,\dots,X_n]/g^{-1}(\underline{a})D[X_1,\dots,X_n] .$$

It follows that the homomorphism $^h(A'[X_1,\dots,X_n,Y_1,\dots,Y_m], h^{-1}(\underline{b})) \to B$ decomposes in $^h(D[X_1,\dots,X_n],\underline{d}) \to {}^h(A[X_1,\dots,X_n], f^{-1}(\underline{b})) \to B$ and both factors are surjective.

4) With the previous notations $(B,\underline{b})$ is of finite type over A' and we must prove that $(B,\underline{b})$ is of finite type over A. There exists an homomorphism $f' : A'[X_1,\dots,X_n] \to B$

s.t. ${}^h(A'[X_1,\dots,X_n],f'^{-1}(\underline{b}))\to B$ is surjective. If we consider the induced homomorphism $f : A[X_1,\dots,X_n]\to B$ we have that ${}^h(A'[X_1,\dots,X_n],f'^{-1}(\underline{b}))\to B$ factors through ${}^h(A[X_1,\dots,X_n],f^{-1}(\underline{b}))\to B$ hence this last homomorphism is surjective.

5) We easily reduce to the affine case; hence we have H-couples and morphisms $(C,\underline{c})\to(A,\underline{a})\to(B,\underline{b})$ and $(C,\underline{c})\to$ $\to(C',\underline{c}')$ s.t. $(B,\underline{b})$ is of finite type over A. So we have an homomorphism $f : A[X_1,\dots,X_n]\to B$ s.t. ${}^h(A[X_1,\dots,X_n],$ $f^{-1}(\underline{b}))\to B$ is surjective. Consider the morphism of couples

$$(D,\underline{d}) = {}^h(A\otimes_C C',\ Im(\underline{a}\otimes_C C') + Im(A\otimes_C \underline{c}'))\to$$
$$\to {}^h(B\otimes_C C',\ Im(\underline{b}\otimes_C C') + Im(B\otimes_C \underline{c}')) = (E,\underline{e})$$

and prove that $(E,\underline{e})$ is of finite type over D. Letting $\underline{m} = Im(\underline{c}'\otimes_C A[X_1,\dots,X_n]) + Im(C'\otimes_C f^{-1}(\underline{b}))\subset C'\otimes_C A[X_1,\dots,X_n]$, as in 4) it is easy to show that

$${}^h(C'\otimes_C A[X_1,\dots,X_n],\underline{m}) \simeq {}^h(D[X_1,\dots,X_n],\underline{m}D[X_1,\dots,X_n]) \simeq$$
$$\simeq {}^h(C'\otimes_C {}^hA[X_1,\dots,X_n],\underline{m}^e)$$

where $\underline{m}^e$ is the extension of $\underline{m}$ in $C'\otimes_C {}^hA[X_1,\dots,X_n]$ and ${}^hA[X_1,\dots,X_n]$ is the henselization of $(A[X_1,\dots,X_n],$ $f^{-1}(\underline{b}))$. Hence we have a surjective homomorphism ${}^h(C'\otimes_C {}^hA[X_1,\dots,X_n],\underline{m}^e)\to E$ and an induced homomorphism $g : D[X_1,\dots,X_n]\to E$ s.t. $\underline{m}E = \underline{e}$.

6) and 7) follow easily from the definition.

2.5. Theorem. Let $\varphi : X\to Y$ be a morphism of affine Hensel schemes and let $X = Sph(B,\underline{b})$, $Y = Sph(A,\underline{a})$; then φ is locally of finite type $\Leftrightarrow$ $(B,\underline{b})$ is of finite type over A.

Proof. First we suppose that there exists an affine open

covering $Sph({}^{h}B_{f_j},\underline{b}\,{}^{h}B_{f_j})$ of X s.t. $({}^{h}B_{f_j},\underline{b}\,{}^{h}B_{f_j})$ is of finite type over A, $f_j \in B$. Write $(B,\underline{b})$ as a direct limit of couples $(B_\alpha,\underline{b}_\alpha)$ with B_α of finite type over A and $\underline{b}_\alpha$ inverse image of $\underline{b}$ in B_α.

Then $({}^{h}B_{f_j},\underline{b}\,{}^{h}B_{f_j})$ is the direct limit of $({}^{h}B_{\alpha_{f_j}},\underline{b}_\alpha\,{}^{h}B_{\alpha_{f_j}})$ and $\underline{b}_\alpha\,{}^{h}B_{\alpha_{f_j}}$ is the inverse image of $\underline{b}\,{}^{h}B_{f_j}$ in ${}^{h}B_{\alpha_{f_j}}$ because of the inclusion $(B_\alpha/\underline{b}_\alpha)_{f_j} \hookrightarrow (B/\underline{b})_{f_j}$.

Since $({}^{h}B_{f_j},\underline{b}\,{}^{h}B_{f_j})$ is of finite type over A there exists an homomorphism $g : A[X_1,\dots,X_n] \to {}^{h}B_{f_j}$ s.t. ${}^{h}(A[X_1,\dots,X_n],g^{-1}(\underline{b}\,{}^{h}B_{f_j})) \to {}^{h}B_{f_j}$ is surjective; then for some α, ${}^{h}(B_{\alpha_{f_j}},\underline{b}_\alpha B_{\alpha_{f_j}}) \to {}^{h}B_{f_j}$ is surjective.

Put $(C_\alpha,\underline{c}_\alpha) = Im({}^{h}B_\alpha,\underline{b}_\alpha\,{}^{h}B_\alpha) \subset (B,\underline{b})$. From the exact sequences

$$\begin{array}{ccccc} 0 \to C_\alpha & \to & \Pi\, {}^{h}C_{\alpha_{f_j}} & \to & \Pi\, {}^{h}C_{\alpha_{f_j f_k}} \\ \downarrow & & \wr\vert & & \wr\vert \\ 0 \to B & \to & \Pi\, {}^{h}B_{f_j} & \to & \Pi\, {}^{h}B_{f_j f_k} \end{array}$$

we have $C_\alpha = B$.

The general case follows easily from the previous one using Th. 2.4.

3. In this section we prove the main results of the paper. Let us start with some preliminary results.

3.1. Lemma. Let $f : B \to C$ be a ring homomorphism, C noetherian, $\underline{c} \subset Rad\ C$ be an ideal of C. For every $n > 0$ put $b^{(n)} = f^{-1}(\underline{c}^n)$ and assume that $B/\underline{b}^{(n)} \simeq C/\underline{c}^n$. Then $\underline{c} = \underline{b}C$

where $\underline{b} = f^{-1}(\underline{c})$.

Proof. From the isomorphisms $B/\underline{b}^{(n)} \simeq C/\underline{c}^n$ we get $\underline{b}(C/\underline{c}^n) = \underline{c}/\underline{c}^n$, hence $\underline{b}C + \underline{c}^n = \underline{c}$ and $\underline{b}C = \bigcap_n (\underline{b}C + \underline{c}^n) = \underline{c}$.

Recall that an Hensel scheme X is said *locally noetherian* if it can be covered by affine open subsets of the form $Sph(B,\underline{b})$ with B noetherian ring. It is easy to see that if $X = Sph(B,\underline{b})$ is affine then X is locally noetherian $\Leftrightarrow$ B is noetherian.

Now let $\varphi : X \to Y$ be a morphism of Hensel schemes locally of finite type, X locally noetherian, Y affine. Let I, J be ideals of definition of X , Y respectively s.t. $\varphi^*(J) \cdot O_X \subset I$. Put $A = \Gamma(Y,O_Y)$, $\underline{a} = \Gamma(Y,J)$ and suppose that Y is the union of two affine open subsets $Y_i = Sph({}^hA_{f_i}, \underline{a}\,{}^hA_{f_i})$, $f_i \in A$, $i = 1,2$ s.t. the inverse images $X_i = \varphi^{-1}(Y_i)$ are affine i.e. $X_i = Sph(B_i,\underline{b}_i)$, $B_i = \Gamma(X_i,O_X)$, $\underline{b}_i = \Gamma(X_i,I)$.

3.2. Proposition. Under the above hypotheses and notations the sequence

$$0 \to \Gamma(X,I) \to \Gamma(X,O_X) \to \Gamma(X,O_X/I) \to 0$$

is exact.

Proof. Put $B = \Gamma(X,O_X)$, $\underline{b} = \Gamma(X,I)$, $B_{12} = \Gamma(X_1 \cap X_2, O_X)$, $\underline{b}_{12} = \Gamma(X_1 \cap X_2, I)$. We have $B_{12} = {}^h(B_{1f_2}, \underline{b}_1 B_{1f_2}) = {}^h(B_{2f_1}, \underline{b}_2 B_{2f_1})$ and $\underline{b}_{12} = \underline{b}_1 B_{12} = \underline{b}_2 B_{12}$. We have a commutative diagram with exact rows

$$(\circ) \qquad \begin{array}{ccccccc} 0 \to & \underline{b} & \to & \underline{b}_1 \times \underline{b}_2 & \to & \underline{b}_{12} \\ & \downarrow & & \downarrow & & \downarrow \\ 0 \to & B & \to & B_1 \times B_2 & \to & B_{12} \end{array} \quad .$$

Moreover since X/I is an affine scheme, there exists an

A-algebra $\bar{B}$ s.t. $B_i/\underline{b}_i = \bar{B}_{f_i}$, $B_{12}/\underline{b}_{12} = \bar{B}_{f_1 f_2}$ and an inclusion $B/\underline{b} \hookrightarrow \bar{B}$.

By tensoring the diagram (°) with ${}^h B_{f_1 f_2}$ over B we have

$$\begin{array}{ccccccccc} 0 \to & \underline{b}\,{}^h B_{f_1 f_2} & \to & \underline{b}_1 \otimes_B {}^h B_{f_1 f_2} \times \underline{b}_2 \otimes_B {}^h B_{f_1 f_2} & \to & \underline{b}'_{12} & \to 0 \\ & \downarrow & & \downarrow & & \downarrow & \\ 0 \to & {}^h B_{f_1 f_2} & \to & B_1 \otimes_B {}^h B_{f_1 f_2} \times B_2 \otimes_B {}^h B_{f_1 f_2} & \to & B'_{12} & \to 0 \end{array}$$

where $\underline{b}'_{12}$ and B'_{12} are the images of the middle terms in $B_{12} \otimes_B {}^h B_{f_1 f_2}$. From this we have the exact sequence

$$0 \to (B/\underline{b})_{f_1 f_2} \to \bar{B}_{f_1 f_2} \times \bar{B}_{f_1 f_2} \to B'_{12}/\underline{b}'_{12} \to 0 .$$

The inclusion $B'_{12} \hookrightarrow B_{12} \otimes_B {}^h B_{f_1 f_2}$ induces a surjective homomorphism $B'_{12}/b'_{12} \to \bar{B}_{f_1 f_2}$ and a commutative diagram with exact rows

$$(°°) \qquad \begin{array}{ccccccc} 0 \to & (B/\underline{b})_{f_1 f_2} & \to & \bar{B}_{f_1 f_2} \times \bar{B}_{f_1 f_2} & \to & B'_{12}/\underline{b}'_{12} & \to 0 \\ & \downarrow & & \wr\| & & \downarrow & \\ 0 \to & \bar{B}_{f_1 f_2} & \to & \bar{B}_{f_1 f_2} \times \bar{B}_{f_1 f_2} & \to & \bar{B}_{f_1 f_2} & \to 0 \end{array}$$

From the commutative diagram

$$\begin{array}{ccc} {}^h B_{f_1 f_2} & \longrightarrow & B_1 \otimes_B {}^h B_{f_1 f_2} \\ \downarrow & & \downarrow \\ B_2 \otimes_B {}^h B_{f_1 f_2} & \longrightarrow & B'_{12} \end{array}$$

we have an induced homomorphism $\bar{B}_{f_1 f_2} \to B'_{12}/\underline{b}'_{12}$ which makes the right square in (°°) commute. So we get an homomorphism $\bar{B}_{f_1 f_2} \to (B/\underline{b})_{f_1 f_2}$ and since the composition $\bar{B}_{f_1 f_2} \to B'_{12}/\underline{b}'_{12} \to$

$\to \overline{B}_{f_1 f_2}$ is the identity, the same is true for the composition $\overline{B}_{f_1 f_2} \to (B/\underline{b})_{f_1 f_2} \to \overline{B}_{f_1 f_2}$ hence $(B/\underline{b})_{f_1 f_2} \simeq \overline{B}_{f_1 f_2}$.

More generally if we consider I^n instead of I we get $(B/\underline{b}^{(n)})_{f_1 f_2} \simeq B_{12}/\underline{b}_{12}^n$ where $\underline{b}^{(n)} = \Gamma(X, I^n)$.

Then we have ${}^h B_{f_1 f_2}/\underline{b}^{(n)}\, {}^h B_{f_1 f_2} \simeq B_{12}/\underline{b}_{12}^n$ and by Lemma 3.1, since B_{12} is noetherian, $\underline{b}B_{12} = \underline{b}_{12}$. From this, applying [6], I,1,5, we obtain that ${}^h B_{f_1 f_2} \to B_{12}$ is surjective. Hence the sequence $0 \to B \to B_1 \times B_2 \to B_{12} \to 0$ is exact and $0 \to \underline{b} \to \underline{b}_1 \times \underline{b}_2 \to \underline{b}_{12} \to 0$ is exact too, because $\underline{b}_{12} = \underline{b}B_{12}$. Hence $0 \to B/\underline{b} \to \overline{B}_{f_1} \times \overline{B}_{f_2} \to \overline{B}_{f_1 f_2} \to 0$ is exact i.e. $B/\underline{b} = \overline{B}$.

3.3. Theorem. Let $\varphi : X \to Y$ be an affine morphism of Hensel schemes, locally of finite type with X locally noetherian. Then if Y is affine X is also affine.

Proof. By definition there exists an affine open covering $\{Y_{f_i}\}_{i \in I}$ of $Y = Sph(A, \underline{a})$ such that $X_i = \varphi^{-1}(Y_{f_i})$ is affine, $X_i = Sph(B_i, \underline{b}_i)$, with $(B_i, \underline{b}_i)$ of finite type over ${}^h A_{f_i}$, hence of finite type over A .

By [3] 4.3 and 4.4 we can suppose the covering Y_{f_i} to be special, hence, by induction on the number r of the f_i's, we can assume $r = 2$. If $B = \Gamma(X, O_X)$, $\underline{b} = \Gamma(X, I)$, I ideal of definition of X , put $Z = Sph(B, \underline{b})$. We first prove that the canonical morphism $X \to Z$ is adic; in fact for every $n \geq 1$, letting $\underline{b}^{(n)} = \Gamma(X, I^n)$ by Prop. 3.2 we have $(B/\underline{b}^{(n)})_{f_i} \simeq B_i/\underline{b}_i^n$ and from Lemma 3.1, $\underline{b}B_i = \underline{b}_i$.

From $(B/\underline{b})_{f_i} \simeq B_i/\underline{b}B_i$, by [6] I,1.5, we have that ${}^h B_{f_i} \to B_i$ is finite and in this case the result follows from [3] Prop. 1.22.

From Theorem 3.3 we easily deduce the following *Serre's criterion* for Hensel schemes. Recall that an Hensel scheme is said *noetherian* if it is quasi-compact and locally noetherian.

3.4. Theorem. Let X be a noetherian Hensel scheme which is of finite type over an affine Hensel scheme Y. If $H^1(X,F)=0$ for every quasi-coherent ideal F of O_X, then X is affine.

4. In this section, on the analogy with the case of schemes, we deduce some corollaries.

Recall that an Hensel scheme is said *reduced* if $O_{X,x}$ is reduced for every $x \in X$. To every Hensel scheme X we can associate a quasi-coherent ideal $N \subset O_X$ s.t. $N_x = Nil(O_{X,x})$ for every $x \in X$; the closed Hensel subscheme X/N is reduced and is denoted by X_{red}. As in the case of schemes we prove the following.

4.1. Proposition. Let X be an Hensel scheme and F a quasi-coherent nilpotent ideal of O_X. Then X is affine $\Leftrightarrow X/F$ is affine.

Proof. If $X = Sph(A,\underline{a})$ is affine then by [3] Th. 1.11 we have $F = \tilde{\underline{c}}$ with $\underline{c}$ ideal of A. Then $X/F = Sph(A/\underline{c},\underline{a}+\underline{c}/\underline{c})$ is also affine.

Conversely suppose X/F affine and $F^n = 0$. By induction, using [3] Cor. 1.15 we can suppose $n = 2$. Put $A = \Gamma(X,O_X)$, $\underline{a} = \Gamma(X,I)$, I ideal of definition of X, $Y = Sph(A,\underline{a})$ and $\varphi : X \to Y$ the canonical morphism.

We prove that φ is an isomorphism. Note that $I \cdot O_X/F$ is an ideal of definition of $X_o = X/F$. From the exact sequences

$$0 \to \Gamma(X,F) \to \Gamma(X,O_X) \to \Gamma(X,O_X/F) \to H^1(X,F)$$

$$0 \to \Gamma(X, I \cap F) \to \Gamma(X, I) \to \Gamma(X, I \cdot \mathcal{O}_X/F) \to H^1(X, I \cap F)$$

since F , $I \cap F$ are quasi-coherent $\mathcal{O}_X$-modules and also quasi-coherent $\mathcal{O}_{X_o}$ -modules we have

$$H^1(X, F) = H^1(X_o, F) = H^1(X, I \cap F) = H^1(X_o, I \cap F) = 0 .$$

If $\underline{c} = \Gamma(X, F)\ A$, we have $\underline{c}^2 = 0$ and $A_o = \Gamma(X_o, \mathcal{O}_{X_o}) = A/\underline{c}$; moreover $X_o = Sph(A/\underline{c}, \underline{a}A/\underline{c})$. Consider the canonical morphisms $X_o \to X \to Y$: since $X_o = X$ and $X_o = Y$ as topological spaces we have $X = Y$.

From the exact sequence of sheaves over $X = Y$

$$\begin{array}{ccccccccc} 0 & \to & \tilde{\underline{c}} & \to & \tilde{A} & \to & \widetilde{(A/\underline{c})} & \to & 0 \\ & & \wr\vert & & \downarrow & & \wr\vert & & \\ 0 & \to & F & \to & \mathcal{O}_X & \to & \mathcal{O}_X/F & \to & 0 \end{array}$$

we have $\tilde{A} = \mathcal{O}_X$.

4.2. Proposition. Let $\varphi : X \to Y$, $\psi : Y \to Z$ be affine morphisms of Hensel schemes and suppose φ locally of finite type and X locally noetherian. Then $\psi \circ \varphi$ is affine.

Proof. Easy from Th. 3.3.

4.3. Proposition. Let $\varphi : X \to Y$, $\psi : X' \to Y'$ be affine morphisms of Hensel schemes over a base S . Suppose φ or ψ locally of finite type and $X \times_S X'$ locally noetherian. Then the morphism $\varphi \times \psi : X \times_S X' \to Y \times_S Y'$ is affine.

Proof. If φ is locally of finite type, factor $\varphi \circ \psi$ as follows $X \times_S X' \to Y \times_S X' \to Y \times_S Y'$ and the proposition follows from Prop. 1.2, Th. 2.4 and Prop. 4.2.

4.4. Proposition. Let $\varphi : X \to Y$, $\psi : Y \to Z$ be morphism of Hensel schemes. Suppose $\psi \circ \varphi$ affine and ψ separated. Then φ is affine.

Proof. As in the case of schemes ([1] I.9.1.16).

4.5. Proposition. Let $\varphi : X \to Y$ be a morphism of Hensel schemes, Then

a) if φ is affine, $\bar{\varphi} : X_{red} \to Y_{red}$ is affine

b) if $\bar{\varphi} : X_{red} \to Y_{red}$ is affine and X is locally noetherian, φ is affine.

Proof. a) easy. b) we can cover Y_{red} by affine open subsets U_i s.t. the Hensel open subschemes U_i of Y are affine and $(Z_i)_{red} = \bar{\varphi}^{-1}(U_i) \subset X_{red}$ are also affine. But $Z_i \subset X$ is quasi-compact and locally noetherian, hence the sheaf $N = Nil(O_{Z_i})$ is nilpotent and we can apply Prop. 4.1.

References

1. A. Grothendieck - J.A. Dieudonné, *Eléments de Géométrie algébrique, I*, Springer-Verlag, Berlin, 1971.

2. A. Grothendieck - J.A. Dieudonné, *Eléments de Géométrie algébrique, II*, I.H.E.S. Publ. Math., 8, Paris, 1961.

3. S. Greco - R. Strano, *Quasi-coherent sheaves over affine Hensel schemes*, Trans. A.M.S., 268 (1981), 445-465.

4. F. Mora, *Ideali di definizione e morfismi di schemi henseliani*, Ann. Mat. Pura e Appl., Ser. 4, 105 (1975), 191-204.

5. H. Kurke - G. Pfister - M. Roczen, *Henselsche Ringe und algebraische Geometrie*, V.E.B. Deutscher Verlag der Wiss., Berlin, 1975.

6. H. Kurke - G. Pfister - D. Popescu - M. Roczen - T. Mostowski, *Die Approximationseigenschaft localer Ringe*, L.N.M. n.634, Springer-Verlag, Berlin, 1978.

BLOWING-UP AND GLUEINGS IN ONE-DIMENSIONAL RINGS

Grazia Tamone

Istituto di Matematica-University of Genoa

Genova

INTRODUCTION. Let C be an affine curve, P a closed point on C, X a non-singular surface containing C as a closed subscheme of codimension 1. The blowing-up X' of X with center P induces the blowing-up C' of C with center P, which is a curve on X' (for the definition and the basic properties of blowing-up see [3], Ch.II,V). It is known that C' is affine. Let R,R' be respectively the coordinate ring of C,C'; if $\underline{m}$ is the maximal ideal in R corresponding to P, R' is called the ring "obtained from R by blowing-up $\underline{m}$". In [5] Lipman gives a more general definition of the ring "obtained from a one-dimensional ring by blowing-up an ideal". Let A be the local ring of C at P. There is a sequence of semilocal rings:

$$(.) \quad A = A_k \subset A_{k-1} \subset \dots \subset A_1 \subset A_o = \bar{A}$$

where $\bar{A}$ is the normalization of A, and each A_i (i = 0,...,k-1) is obtained from A_{i+1} by blowing-up the radical of A_{i+1}.

In [8] we proved that for each ring A with finite normalization $\bar{A}$ there is a sequence:

$$(..) \quad A = B_n \subset B_{n-1} \subset \dots \subset B_1 \subset B_o = \bar{A}$$

where each B_{i+1} (i = 0,...,n-1) is a "glueing of primary ideals of B_i over

This work was supported by C.N.R. (Consiglio Nazionale delle Ricerche)

a prime ideal of A".If $(A,\underline{m})$ is a local ring of dimension 1,every B_{i+1} is local,and coincides with the glueing,over its own maximal ideal $\underline{m}_{i+1}$,of the primary ideals belonging to $\underline{m}_{i+1}B_i$ (more details will be given in section 1) . If,moreover,A is analytically irreducible,in (.) each ring is local,so that A_i is obtained from A_{i+1} by blowing-up the maximal ideal $\underline{n}_{i+1}$ in A_{i+1} $(i=0,\ldots,k-1)$. Under these assumptions on A, in this paper we give conditions on A in order that each A_{i+1} in (.) can be obtained from A_i by glueing of the primary ideals belonging to $\underline{n}_{i+1}A_i$,over the maximal ideal $\underline{n}_{i+1}$. With regard to the rings we are interested in,the question can be answered by studying certain properties of the value-semigroup of the elements of A with respect to the discrete valuation of $\bar{A}$. In particular,if A is the local ring of a monomial curve,we propose an algorithmic method ,as a means to study these properties. As regards this last class of rings,we find also necessary and sufficient conditions on the generators of the semigroup,in order that the sequence (.) is not only as above,but coincides with the sequence (..) of the glueings.

1 . In first,we recall some results and notations,we'll need in the paper. Let $(A,\underline{m})$ be a local Cohen-Macaulay ring of dimension 1. Then,the ring "obtained by blowing-up $\underline{m}$ " is $\bigcup_{n>0}(\underline{m}^n:\underline{m}^n)=\bigcup_{n>0}\{a\in\bar{A}\mid a\,\underline{m}^n\subset\underline{m}^n\}$. (see [5],§ 1,page 651),which is also equal to $A[z_1/x,\ldots,z_t/x]$,where $\{z_1,\ldots,z_t\}$ is a set of generators of $\underline{m}$,x is a suitable element in $\underline{m}$ ([5],prop.1.1,def.1.7,lemma1.8). One can prove that this ring coincides with the "first neighbourhood of A" defined by Matlis in [6]:

$\Lambda = \{ b/a \mid b \in \underline{m}^s, a$ is a superficial element of degree s $\}$, (see [6], chap.XII and also [2],cor.3.5). From now onwards,we will denote the "embedding dimension" of A and the "multiplicity" of A respectively by: emdim(A) and e(A).

LEMMA 1.1 *Let* $(A, \underline{m})$ *be a local Cohen-Macaulay ring of dimension* 1, Λ *be the ring obtained from* A *by blowing-up* $\underline{m}$. *The following conditions are equivalent:*

1) emdim(A) = e(A)
2) $\Lambda = \underline{m} : \underline{m}$
3) *there is an element* x *in* $\underline{m}$ *such that:*
   i) $x\,\underline{m} = \underline{m}^2$
   ii) $\Lambda = \underline{m}\, x^{-1}$

Pf. See [5],lemma 1.8 (i),cor.1.10,lemma1.11.

As regards the definitions and the properties of the "glueings of primary ideals over a prime ideal",as well as the construction of the sequence (..),we refer to [8]. Here,we notice some results concerning 1-dimensional rings,we'll mainly use.

LEMMA 1.2 *Let* $A \subset B$ *be rings such that* B *is finite over* A , $\underline{m}$ *be a maximal ideal of* A . *If* A' *is the glueing over* $\underline{m}$ *of the primary ideals belonging to* $\underline{m}$ B, *we have:*

1) $A' = A + \underline{m}B$, *and* $\underline{m}' = \underline{m}B$ *is a maximal ideal of* A'
2) A' *coincides with the glueing of the primary ideals belonging to* $\underline{m}'B$, *over* $\underline{m}'$.

In particular, if $(A, \underline{m})$ *is local, we have also*

3) A' *is local, with maximal ideal* $\underline{m}' = \underline{m}B$.

Pf. Let $\underline{q}_1,\ldots,\underline{q}_n$ be the primary ideals belonging to $\underline{m}B$. Then we have:

1) $\cap\underline{q}_i = \underline{m}B$, and $\underline{m} = \underline{q}_i \cap A$ ([8], lemma 3.1) $\supset (\cap\underline{q}_i) \cap A = \underline{m}B \cap A = \underline{m}$, then $(\cap\underline{q}_i) \cap A = \underline{m}$, so that $B/\cap\underline{q}_i$ is integral over $A/\underline{m}$. Moreover, $A/\underline{m}$ is integrally closed (it is a field). So, $A' = A + \underline{m}B$, owing to the remark after prop.1.7 in [8]. Now, it is easy to see that $\sqrt{\underline{q}_1},\ldots,\sqrt{\underline{q}_n}$ are the only prime ideals of B over $\underline{m}$; moreover, A' is integral over A. Then, $\underline{m}' = \cap\underline{q}_i = \underline{m}B$ is the only prime ideal in A' over $\underline{m}$ ([8], osserv.II); therefore, $\underline{m}'$ is a maximal ideal of A' (because of the integral dependence of A' over A, and the maximality of $\underline{m}$).

2) The glueing over $\underline{m}'$ of the primary ideals belonging to $\underline{m}'B$ is $A'' = A' + \underline{m}'B$ (see [8], prop.1.7 and the following remark) $= (A + \underline{m}B) + \underline{m}'B = (A + \underline{m}B) + \underline{m}B$ (according to 1)) $= A + \underline{m}B = A'$.

3) Now, let us suppose $(A, \underline{m})$ local. Let $\underline{n} \in \mathrm{Spm}(A')$. Then, $\underline{n} \cap A = \underline{m}$ (because of the integral dependence); on the other hand, $\underline{m}' = \underline{m}B$ is the only prime ideal in A' over $\underline{m}$ (see the above proof of 1)), so $\underline{n} = \underline{m}'$.

The above lemma and th.3.2 in [8] show that each B_{i+1} in the sequence (..) is equal to $A + \underline{m}B_i$, is local, and coincides with the glueing, over its own maximal ideal $\underline{m}'_{i+1}$, of the primary ideals belonging to $\underline{m}'_{i+1}B_i$. Now, let $(A, \underline{m})$ be analytically irreducible. It means that $\hat{A} = (A, \underline{m})\hat{}$ is an integral domain, and we note that this condition is equivalent to the following one: $\bar{A}$ is finite over A, and is a discrete valuation ring (see [4], page 748). Because of these facts, each ring A_{i+1} is local. Let $\underline{n}_{i+1}$ be the maximal ideal in A_{i+1}. So, A_i is the ring obtained from A_{i+1} by blowing-up $\underline{n}_{i+1}$ $(i = 0,\ldots,k-1)$. Our purpose is now to find conditions on

A in order that each A_{i+1} is also the glueing, over $\underline{n}_{i+1}$, of the primary ideals belonging to $\underline{n}_{i+1}A_i$.

REMARKS. In general, (A, $\underline{m}$) doesn't have the above property. We get a simple example. Let $A = k[[t^3,t^5]]$. Then, $\bar{A} = k[[t]]$. The sequence (.) is: $A = k[[t^3,t^5]] \subset k[[t^2,t^3]] \subset k[[t]] = \bar{A}$. In fact, the ring obtained from A blowing-up (t^3,t^5) is $k[[t^3/t^3,t^5/t^3]]$([5], prop.1.1, def.1.7, lemma 1.8) $= k[[t^2,t^3]]$, and the ring obtained from $k[[t^2,t^3]]$ by blowing-up (t^2,t^3) is $(t^2,t^3)k[[t^2,t^3]](t^2)^{-1} = k[[t]]$ (lemma1.1). The ring A is not obtained from $k[[t^2,t^3]]$ by glueing of the primary ideals belonging to $(t^3,t^5)k[[t^2,t^3]]$ over (t^3,t^5). In fact, this glueing gives $A+(t^3,t^5)k[[t^2,t^3]]$ (lemma 1.2,1)) $= k[[t^3,t^5,t^7]] \neq A$.

If A is seminormal, the problem is trivially solved, for in this case both the sequences (.) and (..) coincide with: $A \subset \bar{A}$. In fact, the conductor of A in $\bar{A}$ is $\underline{m}$, then $B_1 = A + \underline{m}\bar{A}$ (lemma 1.2,1)) $= A$. On the other hand, $\bar{A}$ is obtained by blowing-up $\underline{m}$ in A. In fact, emdim(A) = e(A) because of the seminormality of A (see [1], th.1), so the ring obtained by blowing-up $\underline{m}$ is $\Lambda = \underline{m} : \underline{m}$ (lemma 1.1) $= \{a \in \bar{A} \mid a\underline{m} \subset \underline{m}\} = \Lambda$, since $\underline{m}$ is the conductor.

PROPOSITION 1.3 *Let* (A, $\underline{m}$), Λ *be as in lemma* 1.1, *and let* A' *be the glueing of the primary ideals belonging to* $\underline{m}\Lambda$, *over* $\underline{m}$. *The following conditions are equivalent:*

1) *the rings* A, A' *coincide*

2) emdim(A) = e(A)

3) *the conductor of* A *in* Λ *is* $\underline{m}$.

Pf. 1) ⇒ 2) We have: $A' = A + \underline{m}\Lambda$ (lemma 1.2,1)); so, $A = A'$ implies $\underline{m}\Lambda \subset A$, and then $\underline{m}\Lambda = \underline{m}\Lambda \cap A = \underline{m}$. Therefore, $\Lambda = \underline{m} : \underline{m}$ ([5], def.1.3), that is $\mathrm{emdim}(A) = e(A)$ (lemma 1.1).

2) ⇒ 3) If $\mathrm{emdim}(A) = e(A)$, we have: $\Lambda = \underline{m} : \underline{m}$ (lemma 1.1), then $\underline{m}\Lambda \subset \underline{m} \subset A$; so, $\underline{m}$ is contained in the conductor $\underline{b}$ of A in Λ, and therefore $\underline{m} = \underline{b}$.

3) ⇒ 1) In view of lemma 1.2, $A' = A + \underline{m}\Lambda$. Then, $A' \subset A$ because $\underline{m}$ is the conductor. So, $A' = A$.

From the equivalence of 1), 2) in prop.1.3, it follows:

COROLLARY 1.4 *Let* $(A, \underline{m})$ *be a 1-dimensional analytically irreducible ring. The following conditions are equivalent:*

1) *each* A_{i+1} *in the sequence* (.) *is obtained from* A_i *by glueing, over the maximal ideal* $\underline{n}_{i+1}$, *of the primary ideals belonging to* $\underline{n}_{i+1}A_i$

2) $\mathrm{emdim}(A_{i+1}) = e(A_{i+1})$ *for* $i = 0,\dots,k-1$.

REMARK. From 2) in cor.1.4 it follows obviously: $\mathrm{emdim}(A) = e(A)$. This condition is not sufficient to obtain 1). Let us consider an example. Let $A = k[[t^4, t^{10}, t^{11}, t^{17}]]$. Now, $\mathrm{emdim}(A) = e(A)$, because $\underline{m} = (t^4, t^{10}, t^{11}, t^{17})$ is such that $t^4\underline{m} = \underline{m}^2$, and moreover the blowing-up is $\Lambda = \underline{m}(t^4)^{-1}$ (lemma 1.1). So, $\Lambda = k[[t^4, t^6, t^7, t^{13}]] = k[[t^4, t^6, t^7]]$, and we have: $\mathrm{emdim}(\Lambda) = \dim_k((t^4, t^6, t^7)/(t^4, t^6, t^7)^2) \leq 3$, while $e(\Lambda) = 4$.

It is rather natural for us to wonder if it is possible to verify 1) or

2) in cor.1.4, only in terms of the ring A. Under the assumptions of cor.1.4, the rings A_{i+1} in (.) are local, and for each $i = 0,\dots,k-1$ the maximal ideal $\underline{n}_{i+1}$ is the contraction of the maximal ideal of $\bar{A}$. Therefore, the rings A_{i+1} are the "branch sequence of A along the maximal ideal of $\bar{A}$" (see [5], page 669). According to this, we can give a necessary and sufficient condition on a particular class of analytically irreducible rings, in order to satisfy condition 2) of cor.1.4. We will denote by μ_j the multiplicity $e(A_j)$ for each A_j in (.).

THEOREM 1.5 (Lipman-Du Val) *Let A be a complete local 1-dimensional domain with algebrically closed residue field. The following conditions are equivalent:*

1) *each* A_j *in the sequence* (.) *is such that* $\text{emdim}(A_j) = e(A_j) = \mu_j$ *for* $j = 1,\dots,k$

2) *the value-semigroup of the elements of* A *(for the discrete valuation with valuation ring* $\bar{A}$ *) is equal to:*

$$\{0,\mu_k, \mu_k+\mu_{k-1},\dots,\mu_k+\mu_{k-1}+\dots+\mu_o\}\cup\{n\in\mathbb{N} \mid n>\mu_k+\dots+\mu_o\}$$

Pf. See [5], th.2.2 and cor.3.2.

2. Now, we will apply th.1.5 to the local rings of "monomial curves". In first, we recall some facts concerning semigroups we will use from now onwards. Let S be a semigroup consisting of natural numbers, and let $0 \in S$. We say that S is numerical if the generators $a_1,\dots,a_n$ of S are such that $(a_1,\dots,a_n) = 1$, or equivalently if there exists $s \in S$ such that $s+\mathbb{N} \subseteq S$.

The least of these integers is called the <u>conductor</u> of S, and it will be noted by c. If S is generated by $a_1,\dots,a_n$ we will write: $S = \langle a_1,\dots,a_n \rangle$.

Now, let $A = k[[t^{n_1},\dots,t^{n_p}]]$, k algebrically closed. We assume:

$n_1 < n_2 < \dots < n_p$, and $(n_1,\dots,n_p) = 1$. The integral closure of A in its quotient field is $\bar{A} = k[[t]]$. If v is the discrete valuation with valuation ring $\bar{A}$, the value-semigroup v(A) of the elements of A is $S = \langle n_1,\dots,n_p \rangle$.

Since $(n_1,\dots,n_p) = 1$, S is a numerical semigroup. We will give an algorithmic method to find the multiplicities of the rings A_j in (.) and the value-semigroup S of the elements of A.

REMARK. Since $\mu_o = \mu(\bar{A}) = 1$, it is easy to see that 2) of th.1.5 is equivalent to the following condition:

$$\{0,\mu_k, \mu_k+\mu_{k-1},\dots, \mu_k+\mu_{k-1}+\dots+\mu_1\} = \{ n \in S \mid n \leqslant c \}$$

LEMMA 2.1 *Let* $A = k[[t^{n_1},\dots,t^{n_p}]]$, *where* $n_1 < \dots < n_p$. *The ring obtained from* A *by blowing-up* $\underline{m} = (t^{n_1},\dots,t^{n_p})$ *is* $\Lambda = k[[t^{n_1}, t^{n_2-n_1},\dots,t^{n_p-n_1}]]$

<u>Pf.</u> According to [5], prop.1.1, we have: $\Lambda = A[t^{n_1}/x,\dots,t^{n_p}/x]$, where $x \in \underline{m}$ is such that $x\,\underline{m}^n = \underline{m}^{n+1}$ for a suitable $n > 0$. We prove: $x = t^{n_1}$, showing that the valuation v(x) is equal to n_1. Since $x\,\underline{m}^n = \underline{m}^{n+1}$, we have: $(n+1)n_1 \in \Sigma = \langle v(x)+nn_1,\dots,v(x)+nn_p,\dots,v(x)+k_1n_1+\dots+k_pn_p,\dots \rangle$, where $k_1+\dots+k_p = n$. If $v(x) > n_1$, for each $(k_1,\dots,k_p)$ such that $k_1+\dots+k_p = n$ we have: $(n+1)n_1 = nn_1+n_1 = (k_1+\dots+k_p)n_1+n_1 \leqslant (k_1n_1+\dots+k_pn_p)+n_1 < k_1n_1+\dots+k_pn_p+v(x)$, so that $(n+1)n_1 \notin \Sigma$, which is a contradiction.

It follows: $\Lambda = A[t^{n_1}/t^{n_1}, t^{n_2}/t^{n_1},\dots,t^{n_p}/t^{n_1}] = k[[t^{n_1}, t^{n_2-n_1},\dots,t^{n_p-n_1}]]$.

REMARK. Under the assumptions of lemma 2.1, we have: $e(A)=\min(n_1, n_2-n_1, .., n_p-n_1)$

Using lemma 2.1, we can find a method to compute the multiplicities μ_j.

Let $A = k [\![t^{n_1}, \dots, t^{n_p}]\!]$. We can assume: $n_2 \notin \langle n_1 \rangle$. Let $n_2 = xn_1+r_2$ be the division of n_2 by n_1. Starting from $A = A_k$, we have in (.) a sequence of x rings of multiplicity n_1:

$$A = A_k \subset A_{k-1} = k [\![t^{n_1}, t^{n_2-n_1}, \dots, t^{n_p-n_1}]\!] \subset \dots \subset A_{k-x+1} =$$

$$= k [\![t^{n_1}, t^{n_2-xn_1}, \dots, t^{n_p-xn_1}]\!] = k [\![t^{r_2}, t^{n_1}, \dots, t^{n_p-xn_1}]\!].$$

If $r_2 = 1$, we have: $A_{k-x+1} = k [\![t]\!] = \bar{A}$. If $r_2 > 1$, after A_{k-x+1} there begins a sequence of rings of multiplicity r_2, the first of which is A_{k-x}. We want to find the number of these rings, so we write the generators of the semigroup $S(A_{k-x})$, say $a_1, a_2, \dots, a_p$ in encreasing order, and in such a way that $a_2 \notin \langle a_1 \rangle$: the number of these rings, say y, is equal to the quotient of the division of a_2 by a_1. Using the same argument, we will find the number of the following rings of multiplicity equal to $e(A_{k-x-y})$, and so on, until we reach $k [\![t]\!] = \bar{A}$.

Now, for each h such that A_h is the first in a sequence of rings of the same multiplicity $e(A_h)$, we need to know those generators $a_1, \dots, a_{p_h}$ of the semigroup $S(A_h)$ such that a_2 is not a multiple of a_1, in encreasing order (there is at least one of the generators of $S(A_h)$ which is not a multiple of a_1, because of the assumptions on $n_1, \dots, n_p$ and the structure of the rings following A). We will use the following algorithmic procedure.

Given $n_1, \dots, n_p$, we denote by x the quotient of the division of n_2 by n_1. For each $j = 2, \dots, p$ we put: $r_j = n_j - xn_1$, and we call q_j the quotient of

the division of r_j by r_2. If $r_2 = 1$, the sequence (.) is simply:
$A = A_k \subset A_{k-1} = k[[t^{n_1}, t^{n_2 - n_1}, \ldots, t^{n_p - n_1}]] \subset \ldots \subset k[[t^{r_2}, t^{n_1}, t^{r_3}, \ldots, t^{r_p}]] =$
$= k[[t]] = \bar{A}$. So, we consider the case: $r_2 > 1$.

a) If n_1 is a multiple of r_2, let j be the least integer such that r_j is not a multiple of r_2. Then, we put:

$$n'_1 = r_2\ ,\ n'_2 = r_j,\ n'_3 = r_{j+1}\ ,\ldots, n'_{p-j+2} = r_p\ .$$

Then, after the x rings $A = A_k \subset A_{k-1} \subset \ldots \subset A_{k-x+1}$ of multiplicity n_1, in the sequence (.) we have the ring $A_{k-x} = k[[t^{r_2}, t^{r_j}, t^{r_{j+1}}, \ldots, t^{r_p}]]$, where $r_2 < r_j < \ldots < r_p$ and r_j is not a multiple of r_2; so, in (.) there are q_j rings of multiplicity $r_2 = e(A_{k-x})$ following A_{k-x+1}.

b) If n_1 is not a multiple of r_2, we proceed as follows:

b_1) if $n_1 \leqslant r_3$, we put:

$$n'_1 = r_2\ ,\ n'_2 = n_1\ ,\ n'_3 = r_3\ ,\ldots, n'_p = r_p.$$

So, after A_{k-x+1} in (.) we have $A_{k-x} = k[[t^{r_2}, t^{n_1}, t^{r_3}, \ldots, t^{r_p}]]$; then, if q'_2 is the quotient of the division of n'_2 by n'_1, in (.) there are q'_2 rings of multiplicity r_2 following A_{k-x+1}.

b_2) when $n_1 > r_3$, if there exists a $j \geqslant 3$ such that r_j is not a multiple of r_2, let $\bar{j}$ be the least of these integers. Then we consider the elements $r_2, r_{\bar{j}}, r_{\bar{j}+1}, \ldots, r_p, n_1$ and put: $n'_1, \ldots\ldots, n'_{p-\bar{j}+3}$ these elements, written in encreasing order. Then, $A_{k-x} = k[[t^{n'_1}, \ldots\ldots, t^{n'_{p-\bar{j}+3}}]]$, where $n'_1 = r_2$; then, if q'_2 is the quotient of the division of n'_2 by n'_1, after A_{k-x+1} there are in (.) q'_2 rings of multiplicity r_2. On the contrary, if such a j doesn't exist, we put: $n'_1 = r_2$, $n'_2 = n_1$. In this case, the ring following A_{k-x+1} is $A_{k-x} = k[[t^{r_2}, t^{n_1}]]$; so, if q'_2 is the quotient of the division of n'_2 by n'_1, in (.) we have q'_2 rings of multiplicity r_2 following A_{k-x+1}.

Now, starting from the generators n'_j of $S(A_{k-x})$, we repeat the argument, and we continue in this way until we find a ring $A_h = k[[t^{a_1}, \ldots, t^{a_{p_h}}]]$ in

(.) such that the remainder of the division of a_2 by a_1 is equal to 1. Then, after the sequence of the rings of multiplicity $a_1 = e(A_h)$, in (.) there is $k[\![t]\!] = \bar{A}$, so the procedure ends.

EXAMPLES. 1) $A = k[\![t^6, t^{21}, t^{23}, t^{25}, t^{26}, t^{28}]\!]$. We have:

$x = 3, r_2 = 3, r_3 = 5, r_4 = 7, r_5 = 8, r_6 = 10$. Therefore, n_1 is a multiple of r_2, so it falls within a). The least j such that r_j is not a multiple of r_2 is $j = 3$; therefore, we put: $n'_1 = 3, n'_2 = 5, n'_3 = 7, n'_4 = 8, n'_5 = 10$.

The quotient of the division of n'_2 by n'_1 is 1; the remainder of this division is greater than 1, so we start again from $n'_1, \ldots, n'_5$. We have:

$r'_2 = 2, r'_3 = 4, r'_4 = 5, r'_5 = 7$; so, n'_1 is not a multiple of r'_2, and $n'_1 < r'_3$. Therefore, as in b_1), we put:

$n''_1 = 2, n''_2 = 3, n''_3 = 4, n''_4 = 5, n''_5 = 7$.

The remainder of the division of n''_2 by n''_1 is 1; then, in (.) we have:

3 rings of multiplicity 6

1 ring of multiplicity 3

1 ring of multiplicity 2

and, then, $\bar{A} = k[\![t]\!]$.

2) $A = k[\![t^5, t^{27}, t^{29}, t^{31}, t^{33}]\!]$. We have:

$x = 5, r_2 = 2, r_3 = 4, r_4 = 6, r_5 = 8$, then n_1 is not a multiple of r_2, and $n_1 > r_3$. So, it falls within b_2); moreover, each r_j is a multiple of r_2. Then, we put: $n'_1 = 2, n'_2 = 5$. Since the remainder of the division of n'_2 by n'_1 is 1, in (.) we have:

5 rings of multiplicity 5

2 rings of multiplicity 2

and finally $\bar{A} = k[\![t]\!]$.

This algorithmic method can be translated into a suitable algorithm for a computer. Using this procedure, we can compute the set: $\{0, \mu_k, \mu_k + \mu_{k-1}, \ldots, \mu_k + \mu_{k-1} + \ldots + \mu_1\}$. Now, we show how we can compute $\{n \in S \mid n \leq c\}$.

Since the rings we are interested in are necessarily such that emdim(A) = = e(A), we can assume the generators of S satisfy some other conditions, as in the next lemma. We first recall a definition. Let $S \subset \mathbb{N}$ be a numerical semigroup whose least positive integer is m. The elements $a_1, \ldots, a_p$ are a <u>standard basis</u> for S if:

i) $a_1 = m$

ii) $p = a_1$

iii) for $1 \leq i \leq m-1$, a_{i+1} is the least integer is S, having a_1-residue class distinct from those of $a_1, \ldots, a_i$.

(see [7], pages 455-456)

LEMMA 2.2 *Let* $A = k[[t^{n_1}, \ldots, t^{n_p}]]$ *where* $n_1 < \ldots < n_p$ *generate minimally* $S = < n_1, \ldots, n_p >$. *If* emdim(A) = e(A), *the elements* $n_1, \ldots, n_p$ *are a standard basis for* S.

<u>Pf.</u> It is known that $e(A) = n_1$ (see [7], prop. 2.1). Moreover, in this case, emdim(A) = p, so $n_1 = e(A) = \text{emdim}(A) = p$. Now, we prove: if $j < h$, then n_j, n_h have distinct n_1-residue classes. Let $n_j = q_j n_1 + r_j$, $n_h = q_h n_1 + r_h$ be the divisions of n_j and n_h by n_1. If $r_j = r_h$, then $n_h = (q_h - q_j) n_1 + n_j$ (where $q_h - q_j > 0$), so that $n_h \in < n_1, n_j >$, absurd. This shows that for each i, $1 \leq i \leq p-1$, n_{i+1} has n_1-residue class distinct from those of $n_1, \ldots, n_i$. Now, we prove that n_{i+1} is the least integer in S having this property. If $a \in S$ is such that $a < n_{i+1}$, $a(\text{mod}\, n_1) \neq n_j(\text{mod}\, n_1)$ for $j = 1, \ldots, i$ we have: $a = \sum_{k \leq i} a_k n_k$ $(a_k \in \mathbb{N})$, and $a(\text{mod}\, n_1) = n_h(\text{mod}\, n_1)$ for some $h \geq i+1$. Then, the same argument as before shows we have: $n_h \in < n_1, \ldots, n_i >$, which is a contradiction.

If $n_1, \ldots, n_p$ are a standard basis for S, the conductor c of S is $n_p - n_1 + 1$

(see [7],prop.2.2,i)).Moreover,the elements $n \in S$ such that $n \leqslant c$ (i.e. the ones we must compare with the elements of $\{0, \mu_k, \ldots, \mu_k + \mu_{k-1} + \ldots + \mu_1\}$) can be written in a very useful way,as shown in the following

PROPOSITION 2.3 *Let* A,S *be as in lemma* 2.2.*If* $n_1,\ldots,n_p$ *are a standard basis for* S,*each non-zero element of* S *can be written like this:*

$\lambda n_1 + n_k$, $\lambda \geqslant 0$, $k \geqslant 1$.

Pf. For each $n \in S-\{0\}$ we have: $n = \sum_{i=1}^{p} a_i n_i$,where $a_i \in \mathbb{N}$,$a_i \neq 0$ for some i.So,it is enough to prove that each $n = a_i n_i + a_j n_j$ (where $(a_i,a_j) \neq (0,0)$) can be written like this: $\lambda n_1 + n_k$,for a suitable $\lambda \geqslant 0$, $k \geqslant 1$.
Let us divide $a_i n_i + a_j n_j$ by n_1: $a_i n_i + a_j n_j = a n_1 + s$, $0 \leqslant s < n_1$.
If $s = 0$,the statement is trivial. If $s > 0$,since $n_1,\ldots,n_p$ are a standard basis we can find $k \in \{2,\ldots,p\}$ such that $n_k (\mathrm{mod}\, n_1) = s$. Let $n_k = b n_1 + s$.
So,we have: (+) $a_i n_i + a_j n_j = a n_1 + n_k - b n_1 = (a-b) n_1 + n_k$. Now,it is enough to prove: $a-b \geqslant 0$. If $n_k \neq n_i, n_j$,then $a-b \geqslant 0$,otherwise from (+) it follows: $n_k \in \langle n_1, n_i, n_j \rangle$. If $n_k = n_i, n_k \neq n_j$, (+) becomes:
$(a_i - 1) n_i + a_j n_j + (b-a) n_1 = 0$,so if $a-b < 0$ we have a contradiction , because $a_j \geqslant 0$ and $a_i - 1 \geqslant 0$,otherwise $n_i \in \langle n_1, n_j \rangle$.The same argument shows that $b-a < 0$ gives a contradiction when $n_k = n_j, n_k \neq n_i$. Finally, if $n_k = n_i = n_j$, then $a-b \geqslant 0$,otherwise (+) gives:
$(a_i + a_j - 1) n_i + (b-a) n_1 = 0$,when $b-a > 0$, $a_i + a_j - 1 > 0$.

So,prop.2.3 shows that $\{n \in S \mid n \leqslant c\} = \{n = \lambda n_1 + n_k,\ \lambda \in \mathbb{N}, 1 \leqslant k \leqslant n_1, n \leqslant c\}$.
Now,given $A = k[\![t^{n_1},\ldots,t^{n_p}]\!]$,where $n_1,\ldots,n_p$ are a standard basis for S,we construct the two sets: $\{n \in S \mid n \leqslant c\}$,and $\{0, \mu_k, \ldots, \mu_k + \ldots + \mu_1\}$, and compare them,applying th.1.5. We show some examples,using the rings already considered in the above examples 1),2).

1) Let $A = k[\![t^6,t^{21},t^{23},t^{25},t^{26},t^{28}]\!]$. We have: $\{0, \mu_k,\ldots, \mu_k+\mu_{k-1}+..\mu_1\}$ $= \{0,6,12,18,21,23\}$. On the other hand, the conductor c of S is 28-6+1=23, so $\{n \in S \mid n \leqslant c\} = \{0,6,12,18,21,23\}$. So, A satisfies condition 2) of th.1.5 .

2) Let $A = k[\![t^5,t^{27},t^{29},t^{31},t^{33}]\!]$. Then, $\{0, \mu_k,\ldots, \mu_k+\ldots+\mu_1\} =$ $= \{0,5,10,15,20,25,27,29\}$. The conductor c of S is 33-5+1 = 29. So, $\{n \in S \mid n \leqslant c\} = \{0,5,10,15,20,25,27,29\}$. Then, A satisfies condition 2) of th.1.5.

3. In some particular cases, there is also an easy complete characterization of the generators $n_1,\ldots,n_p$ of S in order to get the equality of the two sets $\{n \in S \mid n \leqslant c\}$ and $\{0, \mu_k,\ldots, \mu_k+\ldots+\mu_1\}$.
A characterization in the general case, even if possible, would result hard reading, and not very useful; one can easy realize it, studying some particular situations, as p = 5, p = 6, that in themselves require the investigation of a quite great number of cases. On the other hand, the procedure considered in section 2), allow just as well to verify if a given ring $A = k[\![t^{n_1},\ldots,t^{n_p}]\!]$ has the required property. On the contrary, as we will see, it is rather easy to obtain a characterization of the rings A as before, such that the sequence (.) coincides with the sequence (..) of glueings.

PROPOSITION 3.1 *Let* A,S *be as in prop.*2.3, $n_1,..,n_p$ *be a standard basis for* S.

a) *If* $n_2 \equiv 1 \bmod n_1$, *then condition* 2) *of th.*1.5 *holds if and only if* $n_j = n_{j-1}+1$ *for* $3 \leqslant j \leqslant p$.

b) *If* $n_2 \equiv n_1-1 \bmod n_1$, *we have: condition* 2) *of th.*1.5 *holds if and only if* $n_3 = n_2+2$, $n_j = n_{j-1}+1$ *for* $4 \leqslant j \leqslant p$.

Pf. a) Let $n_2 = q_2n_1+1$. The sequence (.) is:

$A \subset k[\![t^{n_1}, t^{n_2-n_1}, \ldots, t^{n_p-n_1}]\!] \subset \ldots \subset k[\![t^{n_1}, t^{n_2-(q_2-1)n_1}, \ldots, t^{n_p-(q_2-1)n_1}]\!] \subset$

$k[\![t^{n_1}, t, t^{n_p-q_2n_1}]\!] = k[\![t]\!] = \bar{A}$. Then, $\{0, \mu_k, \ldots, \mu_k+\ldots+\mu_1\} =$

$= \{0, n_1, 2n_1, \ldots, q_2n_1\}$, so it is enough to prove:

$\{n \in S \mid n \leqslant c\} = \{0, n_1, 2n_1, \ldots, q_2n_1\}$ if and only if $n_j = n_{j-1}+1$ for $3 \leqslant j \leqslant p$ (th.1.5 and remark). Let us suppose $\{n \in S \mid n \leqslant c\} =$ $\{0, n_1, 2n_1, \ldots, q_2n_1\}$. Then, in particular we have: $n_p-n_1+1 = c = q_2n_1$, so $n_p = q_2n_1+n_1-1 = n_2+(n_1-2) = n_2+(p-2)$; this implies: $n_j = n_{j-1}+1$ for $3 \leqslant j \leqslant p$. Conversely, if $n_j = n_{j-1}+1$, $3 \leqslant j \leqslant p$, we have: $n_p = n_2+p-2$, so $c = n_p-n_1+1 = n_2+p-2-n_1+1 = n_2+n_1-2-n_1+1 = q_2n_1$, then $\{n \in S \mid n \leqslant c\} =$ $= \{0, n_1, 2n_1, \ldots, q_2n_1\}$.

b) Let $n_2 = q_2n_1+n_1-1$. Then, in the sequence (.) we have:

$A \subset k[\![t^{n_1}, t^{n_2-n_1}, \ldots\ldots, t^{n_p-n_1}]\!] \subsetneq \ldots \subset k[\![t^{n_1}, t^{n_2-(q_2-1)n_1}, \ldots, t^{n_p-(q_2-1)n_1}]\!] \subset$

$\subset k[\![t^{n_1}, t^{n_1-1}, t^{n_3-q_2n_1}, \ldots, t^{n_p-q_2n_1}]\!] \subset \ldots\ldots$

Now, we notice: $n_3-q_2n_1 > n_1$, because $n_3-q_2n_1 > n_2-q_2n_1 = n_1-1$, and $n_3-q_2n_1 \neq n_1$, otherwise $n_3 \in \langle n_1 \rangle$. Then, $n_1-1 < n_1 < n_3-q_2n_1 < \ldots < n_p-q_2n_1$ so (.) continues as follows:

$\ldots\ldots \subset k[\![t^{n_1-1}, t^{n_1-(n_1-1)}, \ldots, t^{n_p-q_2n_1-(n_1-1)}]\!] = k[\![t]\!] = \bar{A}$.

Therefore, $\{0, \mu_k, \ldots, \mu_k+\ldots+\mu_1\} = \{0, n_1, 2n_1, \ldots, q_2n_1, q_2n_1+n_1-1\}$

It is enough to prove: $\{n \in S \mid n \leqslant c\} = \{0, n_1, 2n_1, \ldots, q_2n_1, q_2n_1+n_1-1\}$ if and only if $n_3 = n_2+2$, $n_j = n_{j-1}+1$ for $4 \leqslant j \leqslant p$ (th.1.5 and remark).

If $\{n \in S \mid n \leqslant c\} = \{0, n_1, \ldots, q_2n_1, q_2n_1+n_1-1\}$, then $n_p-n_1+1 = c = q_2n_1+n_1-1$ so $n_p = q_2n_1+n_1-1+n_1-1 = n_2+n_1-1$. Owing to this, and because of $n_p > \ldots > n_2$ and $n_j \notin \langle n_1 \rangle$ for $j \geqslant 2$, we have the statement. Conversely, if $n_3 = n_2+2$, $n_j = n_{j-1}+1$ for $4 \leqslant j \leqslant p$, we get: $c = (n_2+n_1-1)-n_1+1 = n_2 = q_2n_1+n_1-1$, so $\{n \in S \mid n \leqslant c\} = \{0, n_1, \ldots, q_2n_1, q_2n_1+n_1-1\}$.

The rings considered in prop.3.1 a) have also the following property

(stronger than 2) of th.1.5): the sequence (.) coincides with the sequence (..). In the following proposition we also show that these rings are the only ones of the above type having this property.

PROPOSITION 3.2 *Under the assumptions of prop.3.1, the following conditions are equivalent:*

1) *the sequence* (.) *coincides with the sequence* (..) *of the glueings between* A *and* $\bar{A}$

2) $n_2 \equiv 1 \bmod n_1$, $n_j = n_{j-1} + 1$ *for* $3 \leqslant j \leqslant p$.

Pf. 1) $\Rightarrow$ 2) According to prop.3.1 a), it is enough to prove: $n_2 \equiv 1 \bmod n_1$. For the assumptions, we have in particular: $A_1 = B_1$, so that $A_1 =$
$= A + (t^{n_1},\dots,t^{n_p})\bar{A}$ (lemma 1.2) $= k[\![t^{n_1},\dots,t^{n_p}]\!] + (t^{n_1})k[\![t]\!] =$
$= k[\![t^{n_1},t^{n_1+1},\dots,t^{n_1+n_1-1}]\!]$ (in fact, $p = n_1$ and $t^a \in k[\![t^{n_1},t^{n_1+1},\dots,t^{n_1+n_1-1}]\!]$ for $a \geqslant n_1$). Therefore, $n_1 = e(B_1) = e(A_1) = \mathrm{emdim}(A_1)$ (cor.1.4) $= \mathrm{emdim}(B_1)$. From these equalities we have: the n_1-residue class of n_2 (say r_2) is equal to 1. Indeed, if $r_2 > 1$, the sequence (.) is:
$A \subset k[\![t^{n_1},t^{n_2-n_1},\dots,t^{n_p-n_1}]\!] \subset \dots \subset k[\![t^{r_2},t^{n_1},\dots]\!] \subsetneq \dots \subset \bar{A}$, so $e(A) \leqslant r_2 < n_1$ which contradicts the above equality.

2) $\Rightarrow$ 1) Let $n_2 = q_2 n_1 + 1$, $n_j = n_{j-1} + 1$ for $3 \leqslant j \leqslant p$. Then, $B_1 =$
$= A + (t^{n_1},\dots,t^{n_p})\bar{A}$ (lemma 1.2) $= k[\![t^{n_1},\dots,t^{n_p}]\!] + (t^{n_1})k[\![t]\!] =$
$= k[\![t^{n_1},t^{n_1+1},\dots,t^{n_1+n_1-1}]\!]$ (see 1) $\Rightarrow$ 2)), and it is easy to see that $B_j = k[\![t^{n_1},t^{jn_1+1},\dots,t^{jn_1+n_1-1}]\!]$ for $j \geqslant 1$. Therefore, $A = B_{q_2}$, and the sequence (..) of glueings is:
$A = k[\![t^{n_1},t^{q_2n_1+1},\dots,t^{q_2n_1+n_1-1}]\!] \subset k[\![t^{n_1},t^{(q_2-1)n_1+1},\dots,t^{(q_2-1)n_1+n_1-1}]\!]$
$\subset \dots \subset k[\![t^{n_1},t^{2n_1+1},\dots,t^{2n_1+n_1-1}]\!] \subset k[\![t^{n_1},t^{n_1+1},\dots,t^{n_1+n_1-1}]\!] \subset k[\![t]\!] =$
$= \bar{A}$, which is also the sequence (.).

REFERENCES

[1] E.D.DAVIS On the geometric interpretation of seminormality
Proc.of Amer.Mat.Soc. 68 (1978) ,1-5

[2] S.GRECO-P.VALABREGA On the theory of adjoints
Lecture Notes in Math. 732 (Algebraic Geometry)
Springer-Verlag (1978) ,98-123

[3] R.HARTSHORNE Algebraic Geometry
Springer-Verlag New York (1977)

[4] E.KUNZ The value-semigroup of a one-dimensional Gorenstein ring
Proc. of Amer.Mat.Soc. 25 (1970) ,748-751

[5] J.LIPMAN Stable ideals and Arf rings
Amer.J.of Mat. 93 (1971) ,649-685

[6] E.MATLIS 1-Dimensional Cohen-Macaulay rings
Lecture Notes in Math. 327 Springer-Verlag (1970)

[7] D.S.RIM-M.A.VITULLI Weierstrass Points and Monomial Curves
J.of Algebra 48 (1977) ,454-476

[8] G.TAMONE Sugli incollamenti di ideali primari e la genesi di certe singolarità
B.U.M.I. (Supplemento) Algebra e Geometria Suppl.Vol.2 (1980) ,243-258.

Rational singularities with k*-action

Kei-ichi WATANABE

Department of Mathematics
Nagoya Institute of Technology
Nagoya, Japan

Introduction.

Let k be a field of characteristic 0, Y be a normal variety over k and $\Phi : Y' \longrightarrow Y$ be a resolution of singularities of Y. A point y of Y is called a rational singularity if the higher direct image sheaves $R^q\Phi_*(O_{Y'})$ vanish at y for $q > 0$. It is known that $R^q\Phi_*(O_{Y'})$ is independent of the resolution Φ ([6]). So we don't specify the resolution Φ when we write $R^q\Phi_*(O_{Y'})$.

Also, the modules $R^q\Phi_*(O_{Y'})_y$ reflects some ring-theoretic properties of the local ring $O_{Y,y}$. For example, if y is an isolated singularity, $R^q\Phi_*(O_{Y'})_y$ is isomorphic to the local cohomology group $H_y^{q+1}(O_Y)$ $(0 < q < \dim(O_{Y,y}) - 1)$ and, consequently, $O_{Y,y}$ is a Cohen-Macaulay ring if and only if $R^q\Phi_*(O_{Y'})_y = 0$ for $0 < q < \dim(O_{Y,y}) - 1$.

The aim of this paper is to describe the highest direct image sheaf $R^{d-1}\Phi_*(O_{Y'})_y$ $(d = \dim(O_{Y,y}))$ by purely ring-theoretic data of $O_{Y,y}$ when $O_{Y,y}$ has a k*-action and to give criteria for $O_{Y,y}$ to be a rational singularity or a canonical singularity (which was defined by M. Reid [10]).

If (Y,y) has a k*-action, we can attach a normal graded ring R to (Y,y). Hereafter, we put Y = Spec(R) and we will use the language of graded rings and graded modules. To get our result, we observe the space $\mathrm{Proj}(R^{\natural})$ defined in [E.G.A.], Chapter II, § 8. (This space is called an "affine cylinder" in [2].) There is a canonical proper birational morphism $\Psi : \mathrm{Proj}(R^{\natural}) \longrightarrow \mathrm{Spec}(R)$. So, if the singularities of $C = \mathrm{Proj}(R^{\natural})$ are rational singularities, we have (taking Φ as the composition of a resolution of C and Ψ), an isomorphism $R^q\Phi_*(O_{Y'}) \cong H^q(C, O_C)$ for every q and the latter cohomology group can be described by the local cohomology group of R. Thus the key point of our argument is to show that the singularities of $C = \mathrm{Proj}(R^{\natural})$ are rational singularities if the singularities of Spec(R) - {y} are rational singularities. We can prove this by the aid of Boutot's theorem which states that a direct summand of a rational singularity is also a rational singularity.

This procedure was achieved by Pinkham [9] in the case dim R = 2 and by the author [11] in some special cases.

It was shown by Demazure [2] that a normal graded ring R is characterized by the pair of X = Proj(R) and a "rational coefficient Weil divisor" D on X. We will write R = R(X,D) the normal graded ring determined by the pair (X,D). This expression is very useful in studying the graded ring or to construct examples of normal graded rings. In § 3, we will give a condition for R(X,D) to be a canonical singularity (defined by Reid [10]) in terms of (X,D).

By our theorem, R(X,D) is a rational singularity if and only if $H^q(X, O_X(nD)) = 0$ for $q > 0$ and $n \geqq 0$ once we know that Spec(R) - {y} has at most rational singularities. But the condition "Spec(R) - {y} has at most rational singularities" is rather a delicate one if the fractional part of D has singular points, even if X is smooth. We will analyze this problem in § 4 and will determine the condition for Spec(R) - {y} to have at most rational singularities in terms of (X,D) in some cases.

After the first typescript of this paper was complete, the author knew that H. Flenner also proved a theorem equivalent to Theorem (2.2) of this paper independently in [14].

Throughout this paper, we use the following notations.

k is a field of characteristic 0.

$R = \bigoplus_{n \geq 0} R_n$ is a Noetherian normal graded ring with $R_0 = k$.

$\underline{m} = R_+ = \bigoplus_{n > 0} R_n$ is the unique graded maximal ideal of R.

$M|_n = \bigoplus_{i \geq n} M_i$ for a graded R-module M.

$Y = \mathrm{Spec}(R)$ and $y = \underline{m} \in Y$.

$U = Y - \{y\}$.

$\Phi : Y' \longrightarrow Y$ is a resolution of the singularities of Y. As we are interested in $R^q\Phi_*(O_{Y'})$ and not in Φ itself, we may change Φ according to circumstances.

$X = \mathrm{Proj}(R)$ and $O_X(n) = \tilde{R}(n)$ for every integer n. Note that X is a normal projective space over k since R is a normal graded ring finitely generated over k.

$H^q_{\underline{m}}(R)$ is the q-th local cohomology group of R with respect to $\underline{m}$. Note that $H^q_{\underline{m}}(R)$ has the natural structure of a graded R-module ([5], p. 187).

$K_R = (H^d_{\underline{m}}(R))^*$ (d = dim R) is the canonical module of R. (See [5], p. 184 for the definition of the functor $(\ )^*$.)

$a(R) = \max \{ n \mid (H^d_{\underline{m}}(R))_n \neq 0 \} = - \min \{ n \mid (K_R)_n \neq 0 \}$.

In the following, let Z be a normal quasi-projective variety over k.

Div(Z) is the group of Weil divisors on Z.

$\mathrm{Div}(Z,\mathbb{Q}) = \mathrm{Div}(Z) \otimes_{\mathbb{Z}} \mathbb{Q}$ is the group of rational coefficient Weil divisors.

If $E, E' \in \mathrm{Div}(Z,\mathbb{Q})$, we say E and E' are linearly equivalent and denote $E \equiv E'$ if $E - E' = \mathrm{div}(f)$ for some $f \in k(Z)$ (k(Z) is the rational function field of Z.)

$\llcorner E \lrcorner = \sup \{ W \in \mathrm{Div}(Z) \mid W \leq E \}$ and $O_Z(E) = O_Z(\llcorner E \lrcorner)$ for $E \in \mathrm{Div}(Z, \mathbb{Q})$.

ω_Z is the dualizing sheaf of Z. Recall that if $i : Z' = \mathrm{Reg}(Z) \longrightarrow Z$ is the inclusion map, we may put $\omega_Z = i_*(\Omega^d_{Z'})$. Also, we put $\omega_Z^{[r]} = i_*((\Omega^d_{Z'})^{\otimes r})$ for a positive integer r. (See [10], Appendix to § 1 for the detail.)

K_Z is the canonical divisor on Z defined by $\omega_Z \cong O_Z(K_Z)$. Note that K_Z is a Weil divisor on Z determined up to linear equivalence. We have $\omega_Z^{[r]} \cong O_Z(rK_Z)$.

Cl(Z) is the divisor class group of Z.

§ 1. The canonical sheaf of $C(X,D) = \mathrm{Proj}(R^{\natural})$.

In this section, we investigate the canonical sheaf of $\mathrm{Proj}(R^{\natural})$ for $R = R(X,D)$. First, we recall the following

Theorem (1.1) ([2], 3.5). If we take a non-zero homogeneous element T of degree 1 in the quotient field of R, then there exists a unique $D \in \mathrm{Div}(X, \mathbb{Q})$ such that ND is an ample Cartier divisor for some positive integer N and that $R_n = H^0(X, O_X(nD)).T^n$ for every $n \geq 0$.

In this case, we put $R = R(X,D)$. (D is determined by R up to linear equivalence.) We put $D = \sum p_V/q_V.V$, where p_V, q_V are integers, $(p_V,q_V) = 1$ for every V and $q_V > 0$ (where V moves irreducible closed subvarieties on X). We fix D, p_V, q_V and N for this meaning throughout this paper.

Definition (1.2).

$$C = C(X,D) = \mathrm{Spec}_X(\bigoplus_{n\geq 0} O_X(nD).T^n)$$

$$U = U(X,D) = \mathrm{Spec}_X(\bigoplus_{n\in\mathbb{Z}} O_X(nD).T^n) \cong \mathrm{Spec}(R) - \{\underline{m}\}$$

$$S = S(X,D) = C(X,D) - U(X,D).$$

Let $\pi : C \longrightarrow X$ be the structure morphism. (By abuse of notation, we will use the same letter π for the structure morphism $U \longrightarrow X$.) It is easy to see that π induces an isomorphism of S onto X. Note that our notation differs from that of [2].

Remark (1.3). By [2], 3.3 and 3.4, we have the following commutative diagram;

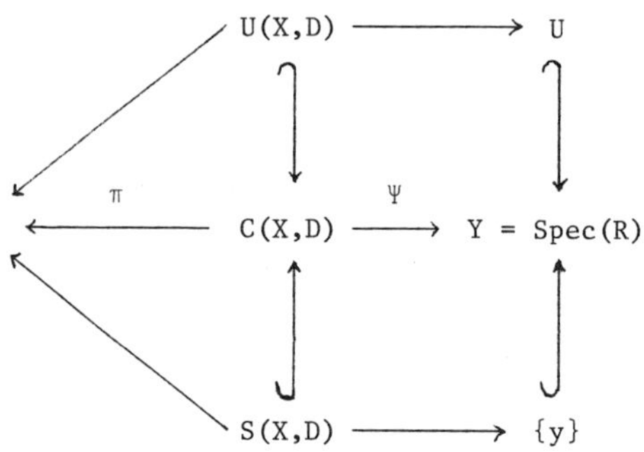

where the morphism Ψ contracts S(X,D) to the point y and induces an isomorphism of $U(X,D)$ onto $U = \mathrm{Spec}(R) - \{\underline{m}\}$.

Remark (1.4). If $R = R(X,D)$, we have a natural isomorphism

$$C(X,D) \cong \mathrm{Proj}(R^{\natural}).$$

Recall that ([E.G.A.], Chap. II, §8), for a graded ring R, $R^{\natural}$ is defined by $R^{\natural} = \bigoplus_{n\geq 0} R|_n$, where the grading of $R^{\natural}$ is given by $(R^{\natural})_n = R|_n$.

(Proof) By [11], (2.1), we have a canonical isomorphism $O_X(n) \cong O_X(nD)$ for every integer n. So, if f is a homogeneous element of R, the restriction of $C(X,D)$ to the open set $D_+(f)$ of $X = \mathrm{Proj}(R)$ is isomorphic to $\mathrm{Spec}(\bigoplus_{n\geq 0} (R_f)_n)$. On the other hand, the formula (8.2.7.3) and the statement (iv), (8.2.10) of [E.G.A], Chap. II show that $\mathrm{Proj}(R^{\natural})$ is covered by affine open sets as above.

Remark (1.5). (i) ([2], 2.9) $\mathrm{div}(T) = \pi^*(D) + S$ in $\mathrm{Div}(C)$.

(ii) $O_C(-S) = \bigoplus_{n>0} O_X(nD).T^n$ being the defining ideal of S. In general, $O_C(-rS) \cong \bigoplus_{n\geq r} O_X(nD).T^n$ as graded O_C-Modules.

(iii) If $E \in \mathrm{Div}(X,\mathbb{Q})$ and if $\pi^*(E)$ is an integral divisor of C, then $O_C(\pi^*(E)) \cong \bigoplus_{n\geq 0} O_X(E + nD).T^n$ as graded O_C-Modules.

Proposition (1.6). On $C = C(X,D)$, we may put

$$K_C = \pi^*(K_X + D') - S$$

where $D' = \sum(q_V - 1)/q_V.V$. (Note that though D' is not an integral divisor on X in general, $\pi^*(D')$ is an integral divisor on C ([2], 2.8, Cor. 2).) In other words,

$$\omega_C \cong \bigoplus_{n>0} O_X(K_X + D' + nD).T^n$$

as graded O_C-Modules.

(Proof) If D is an integral Cartier divisor, then C is an $\underline{\underline{A}}^1$-bundle over X, S is a Cartier divisor on C, $D' = 0$ by definition and $\mathrm{Cl}(C)$ is isomorphic to $\mathrm{Cl}(X)$ via π^*. So, we may put $K_C = \pi^*(E)$ for some $E \in \mathrm{Div}(X)$. Using the adjunction formula $K_S = (K_C + S).S$ and identifying S with X via π (we identify $\pi^*(E).S$ with E in this case), we have $K_X \equiv E - D$, since $S \equiv -\pi^*(D)$ on C. This proves that $K_C \equiv \pi^*(K_X + D) \equiv \pi^*(K_X) - S$. The other part of (1.6) follows from (1.5).

In general case, recall that ND is an integral Cartier divisor. We put

$$C' = \mathrm{Spec}_X(\bigoplus_{n>0} O_X(nND).T^{nN}) \cong C(X, ND).$$

Then there is a canonical finite covering $C \longrightarrow C'$ and we have

$$(*)\ \omega_C \cong \underline{\mathrm{Hom}}_{O_{C'}}(O_C, \omega_{C'}) \cong \underline{\mathrm{Hom}}_{O_{C'}}(\bigoplus_{n\geq 0} O_X(nD).T^n, \bigoplus_{n\geq 0} O_X(K_X+nND).T^{nN})$$

by general duality theory and the previous part of the proof. The right hand side of (*) is a graded O_C-module whose homogeneous part of degree n is 0 for $n \leqq 0$ and $\underline{\mathrm{Hom}}_{O_X}(O_X(-nD), O_X(K_X)) \cong O_X(K_X - \lfloor -nD \rfloor).T^n = O_X(K_X + D' + nD).T^n$ for $n > 0$ (cf. [11], (2.8)). This concludes the proof of (1.6).

§ 2. A criterion for rational singularities.

In this section, we will give a necessary and sufficient condition for R to be a rational singularity.

Definition (2.1). Let Z be a normal scheme of finite type over k and let $f : Z' \longrightarrow Z$ be a resolution of singularities of Z. We say that Z has only rational singularities if $R^i f_*(O_{Z'}) = 0$ for every $i > 0$. Note that, in our terminology, Z has only rational singularities if Z is smooth. We say that a normal ring A, essentially of finite type over k, is a rational singularity if Spec(A) has only rational singularities.

Theorem (2.2). If R is a normal graded ring finitely generated over k, then R is a rational singularity if and only if R satisfies the following two conditions;

(i) $U = \mathrm{Spec}(R) - \{\underline{m}\}$ has only rational singularities.

(ii) R is a Cohen-Macaulay ring and $a(R) < 0$.

To prove this theorem, we will investigate some more properties of C.

Lemma (2.3). (i) $H^q_{\underline{m}}(R) = 0$ $(q = 0,1)$.

(ii) $H^q_{\underline{m}}(R) \cong \bigoplus_{n\in\mathbb{Z}} H^{q-1}(X, O_X(n))$ $(q \geqq 2)$. In particular, R is a Cohen-Macaulay ring if and only if $H^q(X, O_X(n)) = 0$ for every integer n and $1 \leqq q < \dim X$.

(iii) $H^q(C, O_C) \cong (H^{q+1}_{\underline{m}}(R))|_0$ $(q \geqq 1)$.

(Proof) As R is normal, depth $R \geqq 2$ and (i) is clear. Combining (2.1) and (2.2) of [11], we get (ii). As $\pi : C \longrightarrow X$ is an affine morphism, we have $H^q(C, O_C) \cong H^q(X, \pi_*(O_C)) \cong \bigoplus_{n\leqq 0} H^q(X, O_X(n)) \cong H^{q+1}_{\underline{m}}(R)|_0$ (the last isomorphism is due to (ii)).

Lemma (2.4). If U has only rational singularities, so does C and X.

(Proof) As the conditions are local, it suffices to prove the following

Lemma (2.5). If $A = \bigoplus_{n\in\mathbb{Z}} A_n$ is a graded ring essentially of finite type over k and if A is a rational singularity, then so are A_0 and $A|_0 = \bigoplus_{n\geq 0} A_n$.

To prove this lemma, we use the following

Theorem (Boutot, [13]) Let B be a k-algebra essentially of finite type over k and let A be a subalgebra of B which is a direct summand of B as an A-module. In this case, if B is a rational singularity, so is A.

(Proof of (2.5)). As A_0 is a direct summand of A as A_0-module, A_0 is a rational singularity by Boutot's theorem. If we take a variable t over A and if we put deg(t) = -1 in the graded ring A[t], it is easy to see that $A|_0$ is isomorphic to $(A[t])_0$. As A[t] is a rational singularity ([3], Th. 2), so is $A|_0$.

(Proof of (2.2)). We may assume that U has only rational singularities and so does C by (2.4). If we take the resolution Φ to be a composition of a resolution Θ of C and Ψ, by the Leray spectral sequence for $\Phi = \Psi\circ\Theta$, we have $R^q\Phi_*(O_{Y'}) \cong R^q\Psi_*(O_C) \cong H^q(C,O_C) \cong H^{q+1}_{\underline{m}}(R)|_0$. Also, by the definition of a(R), $H^d_{\underline{m}}(R)|_0 = 0$ if and only if $a(R) < 0$ (d = dim R). If R satisfies the conditions (i) and (ii) of (2.2), then $H^q_{\underline{m}}(R)|_0 = 0$ for every q and R is a rational singularity. Conversely, if R is a rational singularity, then R is Cohen-Macaulay ([8], p. 50) and $a(R) < 0$ by the previous argument.

This proof shows the following

Corollary (2.6). If U has only rational singularities, then

$$R^q\Phi_*(O_{Y'}) \cong (H^{q+1}_{\underline{m}}(R))|_0 \quad \text{for every } q \geq 1.$$

Remark (2.7). If X is a normal projective variety over k and if $D \in \mathrm{Div}(X,Q)$ satisfies the conditions; (a) ND is an ample Cartier divisor for some positive integer N and (b) U(X,D) has only rational singularities, then $H^q(X, O_X(nD)) = 0$ for $q < \dim X$ and $n < 0$.

(Proof) As U = U(X,D) has only rational singularities, $H^q_{\underline{m}}(R)$ has finite length for $q < d = \dim R$ (R = R(X,D)). On the other hand, by Grauert-Riemenschneider vanishing theorem and the duality theorem (cf. [3], § II), $R^q\Phi_*(O_{Y'}) \cong H^{q+1}_{\underline{m}}(R)$ for $q < d - 1$. As $H^{q+1}_{\underline{m}}(R) \cong \bigoplus_{n\in\mathbb{Z}} H^q(X, O_X(nD))$ for $q \geq 1$ by (2.3) and as $R^q\Phi_*(O_{Y'}) \cong H^q(C, O_C) \cong \bigoplus_{n\geq 0} H^q(X, O_X(nD))$ for $q \geq 1$ by our condition and (2.6), we have the desired conclusion. If X is smooth and D is an ample Cartier divisor, this statement reduces to Kodaira vanishing theorem.

Definition (2.8). If U has only rational singularities, we put

$$p_g(R) = \dim_k R^{d-1}\Phi_*(O_{Y'}) \qquad (d = \dim R)$$

and we will call it the geometric genus of R. If U has only rational singularities, R is a rational singularity if and only if R is Cohen-Macaulay and $p_g(R) = 0$.

Corollary (2.9). If U has only rational singularities, then

$$p_g(R) = \dim_k (H^d_{\underline{m}}(R))|_0 = \dim_k K_R/(K_R|_0).$$

Corollary (2.10). If R is a Gorenstein ring and if U has only rational singularities, then

$$p_g(R) = \sum_{n=0}^{a(R)} \dim_k R_n.$$

(Proof) By [5], (3.1.4), $K_R \cong R(a(R))$ as graded R-modules and the result follows from (2.9).

Remark. If $A = \bigoplus_{n\geq 0} A_n$ is a Cohen-Macaulay graded ring with $A_0 = k$ and if $x \in A_m$ is an A-regular element, then we have $a(A/xA) = a(A) + m$. So, the invariant a(A) can be computed from that of an Artinian graded ring and if A is an Artinian graded ring, then $a(A) = \max\{ n \mid A_n \neq 0 \}$. Also, if $R = k[X_1,\ldots,X_n]$ is a polynomial ring over k, then $a(R) = -(\deg(X_1) + \ldots + \deg(X_n))$. (cf. [5], (2.2.8) and (2.2.10).)

Corollary (2.11). Let $R = k[X_1,\ldots,X_{d+s}]/(f_1,\ldots,f_s)$ be a normal graded ring which is a complete intersection and assume that U has only rational singularities. then R is a rational singularity if and only if

$$\sum_{j=1}^{s} \deg(f_j) < \sum_{i=1}^{d+s} \deg(X_i).$$

(When R is a hypersurface with isolated singularity, Kimio Watanabe computed $p_g(R)$ in [12].)

Definition (2.12). A local ring $(A,\underline{m})$, essentially of finite type over k is called an elliptic singularity if it satisfies the following conditions;

(i) Spec(A) - {$\underline{m}$} has only rational singularities,

(ii) A is a Gorenstein ring,

(iii) $p_g(A) = 1$.

Corollary (2.13). If R is a Gorenstein ring and if U has only rational singularities, then $R_{\underline{m}}$ is an elliptic singularity if and only if

$0 \leq a(R) < \min \{ n \mid n > 0 \text{ and } R_n \neq 0 \}$.

Remark. If U has only rational singularities, then so does X by (2.5). But U may have non-rational singularities even if X is smooth if the support of $D - \lfloor D \rfloor$ has bad singularities. We will treat this problem in § 4.

Remark. To prove that R is a rational singularity, it suffices to show that $R_{\underline{m}}$ is a rational singularity, since "rational singularity" is an open condition and Y has a k*-action.

§ 3. A criterion for canonical singularities.

Definition (3.1) (M. Reid, [10]). If Z is a normal variety over k, a point $z \in Z$ is a canonical singularity if it satisfies the following conditions;

(i) $\omega_{Z,z}^{[r]}$ is an invertible $O_{Z,z}$-module for some positive integer r,

(ii) $\phi_*(\omega_{Z'}^{\otimes r})_z = \omega_{Z,z}^{[r]}$ (where r is as in (i) and $\phi : Z' \longrightarrow Z$ is a resolution of singularities of Z).

R. Elkik [4] and O. Gabber proved that a canonical singularity is, in fact, a rational singularity using a technique of Shephard-Barron, who proved that a canonical singularity of dimension 3 is a rational singularity. Also, H. Flenner proved in [14] that a canonical singularity of index one is a rational singularity. A Gorenstein rational singularity is a canonical singularity and a canonical singularity of dimension 2 is Gorenstein (cf. [10]).

We will give a necessary condition for R to be a canonical singularity (we will say that R is a canonical singularity if $y \in Y$ is a canonical singularity) in terms of X and D, which also is a sufficient condition in some cases. The letters X, D, D', p_V, q_V and N are used as in §1.

Lemma (3.2). $\omega_{Y,y}^{[r]}$ is an invertible $0_{Y,y}$-module if and only if $r.(K_X + D') \equiv a'.D$ for some integer a'.

(Proof) This follows from the calculation of the class group and the canonical class of R (cf. [11], (1.6) and (2.9)).

Lemma (3.3). If $r.(K_X + D') \equiv a'.D$, then $\Psi^*(\omega_Y^{[r]}) \cong 0_C(r.\pi^*(K_X + D') + a'S) \cong \omega_C^{[r]}((a'+r)S)$.

(Proof) As $\Psi^*(\omega_Y^{[r]})$ is isomorphic to $\omega_C^{[r]}$ on U = C - S, $\Psi^*(\omega_Y^{[r]}) \cong \omega_C^{[r]}(bS) \cong 0_C(r.\pi^*(K_X + D') + (b-r)S)$ for some integer b. (Recall that $K_C \equiv \pi^*(K_X + D') - S$ by (1.6).) As $S \equiv -\pi^*(D)$ and $r.(K_X + D') \equiv a'D$, $r.\pi^*(K_X + D') + (b-r).S \equiv (b-r-a')S$. This shows that $\Psi^*(\omega_Y^{[r]}) \equiv 0_C((b-r-a')S)$. On the other hand, $\Psi^*(\omega_Y^{[r]})$ is an invertible 0_C-Module whose restriction to S is trivial. Identifying S with X, the restriction of $\Psi^*(\omega_Y^{[r]})$ to S is isomorphic to $0_X((r+a'-b)D)$. As $(r+a'-b)D$ is an integral divisor and as ND is ample, this leads to the equality $b = a'+r$, which was to be proved.

Lemma (3.4). (i) $\Psi_*(\omega_C) = \omega_Y$ if and only if $a(R) < 0$.

(ii) If $r.(K_X + D') \equiv a'D$, then $\Psi_*(\omega_C^{[r]}) = \omega_Y^{[r]}$ if and only if $a' \leqq -r$.

(Proof) (i) By [11], (2.8), $K_R = H^0(Y, \omega_Y) \cong \bigoplus_{n\in\mathbb{Z}} H^0(X, 0_X(K_X + D' + nD))$ and by (1.6), $\Psi_*(\omega_C) \cong \bigoplus_{n>0} H^0(X, 0_X(K_X + D' + nD)) \cong (K_R)|_1$ as graded R-modules. Thus $\Psi_*(\omega_C) = \omega_Y$ if and only if $a(R) < 0$ by the definition of a(R).

(ii) As $r.(K_X + D') \equiv a'D$, $H^0(Y, \omega_Y^{[r]}) \cong R(a')$ as graded R-modules and by (1.6), $\Psi_*(\omega_C^{[r]}) \cong \bigoplus_{n\geqq r} H^0(X, 0_X(r.(K_X + D') + nD)) \cong \bigoplus_{n\geqq r} H^0(X, 0_X((a'+n).D)$.
Hence $\Psi_*(\omega_C^{[r]}) = \omega_Y^{[r]}$ if and only if $a'+r \leqq 0$.

Proposition (3.5). If R = R(X,D) is a canonical singularity, then $r.(K_X + D') \equiv a'.D$ for some positive integer r and an integer $a' \leqq -r$. If C = C(X,D) has only canonical singularities and if $r.(K_X + D') \equiv a'.D$ for a positive integer r and an integer a' with $a' \leqq -r$, then R is a canonical singularity.

(Proof) Let us take a resolution Φ of Y so that Φ is a composition of a resolution Θ of C and Ψ. As $\Phi_*(\omega_{Y'}^{\otimes r}) \subset \Psi_*(\omega_C^{[r]}) \subset \omega_Y^{[r]}$ in general and

$\Phi_*(\omega_{Y'}^{\otimes r}) = \omega_C^{[r]}$ if C has only canonical singularities, we have the conclusion by (3.4).

Corollary (3.6). If X and D satisfy one of the following conditions and if $r.(K_X + D') \equiv a'.D$ for some positive integer r and an integer $a' \leqq -r$, then $R = R(X,D)$ is a canonical singularity.

(i) X has only canonical singularities and D is an ample Cartier divisor.

(ii) U has only rational singularities, X is a factorial variety and $p_V \equiv 1 \pmod{q_V}$ for every V.

(Proof) By (3.5), it suffices to prove that C has only canonical singularities if X and D satisfy (i) or (ii). In case (i), C is an $\underline{\underline{A}}^1$-bundle over X and has only canonical singularities since so does X. In case (ii), C has only rational singularities by (2.4) and C is a Gorenstein variety since $K_C = \pi^*(K_X + D' + D)$ is a Cartier divisor on C by our condition. As we have noted, a Gorenstein rational singularity is a canonical singularity.

Remark. The condition " $r.(K_X + D') \equiv a'.D$ for $a' \leqq -r$ " is not a sufficient condition for R to be a canonical singularity in general, even if U is smooth. For example, if $X = \mathbb{P}^1$ and $D = \frac{3}{4}.P$ for some point $P \in \mathbb{P}^1$, $K_X \equiv -2P$ and we have $3(K_X + D') \equiv -5D$. But in this case, R is not a canonical singularity since R is not Gorenstein and canonical singularities of dimension two are Gorenstein.

§ 4. Some criteria for U = U(X,D) to have only rational singularities or to be smooth.

If we want to construct an example of a rational singularity by giving X and D, it is important to know the condition for U = U(X,D) to have only rational singularities. So, our problem in this section is; " Find the condition on (X,D) for U(X,D) to have only rational singularities or to be smooth." Although we cannot give the complete answer, we can solve this problem in some useful cases.

Example (4.1). ([11], (3.1)) If (X,D) satisfies the following conditions, then U(X,D) has only rational singularities.

(i) X has only rational singularities.

(ii) If $x \in X$ is a singular point of X, D is an integral Cartier divisor on some neighborhood of x.

(iii) If $x \in X$ is a smooth point of X, $\mathrm{Supp}(D - \llcorner D \lrcorner)$ is normal crossing at x.

Proposition (4.2). Take a point $x \in X$ and put $A = O_{X,x}$. If $D \equiv \frac{p}{q} V$ near x, where V is a Cartier divisor defined by $f \in A$ near x and p, q are relatively prime integers with $0 < p < q$, there is an étale covering from $\mathrm{Spec}(A[t]/(t^q - f)[T,T^{-1}])$ onto $U(X,D) \times_X \mathrm{Spec}(A)$ (where t and T are indeterminates over A).

(Proof) Put $B = \bigoplus_{n \in \mathbb{Z}} O_X(nD)_x = A[f^a T^n \mid a,n \in \mathbb{Z},\ qa + np \geqq 0]$, so that $U(X,D) \times_X \mathrm{Spec}(A) = \mathrm{Spec}(B)$. If we take an integer p' $(0 < p' < q)$ such that $pp' + bq = 1$ for some integer b, it is easy to see that $B = A[f^b T^{p'}, f^p T^q, (f^p T^q)^{-1}] \cong A[u,v,v^{-1}]/(u^q - fv^{p'})$. If we define the A-homomorphism $\psi : B \longrightarrow A[t,T,T^{-1}]/(t^q - f)$ by $\psi(u) = tT^{p'}$ and $\psi(v) = T^q$, ψ is an étale morphism and we get the desired result.

Corollary (4.3). Under the assumptions of (4.2),

(i) $U(X,D) \times_X \mathrm{Spec}(A)$ has only rational singularities if and only if so does $\mathrm{Spec}(A[t]/(t^q - f))$.

(ii) $U(X,D) \times_X \mathrm{Spec}(A)$ is smooth if and only if X and V are smooth at x or X is smooth at x and $q = 1$.

Corollary (4.4). Let us assume that k is algebraically closed and that (X,D) satisfies the conditions of (4.2) at x. If X is smooth at x, $\dim X = 2$, $q > 1$ and if V is singular at x, then $U(X,D) \times_X \mathrm{Spec}(A)$ has only rational singularities if and only if the singularity of V at x is analytically isomorphic to one of the following ones;

Case (i), $q = 2$. $x^2 = y^n$ $(n \geqq 2)$, $x(y^2 - x^n) = 0$ $(n \geqq 2)$, $x^3 = y^n$ $(n = 3,4,5)$ or $y(y^2 - x^3) = 0$.

Case (ii), $q = 3$. $x^2 = y^n$ $(n = 2,3,4,5)$.

Case (iii), q = 4,5. xy = 0 or $x^2 = y^3$.

Case (iv), q ≧ 6. xy = 0.

(Proof) Straightforward from (4.3) and the classification of rational Gorenstein surface singularities (cf. [1]).

REFERENCES.

[E.G.A.] A.Grothendieck and J.Dieudonne, Eléments de Géométrie Algébrique, II. Publ. I.H.E.S. 8 (1961).

[1] M.Artin, On isolated rational singularities of surfaces, Amer. J. Math. 88 (1966), 129-136.

[2] M.Demazure, Anneaux gradués normaux, in Séminaire Demazure-Giruad-Teissier, Singularités des surfaces, Ecole Polytechnique (1979).

[3] R.Elkik, Singularités rationelles et déformations, Invent. Math. 47 (1978), 139-147.

[4] R.Elkik, Rationalité des singularités canoniques, Invent. Math. 64 (1981), 1-6.

[5] S.Goto and K.-i. Watanabe, On graded rings, I, J. Math. Soc. Japan 30 (1978), 179-213.

[6] H.Hironaka, Resolution of singularities of an algebraic variety over a field of characteristic zero, Ann. of Math. 79 (1964), I; 109-203, II; 205-236.

[7] M.Hochster and J.L.Roberts, Rings of invariants of reductive groups acting on regular rings are Cohen-Macaulay, Adv. in Math. 13 (1974), 115-175.

[8] G.Kempf, F.Knudsen, D.Mumford and B.Saint-Donat, Toroidal Embeddings, I. Lecture notes in Math. 339, Springer 1973.

[9] H.Pinkham, Normal surface singularities with $\mathbb{C}^*$-action, Math. Ann. 227 (1977), 183-193.

[10] M. Reid, Canonical 3-folds, in Journées de géométrie algébrique d'Angers, ed. A.Beauville, Sijthoff and Nordhoff, Alphen (1980), 273-310.

[11] K.-i. Watanabe, Some remarks concerning Demazure's construction of normal graded rings, to appear in Nagoya Math. J. 83 (1981).

[12] Kimio Watanabe, On plurigenera of normal isolated singularities, I, Math. Ann. 250 (1980), 65-94.

[13] J.-F. Boutot, Singularités rationelles et quotient par les groupes reductifs, preprint.

[14] H.Flenner, Rationale quasihomogene Singularitäten, Arch. Math. 36 (1981), 35-44.